# Measurements and Syntheses in the Chemistry Laboratory

**Larry Peck**

Texas A & M University

**Kurt J. Irgolic**

Karl-Franzens-Universität Graz

Copyright © 1992 by Macmillan Publishing Company, a division of Macmillan, Inc.

Printed in the United States of America

Macmillan Publishing Company
866 Third Avenue, New York, New York 10022

Macmillan Publishing Company is part
of the Maxwell Communication Group of Companies.

Maxwell Macmillan Canada, Inc.
1200 Eglinton Avenue East
Suite 200
Don Mills, Ontario M3C 3N1

ISBN 0-02-359835-2

Printing: 1 2 3 4 5 6 7 8         Year: 2 3 4 5 6 7 8 9 0 1

# Contents

# To The Student

You are at the beginning of a two-semester journey through the fundamentals of chemistry. Chemistry has a theoretical and a practical side. In lecture the theoretical aspects will be stressed and, perhaps, accentuated with demonstrations. The beauty of laws, theories, hypotheses, and ideas will be laid out for you in eloquent ways. However, the "real" chemistry is experienced in the laboratory where elements, compounds, solutions, mixtures, solids, liquids, and gases are at your disposal. Chemistry is work with substances: the determination of their composition and their properties; the exploration of their transformation into other substances. Chemists and students of chemistry must be familiar with substances (with chemicals). They must know the appearance of substances, their physical state, their color, their odor, their toxic properties, how to handle them, how to store them, and how to dispose of them.

All this knowledge is best obtained in the laboratory. Chemists perform certain operations in their study of substances. Chemists weigh solids, prepare solutions, measure volumes of liquids, determine the pressure of gases, mix reagents, heat reaction mixtures under reflux, separate solids from liquids by filtration or centrifugation, separate and purify liquids by distillation, purify solids by recrystallization, determine the melting points of crystalline substances, measure temperatures, and establish the light-absorbing properties of solutions. In many of these operations chemists use instruments such as analytical balances, centrifuges, spectrophotometers, and filtration devices. To perform chemical operations efficiently and safely, the practitioner of the chemical arts must have acquired certain skills. Just as you never will be able to change the water pump on your car if you don't know anything about screws and wrenches or repair a tear in your pants if you can't locate the thread reservoir in a sewing machine, you will never be able to work with chemicals and understand and appreciate chemistry without the skills you learn in the laboratory. Thirty-five experiments are included in this laboratory manual. A variety of techniques and substances are used in these experiments. Each experiment consists of several sections.

The *Introduction* presents background information about the experiment, connects the subject matter to everyday experiences, discusses applications of the experimental procedure or the substances used, and – in general – attempts to raise your interest in the experiment.

The *Concepts of the Experiment* section provides the theoretical framework for the experiment. Chemical equations for reactions are presented and discussed; mathematical equations are derived that are needed for the treatment of the experimental data, and advice is given to avoid mistakes during the performance of the experiment. To obtain the greatest benefit from an experiment, you must study this section in detail before coming to the laboratory.

The *Activities* section summarizes the work you will have to perform during the laboratory period. You can use this summary as a guide for the study of the operational aspects of the experiment and as a checklist while you work in the laboratory.

The *Safety* section identifies potential hazards you might encounter during your work in the laboratory. Pay close attention to the statements in this section. If you have any questions about safe procedures, proper handling of chemicals or use of instruments, ask your instructor for advice. Keep in mind that it is easier to answer questions and avoid accidents than to treat injuries and repair equipment.

The *Procedures* section gives a detailed, step-by-step description of each operation needed for the experiment. The steps are numbered. The text provides the detailed directions about the quantities of substances to be used, what to do with them, and what to watch out for. Drawings on the left part of the page illustrate the text and show how equipment should be assembled, how solutions should be mixed, how thermometers should be read, how filtrations should be performed, etc. One picture is said to be worth 1,000 words. Use these wordy, pictorial help sources to your advantage.

The first tear-out page of each experiment has the title *PreLab Exercises*. The work requested must be done and the form completed before coming to the laboratory. The questions and problems in these exercises relate to the experiment and will give you an opportunity to test your knowledge about the subject matter of the experiment and calculations associated with it. Should you have problems with the problems, take the textbook and laboratory manual and study the pertinent topics. If you still have problems, obtain help.

The second set of tear-out pages constitutes the *Report Form*. This form must be completed after the experiment has been finished. Do not keep notes on this form. The completed Report Form will be graded by your instructor.

While you do the experiments, observations, masses, buret readings, and many other data must be recorded. These records must be entered into a permanently bound notebook. Loose pages have a tendency to disperse into the environment with a guarantee of unavailability at the time the Report Form has to be completed. Generally, notebook records can never be detailed enough. Your instructors will give you their recommendations about the proper way of keeping a good notebook.

The laboratory manual has several *Appendices*. These appendices describe common laboratory operations such as weighing, filtration, and working with burets. One of the appendices provides an introduction to the statistical treatment of quantitative data, a topic of great importance. Refer to these appendices whenever you need more information than is given in the *Procedures* for an experiment.

The experiments that you will be conducting should introduce you to the world of chemistry, teach you laboratory skills, let you experience the joy of having successfully synthesized a substance, found the composition of an ore sample, and checked the effectiveness of an antacid. Often, students approach chemistry with great trepidation, with a fear of finding the subject too difficult. Chemistry is no more difficult than many other subjects you will study. Approach chemistry with an open mind and you will learn exciting things and even have fun. Naturally, chemistry is not everyone's favorite. Should chemistry not be your favorite subject, please don't consider the lectures and the laboratory to be an abominable obligation but look upon them as a unique opportunity to learn more about the chemical processes that pervade animate and inanimate nature.

We wish you much success in your studies.

M. Larry Peck
College Station, TX

Kurt J. Irgolic
Gras, Austria
1991

# Laboratory Safety

A chemical laboratory is not necessarily the safest of all places in which to work. Neither, however, is it something of which one has to be afraid. With proper understanding of what you are doing, careful attention to safety precautions, and adequate supervision, you will find the chemical laboratory to be a safe place in which you can learn much about chemistry.

Laboratory accidents belong to two general categories of undesirable events: mishaps caused by your own negligence and accidents beyond your control. Although accidents in the laboratory fortunately are rather rare events, you nevertheless must be familiar with all safety rules and emergency procedures. If you know and follow safe working practices, you will pose no threat of serious harm to yourself or others.

Every laboratory system will have some unique features for the prevention of accidents or for handling emergencies. It is important that you become thoroughly familiar with the special safety aspects of your own laboratory area. Some general precautions and procedures applicable to any chemical laboratory are summarized below.

**SAFETY**

1. **WEAR APPROVED EYE PROTECTION AT ALL TIMES.** Laboratory accidents, such as splattering of ammonia and explosions of small glass containers have caused permanent eye damage. This damage could have been prevented by wearing simple laboratory goggles.

2. **NEVER EAT, DRINK, OR SMOKE IN A CHEMICAL LABORATORY.** Tiny amounts of some chemicals may cause toxic reactions. Many solvents are easily ignited by a lighted cigarette.

3. **NEVER WORK IN A CHEMICAL LABORATORY WITHOUT PROPER SUPERVISION.** Your best protection against accidents is the presence of a trained, conscientious supervisor, who is watching for potentially dangerous situations and who is capable of handling an emergency properly.

4. **NEVER PERFORM AN UNAUTHORIZED EXPERIMENT.** Many "simple" chemicals form explosive or toxic products when mixed.

5. **NEVER PIPET BY MOUTH AND NEVER INHALE GASES OR VAPORS.** Always use a mechanical suction device for filling pipets. Reagents may be more caustic or toxic than you expect. If you must sample the odor of a gas or vapor, use your hand to waft a small sample toward your nose.

6. **EXERCISE PROPER CARE IN HEATING OR MIXING CHEMICALS.** Be sure of the safety aspects of every situation in advance. For example, never heat a liquid in a test tube that is pointed toward you or another student. Never pour water into a concentrated acid. Proper dilution technique requires that the concentrated reagent is slowly poured into water while stirring to avoid localized overheating.

7. **BE CAREFUL WITH GLASS EQUIPMENT.** Cut, break, or fire polish glass only by approved procedures. When inserting a glass rod or tube through a rubber or cork stopper, lubricate the glass and the stopper, protect your hands with a cloth towel, and use a gentle twisting motion.

**EMERGENCY PROCEDURES**

1. **KNOW THE LOCATION AND USE OF EMERGENCY EQUIPMENT.** Find out where the safety showers, eyewash spray, and fire extinguishers are located. If you are not familiar with the use of emergency equipment, ask your instructor for a lesson.

2. **DON'T UNDER-REACT.** Any contact of a chemical with any part of your body may be hazardous. Particularly vulnerable are your eyes and the skin around them. In case of contact with a chemical reagent, wash the affected area *immediately* and *thoroughly* with water and notify your instructor. In case of a splatter of chemical over a large area of your body, don't hesitate to use the safety shower. Don't hesitate to call for help in an emergency.

3. **DON'T OVER-REACT.** In the event of a fire, don't panic. Small, contained fires are usually best smothered with a pad or damp towel. If you are involved in a fire or serious accident, don't panic. Remove yourself from the danger zone. Alert others of the danger. Ask for help immediately and keep calm. Quick and thorough dousing under the safety shower often can minimize the damage. Be prepared to help, calmly and efficiently, someone else involved in an accident, but don't get in the way of your instructor when he or she is answering an emergency call.

These precautions and procedures are not all you should know and practice in the area of laboratory safety. The best insurance against accidents in the laboratory is thorough familiarity and understanding of what you're doing. Read experimental procedures before coming to the laboratory, take special note of potential hazards and pay particular attention to advice about safety.

Take the time to find out all the safety regulations for your particular course and follow them meticulously. Remember that unsafe laboratory practices endanger you and your neighbors.

If you have any questions regarding safety or emergency procedures, discuss them with your instructor. Then sign and hand in the following safety agreement.

# Safety Agreement

I studied, understood, and agreed to follow the safety regulations required for this course. I located all emergency equipment and now know how to use it. I understand that I may be dismissed from the laboratory for failure to comply with stated safety regulations.

Signature_____

Print name_____

Date_____

Course_____

Section_____

Instructor_____

Person(s) who should be notified in the event of an accident:

_____

_____

_____

_____

May we post your grade in this course at mid–semester and again at the end of the semester?

yes____          no____

Signature_____

Local Address                    Permanent Address

_____          _____

_____          _____

_____          _____

Phone Number_____        Phone Number_____

# Investigation 1

## TRUTH IN LABELING - A BALANCE EXPERIMENT

### INTRODUCTION

Chemistry has a long history. Several thousand years ago man learned to smelt sulfidic ores and to reduce metal oxides to metals. Periods of human history were named after the products of such reduction processes: the Bronze-age after the alloy that results when mixtures of copper oxides and tin oxides are reduced, and the Iron-age after the metal that is formed from iron(III) oxide. The fermentation of sugars is an ancient chemical art, and the product of this enzyme-catalyzed reaction, ethanol – commonly called alcohol ($C_2H_5OH$) – is still a much cherished constituent of alcoholic beverages of dubious value. The harnessing of fire led to the invention of cooking, a very beneficial chemical process. During the medieval ages alchemists were working hard and very secretly to discover the "stone of the wise" that would change metals into gold and the "elixir of life" that – taken according to recommendations – would keep a person eternally young. The alchemists never succeeded in achieving their goals; instead they learned much about chemistry and discovered, for instance, methods for the preparation of nitric acid and sulfuric acid. Most of these experiments were carried out in a qualitative manner; little attention was given to the amounts of the substances that were reacted or the amount of product obtained.

In 1789 the French chemist Antoine Lavoisier (1742-1794) published his text in which he stressed the quantitative aspects of chemistry. The subsequent use of balances in chemical investigations advanced chemical knowledge tremendously, led to the formulation of the laws of constant composition and multiple proportions, and was ultimately responsibie for Dalton's atomic theory. Since then, the Lavoisier balance has been an indispensable laboratory tool.

This experiment will familiarize you with the proper use of analytical balances.

### CONCEPTS OF THE EXPERIMENT

An important extensive property of a piece of matter is its mass, in every day language called "weight." In physics and chemistry mass and weight are not the same. Weight is the force exerted by a piece of matter on the support it is resting on. This force (weight) is caused on earth by the gravitational attraction between earth and the piece of matter. In other environments, for instance, the moon, in which the gravitational attraction is smaller than on earth, a piece of matter will have a smaller weight than on earth. In a "weightless" environment, for instance in an orbiting space station, the gravitational attraction is zero and all objects in the station have zero weight; however, their mass is not zero. The property "mass" of an object is a measure of the amount of matter and is independent of any gravitational interactions. The unit of mass in the International System of Units is the kilogram (kg), the unit of weight (force) the Newton (kg m s$^{-2}$). Mass and weight (force) are connected (equation 1) by a proportionality constant "g", the gravitational acceleration.

$$F = m \times g \qquad (1)$$

On the surface of the earth "g" has the value 9.81 m s$^{-2}$. Therefore, an object with a mass of 1 kg has a weight of 9.81 Newtons. (1 Newton = 1kg x 1m/1s$^2$)

1

In chemical experiments the amounts of matter of reacting substances is of importance and not their weights. Therefore, the determination of masses is one of the fundamental operations of chemistry. Because languages do not develop according to scientific principles, the process of determining masses is called "weighing" and not "massing." The mass of an object is determined (the object is weighed) on a balance. The weighing process compares the mass of an object with mass standards that are supplied with each balance. Several types of balances are used in a chemical laboratory. These balances differ with respect to the smallest mass that can be reliably determined and the largest mass that can be placed on them. The most common types of analytical balances and their principles of operation are described in Appendix B. Your instructor may provide you with a handout containing information about the particular type of balance in your laboratory. Most balances operate on the lever-principle and exploit the gravitational attractions between earth and the mass standards and between earth and the object to be weighed. Common laboratory balances will not work in a "weightless" environment.

In this experiment an "analytical" balance will be used. An "analytical" balance allows the mass of an object to be determined to 0.1 milligram. One milligram is one-thousandth of one gram and one-millionth of one kilogram. The largest mass an analytical balance can handle (called the maximal load) is generally 159 grams. Check with your instructor to learn the maximal load of the balances in your laboratory. To provide you with the opportunity to become proficient in determining masses (in weighing) on the analytical balance, four small packages of sugar, artificial sweetener, or salt will be weighed to 0.1 milligram. Then the packages will be opened, their contents quantitatively transferred into a small, preweighed beaker, the package materials weighed, and the combined contents weighed. From the masses of the packages, the packaging materials, and the contents, you will be able to ascertain whether or not the manufacturer delivers on the average the amount of sugar, artificial sweetener, or salt stated on the package. Take this opportunity to learn the proper way of determining masses. What you learn in this experiment will benefit you throughout this course and allow you to determine masses confidently and quickly.

## ACTIVITIES

- Before coming to the laboratory study how analytical balances work and how masses are determined (Appendix B).

- Weigh four small packages to 0.1 mg.

- Weigh a small beaker to 0.1 mg.

- Empty the four packages; collect their contents in a small beaker.

- Weigh the empty packages to 0.1 mg.

- Weigh the small beaker with the contents of the four packages to 0.1 mg.

- Calculate the mass of the contents of each package.

- Calculate the average mass of the contents of the four packages from the differences (mass of package – mass of packing material). Express this average in significant figures only.

- Calculate the average mass of the contents from the difference [(mass of beaker plus contents) – mass of beaker]. Express the average in significant figures.

- Compare the two averages with the mass of the contents printed on the packages.

## SAFETY

*Do not taste any of the products used in this or any other experiment. They may have been contaminated by storage in the chemical stockroom. If you are working in a laboratory in which chemical reagents are set out, wear approved eye protection.*

## IMPORTANT GUIDELINES FOR THE USE OF ANALYTICAL BALANCES

Analytical balances are very likely the most expensive instruments you will use in the introductory chemistry laboratory. Analytical balances are delicate instruments and must be used with care. The following guidelines will help to keep the balances clean, operating properly, and providing correct masses.

- Keep the balance area spotlessly clean. If the area is not clean (spilled chemicals, paper), clean it.
- Keep the pan compartment of the balance clean. If the pan compartment is not clean, notify your instructor and clean the compartment according to your instructor's recommendation.
- Keep the balance doors closed. The doors are opened only when adding or removing an item to be weighed.
- Never place a chemical directly on the pan. Use a weighing paper, watchglass or beaker to hold the chemical.
- Always re-zero a balance before using it.
- Add and remove items from the pan very gently. On mechanical balances arrest the pan before removing or adding an item. However, it is proper to add small increments of chemicals to achieve a certain mass with the arrest knob in the "release" position.
- Read the scale in the light window with your eyes at the height of the indicator gap or line to avoid parallax errors.
- Leave the balance spotlessly clean with the doors closed and all dials at zero.
- Consult your instructor when you encounter difficulties with a balance. Do not attempt to "fix" a balance.
- Never move a balance (not even a short distance).

## PROCEDURES

1. Obtain four small packages of a product from your instructor. Shake the contents into the upper half of the packages. Number the packages by writing with pencil "1", "2" ... on the lower half of the packages.

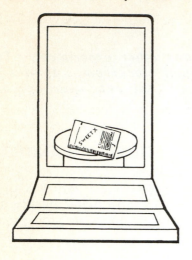

**2.**  Weigh each of the four packages on an analytical balance to 0.1 mg.  Follow the guidelines for mass determinations appropriate for the type of analytical balance in your laboratory.  Record the masses of the packages and identify the balance.  You must use the same balance for all your weighings.

**3.**  Weigh a clean and dry 50-mL or 100-mL beaker.  Record the mass of the beaker.

**4.**  Shake the contents of your four packages to the lower third of the package.  Open each package by tearing off the top.  Carefully empty the contents of each package into the weighed beaker.  Do not lose any of the contents by spillage.  Make sure that none of the contents remains in the package.  Put the torn off piece into the package.

**5.**  Weigh each of the four empty packages and the beaker with the contents of the four packages on the analytical balance.  Record the masses.

**6.**  Calculate the mass of the product in each of the four packages as the difference between the mass of (package + content) and the mass of the packaging material.

**7.**  Weigh the beaker plus the contents of your four packages.  Calculate the average mass of the contents of one package from the results of the weighings of the beaker and the beaker containing the contents from four packages.

**8.**  Express the two average masses of the contents of the packages using significant figures only.

# EXPRESSING THE RESULTS OF THIS EXPERIMENT WITH SIGNIFICANT FIGURES

The numerical results of an experiment must always be expressed in significant figures only. Significant figures are identified with the help of the standard deviation calculated from the results of a series of measurements. The exact method for the identification of significant figures is described in Appendix H and is introduced in Investigation 2. For this experiment the goal is to familiarize you with the use of an analytical balance so a less rigorous method for the identification of significant figures will be used.

When you inspect the masses of your four packages, you will notice that the individual masses cluster around a certain value. For instance, the contents of your four packages could have the following measured masses:

   1.0644 g            1.0851 g         0.9927 g         1.0016 g
     ↑↑                   ↑↑               ↑↑               ↑↑

These masses cluster around one gram. Because the tenths (0.1 g) and the hundredths (0.01 g) digits change from sample to sample, the digits in the remaining decimal places are not very meaningful and are not significant. (The difference between 1.00 and 0.99 is 0.01. Therefore, the third value in the set above is considered to vary in the hundreths digit.) The third and the fourth decimal places in these results should be dropped, because the first two decimal places represent the first two digits that vary. In the data treatment being used for this experiment, the average should be calculated first and then the average should be rounded to the second digit that varies.

The calculated average is 1.03595 g, rounded to 1.04 g.
                           ↑↑

If the calculated average had been: 1.03495 g, it would have been rounded to 1.03 g.
                                       ↑↑

If all the packages weigh less than one gram, in your final result keep the digits that do not change in your individual masses and the next two that do change; for example:

   0.5231 g   0.5110 g   0.5301 g,   0.5095 g      Average: 0.518425 g      Rounded to: 0.518 g
     ↑↑         ↑↑         ↑↑           ↑↑

---

RULES FOR ROUNDING:  If the digit is 0, 1, 2, 3 or 4, drop it and all other digits following:

1.03495; round it to 1.03. If the digit is 5, 6, 7, 8 or 9, increase the digit to the left by one and
  ↑↑

drop it and all digits to the right: 1.03595; round it to 1.04.

# PRELAB EXERCISES

## INVESTIGATION 1: TRUTH IN LABELING - A BALANCE EXPERIMENT

Name:_____

Instructor:_____          ID No.:_____

Course/Section:_____      Date:_____

1. Define mass.


2. In the International System of Units the unit of mass is: _____.


3. What do the prefixes milli and kilo signify?


4. Calculate the average of the following numbers.

    1.5532 g, 1.5478 g, 1.5163 g, 1.5601 g


5. The above numbers each contain 5 significant figures but their average should be rounded to only 4 significant figures. Why?


(over)

6. When may the doors be open on the analytical balance?

7. Give 3 guidelines for proper balance use that you can now add to those given before the Procedures section of this experiment.

8. How might someone leaning on the table near a balance affect its reading?

9. What is the weight, in Newtons, of a 1.0 gram sample on the surface of the earth?

# REPORT FORM

### INVESTIGATION 1: TRUTH IN LABELING - A BALANCE EXPERIMENT

Name: _____

Instructor:_____     ID No.: _____

Course/Section:_____     Date:_____

---

Balance No: _____     Product name: _____     Label mass: _____ g

Complete the table below from your notebook records.

Submit the calculations on a separate page.

| Package No. | Mass (g) | | |
| --- | --- | --- | --- |
| | Package with content | Package without content | Content |
| 1 | | | |
| 2 | | | |
| 3 | | | |
| 4 | | | |

1. Average mass of content: _____ g (from Table)
2. Average mass (significant figures only): _____ g
3. Mass of beaker + contents from four packages: _____ g
4. Mass of beaker: _____ g
5. Mass of contents: _____ g
6. Average mass of contents: _____ g
7. Average mass with significant figures only: _____ g
8. "Beaker average" – "individual package average" (#7 – #2): _____ g
9. Mass of content from label – "individual package average": _____ g

9

Is the "label" mass true? On the average, did the manufacturers fill the packages just right? Did they overfill? Did they add less than claimed on the label? If a ±5% variation is acceptable, is the "label" mass true?

Answer any assigned problems on an attached sheet.

Date _____ Signature _____

# Investigation 2

## PRECISION - ACCURACY - CALIBRATION

### INTRODUCTION

We live in a "numerical" society. Numbers are used for the identification of persons (Do you know your social security number?), are needed when driving a car (Remember the 55 miles per hour speed limit.), when shopping (Did you get the correct change?), when planning meetings (How many people will a room accommodate?), when completing applications (your age, your zip code), and when managing your finances (Have you succeeded in balancing your checkbook?). Some numbers are exact; however, most of the numbers we use are not exact. Exact numbers are generally limited to arbitrarily defined conversion factors (12 inches per foot, 1000 meters per kilometer) and to results obtained by counting discrete items (11 players on the football field, 326 jelly beans in the candy jar). Numbers generated by measuring a property of a piece of matter (mass, volume, temperature, electrical resistance, ...) with a measuring device (balance, graduated cylinder, thermometer, ohm-meter, ...) are inexact to varying degrees because of limitations inherent in instruments and errors associated with the human experimenter. Even numbers obtained by counting will become inexact when the number of items to be counted is very large or when the number of items changes during the counting process. For instance, when a large jar filled with 28,938 jelly beans is given in turn to several persons with the assignment to count the jelly beans, chances are very good that none of the numbers will be the same. The human counting instrument becomes imprecise because the counter becomes tired, may skip a number from time-to-time, or may advance the count by more units than warranted by the transfer of jelly beans from the uncounted to the counted pile. Although the number of living persons in the United States is certainly an exact number, nobody knows this exact number. No reasonable (in terms of expense, people involved, time expended) counting method will provide this exact number because as the counting proceeds the number changes (babies are born, people die, immigrants arrive).

An inexact number obtained by a measurement is characterized by its precision, expressed as its standard deviation or its relative standard deviation, and by its accuracy, expressed as the difference between the exact number (the true value) and the inexact number (the result of measurements). "Accuracy" and "Precision" are discussed in Appendix H. The result of a correctly performed measurement should be precise and accurate. Good precision (low standard deviation) does not necessarily imply accuracy. The precision of a measurement can always be determined. The measurement is repeated n-times, the "n" results are averaged to give the average or mean (equation 1). The standard deviation (equation 2) and the relative standard deviation (equation 3) can then be calculated.

$$\overline{X} = \frac{X_1 + X_2 + X_3 + \ldots + X_n}{n} \quad (1)$$

$$S = \sqrt{\frac{\sum_{1}^{n}\left(X_i - \overline{X}\right)^2}{n-1}} \quad (2)$$

$$RSD = \frac{100S}{\overline{X}} \quad (3)$$

$X_1, X_2 \cdots X_n$ : results of repeated measurements

$n$ : number of measurements

$\overline{X}$ : average or mean

$S$ : standard deviation

$RSD$ : relative standard deviation

Accuracy is quantitatively expressed as the difference between the true value, T, and the average, $\overline{X}$ (equation 4).

$$\text{Accuracy} = T - \overline{X} \tag{4}$$

The goal of every measurement is a number as precise and accurate as demanded by the situation. To work for higher accuracy and precision than necessary is wasteful of time, money, and facilities. For instance, when you buy a watermelon, for which you pay by the pound, weighing the watermelon to ±0.1 g would make no sense. When you order six cubic yards of topsoil for your garden, you certainly don't care whether several more or fewer cubic feet were delivered. You won't measure the volume of the topsoil to the liter. However, when you buy a valuable gemstone, the jeweler will determine its mass very likely to ±0.1 milligram. To achieve the required precision and accuracy the job must be done right: with the correct measuring instrument, the appropriate measuring procedure, and always with the requisite care by the experimenter. The experimenter must be able to make the right choices for each job.

The number of reportable significant figures for an experimental set of data is determined by consideration of the calculated standard deviation. Standard deviation is to be reported with only one significant figure and the average value (or mean) must be reported (rounded, if necessary) so that it contains only the precision (same last digit) as that indicated by the standard deviation.

In this experiment the determination of the accuracies of common laboratory "instruments" for measuring volumes will be used to make you familiar with the concepts of precision, accuracy, standard deviation, and relative standard deviation, and provide experience for selecting appropriate measuring devices.

## CONCEPTS OF THE EXPERIMENT

Beakers, with graduation marks, graduated cylinders, and burets are very frequently used in the laboratory to measure volumes of liquids and liquid solutions. Manufacturers provide these volume-measuring devices with graduation marks and numbers indicating 0.1 mL, 10 mL and similar volumes. A careful experimenter will check how accurate these marks are. Such a checking procedure is called "calibration" and is an important step in any measuring procedure. To calibrate, for instance, a graduated beaker, a volume "standard" is needed. For our purposes a standard is a piece of matter, a property of which (in our case the volume) is well known and preferably certified by an organization such as the U.S. National Institute for Standards and Technology. In this experiment distilled water at room temperature is used as the volume standard. The density of distilled water at room temperature is 0.9973 g mL$^{-1}$ (a value at a temperature other than 24°C can be found in the table in Investigation 3). When a beaker is filled with distilled water, for instance, to the 80-mL mark, the mass of the water in the beaker can be determined by weighing with a triple-beam balance to ±0.01 g or with an analytical balance to ±0.1 mg. From the mass of the water and its density the volume can be calculated. Repeated filling of the beaker and weighings produce several results, for which the average, the standard deviation, and the relative standard deviation are calculated.

The result of a measurement, an "inexact" number, must always be given in such a manner, that the last digit in the number is doubtful. A number written in that way is said to have "significant figures" only. Which one of the digits in a number is the doubtful one is determined by the magnitude of the standard deviation. This magnitude expresses also the degree of inexactness of the number. Significant figures are discussed in Appendix H.

As an example, the average mass of water needed to fill a beaker to the 60-mL mark was determined with an analytical balance to be 68.3550 g. The standard deviation from four measurements was ±0.783 g. The result is already uncertain in the 0.1 g digit. Therefore, the result must be given as 68.4 ±0.8 g after appropriate rounding of the numbers. When the same experiment is carried out with the triple-beam balance, with which masses can be determined to 0.01 g, the same result would have been obtained with a much less expensive and delicate instrument. Although the analytical balance gave us four significant digits to the right of the decimal point for each weighing, we had to discard three of these digits when the average was expressed showing significant figures only. Why work to obtain the additional digits, when their ultimate fate is discard? Proper choice of instruments prevents unproductive work.

The accuracy of the marks on beakers, graduated cylinders, burets and other volume-measuring devices can be expressed as the difference between the true value (obtained from the weighings of water) and the manufacturer's label.

In these experiments the inexactness of the results are caused by errors associated with the instruments (the balances, the beaker) and with the experimenter. Analyze your procedure to determine whether the instrument or the experimenter is responsible for the inexactness of your results. The standard deviations associated with the various weighings will help you make these decisions.

## ACTIVITIES

- Weigh a 100-mL beaker five times on a triple-beam balance and five times on an analytical balance.

- Repeat the weighings using 40.00 mL of water obtained from a buret.

- Calculate averages, standard deviations, and relative standard deviations.

- Report values using only significant figures and use your results to discuss accuracy and to plan other experiments.

## SAFETY

*Always be aware that sharp edges will exist should any piece of glass equipment break.*

## PROCEDURES

## A. CALIBRATION OF A GRADUATED BEAKER

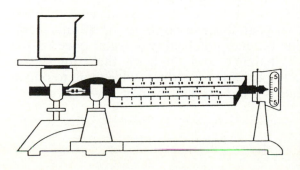

**A-1**. Obtain a clean and dry 100-mL beaker with graduation marks to 80 mL. Weigh this beaker four times on the triple-beam balance and four times on the analytical balance. To prevent observer bias, move the 1-g rider on the triple-beam balance to zero after each weighing and do not look at the rider when you return the balance to equilibrium. You may want to block your line of sight with one of your hands or a piece of paper. On the analytical balance arrest the pan after each weighing, take the beaker from the pan, turn the 0.1-g dial to zero (if possible), return the beaker to the pan, and determine the mass of the beaker. *Use the same balances throughout this experiment.*

**CAUTION**: *Exercise extra care when adding or removing items from the pan of a balance, when adding or removing mass standards from the balance, and when turning the mass selection knobs on the analytical balance.* Appendix A discusses the use of the triple-beam balance, Appendix B the use of the analytical balance.

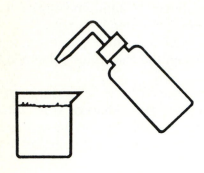

**A-2.** Carefully fill the beaker to the 80-mL mark with distilled water. Use your filled squeeze bottle as the source of water. When the water level approaches the 80-mL mark, add the water drop-wise until, in your judgment, the mark has been reached.

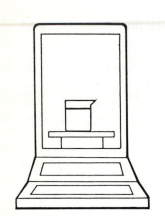

**A-3.** Weigh the beaker filled with water to the 80-mL mark on the analytical balance. Record the mass to 0.1 mg. Leave the filled beaker on the pan for 2 minutes and record the mass every 15 seconds for a total of 8 readings. You will have to explain your observation. Arrest the balance, take the beaker from the pan, turn all mass selection knobs to zero, and close the doors of the balance. (Use the average of these 8 readings as the first reading on the analytical balance in procedures 4 and 5.)

**A-4.** Take the beaker filled with water to the 80-mL mark to the triple-beam balance and determine the mass of the beaker + water. Then remove the beaker from the pan and return all mass-sliders to zero.

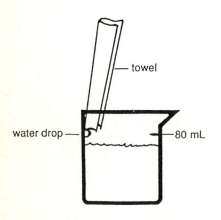

towel

water drop — 80 mL

**A-5.** Pour some water from the beaker to drop the level to approximately the 70-mL mark. Remove all drops of water from the outside of the beaker and from the inside wall above the 80-mL mark with a piece of towel. Fill the beaker again with distilled water from your squeeze bottle to the 80-mL mark. Determine the mass of beaker + water once on the analytical balance and once on the triple-beam balance. Repeat this procedure twice more to obtain four values for the mass of beaker + water on each balance.

**A-6**. Calculate the average mass of the beaker from the four masses obtained on the analytical balance and from the four masses obtained on the triple-beam balance. Calculate the standard deviations and the relative standard deviations for the two averages. Then express the averages using only significant figures. Compare the two averages.

Perform similar calculations for the water samples and determine the accuracy of the 80-mL mark. Complete the appropriate section in the Report Form.

# B. CALIBRATION OF A 50-mL BURET

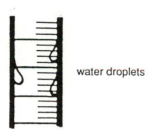

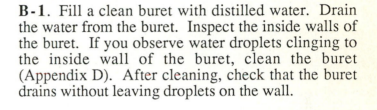

water droplets

**B-1**. Fill a clean buret with distilled water. Drain the water from the buret. Inspect the inside walls of the buret. If you observe water droplets clinging to the inside wall of the buret, clean the buret (Appendix D). After cleaning, check that the buret drains without leaving droplets on the wall.

**B-2**. Clamp a cleanly draining buret to a buret stand. Fill the buret with distilled water. Drain water from the buret until the level has dropped below the zero mark. Read the buret and record the mark just touched by the bottom of the meniscus (Appendix D). Estimate the position of the meniscus to 0.01 mL. Obtain a clean and dry 50-mL or 125-mL Erlenmeyer flask and a tight fitting stopper. Weigh the flask with the stopper on the analytical balance to 0.1 mg and on the triple-beam balance to 0.01 g. Then drain as nearly as practical 40.00 mL of water into the flask. Make sure that water is not deposited on the neck of the flask. Before you read the meniscus again, wait one minute to allow water on the walls of the buret to drain. Should your buret reading indicate that you drained less than 40.00 mL of water, allow a few drops of water to flow into the flask until nearly the desired 40.00-mL volume has been transferred into the flask. Stopper the flask and determine the mass of flask + water + stopper on the analytical balance to 0.1 mg. Record all mass readings. Then take the flask to the triple-beam balance and determine the mass of flask + water + stopper.

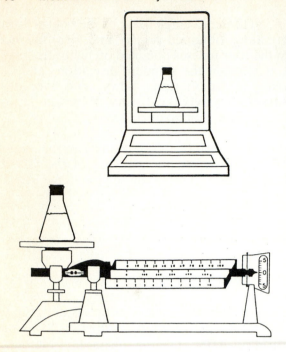

**B-3**. Empty the flask and dry it completely. It is essential that the flask is completely dry. Fill the buret with distilled water and drain 40.00 mL into the flask. Stopper the flask and determine the mass of the flask + water + stopper on the analytical balance and on the triple-beam balance. Repeat this procedure one more time. You should now have three mass values for flask + water + stopper from the analytical balance and three from the triple-beam balance. The mass of the empty flask and stopper was determined only once but it is to be recorded three times on the Report Form.

**B-4**. Express the mass of flask + stopper using only significant figures. Use the standard deviations calculated for the beaker in Procedure A6. Calculate the masses of the two sets of three 40.00 mL-samples, find the averages, the standard deviations, and the relative standard deviations.

## C. PLAN FOR THE CALIBRATION OF OTHER VOLUMETRIC GLASSWARE

**C-1**. A 25-mL graduated cylinder has to be calibrated. Inspect the graduation marks on such a cylinder, consider the capabilities of triple-beam balances and analytical balances and your ability to estimate the smallest volume unit in such a cylinder. Consider a plan for the calibration that would include
  • smallest volume unit that can be estimated.
  • total volume of water to be used.
  • precision of balance required to be responsive to the smallest volume unit that can be estimated.
  • the uncertainty in terms of a relative standard deviation of the calibration.

**C-2**. Repeat this planning exercise for a 5-mL microburet with graduation marks of 0.01 mL.

## CALCULATIONS

The values, which you determined several times, must be averaged and the standard deviations and relative standard deviations calculated. This task is accomplished most efficiently with a calculator that has statistical programs. However, it is strongly recommended that you treat one of the data sets "manually" using the formulas given in the introduction.

The mass determinations you performed in this experiment do not directly give the volumes that are needed for calibration. The masses of the water must be converted to volumes using the density of water given in the "Concepts of the Experiment" section.

Express the accuracies of the calibration marks as the difference between the volumes given by the graduation marks and the volumes determined in this experiment.

# PRELAB EXERCISES

## INVESTIGATION 2: PRECISION - ACCURACY - CALIBRATION

Name:_____

Instructor:_____   ID No.:_____

Course/Section:_____   Date:_____

1. Define each of the following:

    Accuracy

    Precision

    Density

2. Explain the distinction between a measured value and an exact number.

3. How many significant figures in each of the following measured values? (Assume that the last digit shown is uncertain by ± 2 units.)

    a. 150.0

    b. 1.50

    c. 100

    d. $1.0 \times 10^2$

    e. 0.0025

    f. 1,000

4.    Given that the correct value is 1.51, which of the following values is the most precise and which one the most accurate? (Assumption: last digit has an uncertainty of ± 1.)

   a. 1.1831

   b. 1.15

   c. 1.215

5.    A flask was repeatedly weighed on an analytical balance. The following masses were obtained: 50.7727 g, 50.7730 g, 50.7733 g, 50.7735 g, 50.7736 g. Calculate the average, the standard deviation, and the relative standard deviation. Express the average using significant figures only.

# REPORT FORM

## INVESTIGATION 2: PRECISION - ACCURACY - CALIBRATION

Name: _____ ID No.: _____

Instructor: _____ Course/Section: _____

Partner's Name (if applicable): _____ Date: _____

A. Graduated beaker.

| Weighing # | Triple-beam Balance | | | Analytical Balance | | |
|---|---|---|---|---|---|---|
| | Empty Beaker | Beaker & Water | Water | Empty Beaker | Beaker & Water | Water |
| 1 | | | | | | |
| 2 | | | | | | |
| 3 | | | | | | |
| 4 | | | | | | |
| Average | | | | | | |
| Stand. Dev. | | | | | | |
| Rel. Stand. Dev. | | | | | | |
| Average* | | | | | | |
| Volume(mL)** | /////////////////////////// | | | /////////////////////////// | | |

\* Significant figures only.          ** Calculated from significant-figures-only-average.

Compare and comment on the triple-beam and the analytical balance values for average mass and volume of water in the beaker:

Masses of Beaker & Water read every fifteen seconds (procedure A-3):

_____  _____  _____  _____  _____  _____  _____  _____

Explain the results obtained for the weighings over the 2 minute period.

Calculate the accuracy of the 80-mL graduation mark (average volume − 80.0000):

       Accuracy (triple-beam results): _____

       Accuracy (analytical balance results): _____

Manufacturers claim that the graduation marks on the beakers are accurate within ± 5 percent. Calculate the percent accuracy from your results and compare it with the manufacturer's claim.

       Manufacturer's accuracy:     ± 5% 

       Accuracy from your data: _____

## B. Buret

| Trial | Initial Volume Reading | Final Volume Reading | Volume of $H_2O$ |
|---|---|---|---|
| 1 | | | |
| 2 | | | |
| 3 | | | |

Average _____

| Weighing # | Triple-beam Balance | | | Analytical Balance | | |
|---|---|---|---|---|---|---|
| | Flask & Stopper | Flask & Stopper & ~40 mL | Mass of ~40 mL | Flask & Stopper | Flask & Stopper & ~40 mL | Mass of ~40 mL |
| 1 | | | | | | |
| 2 | | | | | | |
| 3 | | | | | | |
| Average | | | | | | |
| Stand. Dev. | * | | | * | | |
| Rel. Stand.Dev. | | | | | | |
| Average** | | | | | | |
| Volume(mL)*** | ///////////////////////////////// | | | ///////////////////////////////// | | |

\* Use standard deviation found for empty beaker in Part A.
\*\* Expressed with significant figures only.
\*\*\* Calculated from mass and density data using significant figures only.

Compare and comment on the triple-beam and the analytical balance averages for the experimentally determined value of the 40.00-mL water sample.

Calculate the accuracy of the 40.00-mL buret volume (average volume from mass values − average volume given by buret readings)

Accuracy (triple-beam results): _____

Accuracy (analytical balance results): _____

Manufacturers claim that the volume delivered by a buret (40.0 mL) is accurate to 0.1 mL. Compare your accuracies with the manufacturers' claimed accuracy.

Compare the precision of the individual volumes (equal to the standard deviation of mass values for the analytical or triple-beam balances) to the precision of the average volumes.

Standard deviation of volume  Procedure A. _____  Procedure B. _____

Explain any differences that are apparent based on the errors ascribable to the instrument and errors caused by the experimenter.

C.  Decide on the basis of these experimental results which balance would be most appropriate for the calibration of

a graduated beaker: _____

a 25-mL graduated cylinder: _____

a 50-mL buret: _____

a microburet with 0.01 mL graduations: _____

Justify your choices.

Date_____  Signature _____

# Investigation 3

## DENSITIES OF LIQUIDS

## INTRODUCTION

At room temperature many substances are liquids. Familiar examples of liquids are water ($H_2O$), tetrachloroethylene ($Cl_2C=CCl_2$, a dry-cleaning agent), and ethanol ($C_2H_5OH$). The determination of the composition and the purity of liquids is one of the jobs of a chemist. This task is especially important when a new liquid compound has been synthesized. The chemist has many methods at his disposal to achieve this task: elemental analyses generally performed by commercial laboratories, nuclear magnetic resonance spectroscopy for organic compounds, determination of the refractive index, determination of the boiling point, determination of the density and many others. When a new compound is being investigated, the properties are determined after each purification step until they do not change any more. When the purity of a known liquid is to be checked, the properties of the liquid are determined and compared with values for the pure liquid reported in the literature or listed in handbooks. The basis of these purity checks is the experimental fact that most properties of a liquid change when impurities are present in the liquid or substances are dissolved in the liquid to obtain a solution. The density of a liquid can be easily determined and can be used to check the purity of a liquid and even the exact composition of a solution. For instance, density measurements are used to ascertain the concentration of sugar in sugar solutions and the composition of ethanol-water solutions. Knowledge about these solutions is very important in the sugar industry and for processes fermenting sugars to alcoholic beverages.

This experiment provides the opportunity to learn how to measure the density of liquids and to apply this knowledge to the identification of an unknown liquid sample.

## CONCEPTS OF THE EXPERIMENT

Density ($\rho$) is defined as the mass expressed in grams of 1.00 mL of a liquid (equation 1). Density is an intensive property because it does not depend on the amount of matter. The formula

$$\rho = \frac{m}{V} \qquad (1)$$

$\rho$: density ($g\,mL^{-1}$)     m: mass (g)     V: volume (mL)

for calculating density (equation 1) suggests that the density of a liquid can be obtained by weighing a known volume of a liquid on a balance. Dividing the mass by the volume produces a value for the density. A glass vessel, called a pycnometer, of exactly known volume can be filled with a liquid. The difference between the mass of the filled pycnometer and the mass of the empty pycnometer provides the mass of the liquid. In this experiment you will construct your own pycnometer from soft-glass tubing and determine its volume from the mass of water that fills the pycnometer and the density of water at the temperature at which you carry out the experiment. Because the density of a liquid changes with temperature (generally the density decreases with increasing temperature), the temperature at which a density was determined must be given. This information is provided as a superscript to the symbol for density; for instance $\rho_{H_2O}^{20} = 0.9982$.

## ACTIVITIES

- Make a pycnometer from soft-glass tubing.

- Determine the volume of your pycnometer.

- Determine the density of "unknown" liquids.

- Calculate the averages and the standard deviations for your results.

## SAFETY

*In this experiment you will be handling glass that has very sharp edges (danger of bleeding cuts) and with hot glass that looks just like cold glass (danger of painful burns). Exercise extreme caution, avoid injuries to yourself and to other students, and prevent damage to property. Never lay hot glass on the bench top or on combustible surfaces. Place all waste glass into the container designated for glass disposal. Whenever practical, avoid sharp edges on glass pieces by fire-polishing.*

## PROCEDURES

### A. CONSTRUCTION OF A PYCNOMETER

blue inner cone

**A-1**. Light and adjust a Bunsen burner to obtain a flame with a distinct bright-blue inner cone.

**CAUTION:** *You will be working with hot flames and hot glass. Hot glass, hot iron rings on the ringstand, and other hot objects look just like cold objects. Avoid burns by being careful what you touch.*

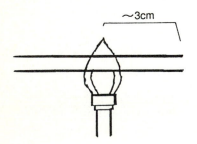

~3cm

**A-2**. Obtain a soft-glass tube that is 15 to 20 cm long and has an inner diameter of 5 to 7 mm. Hold the tube at both ends. Practice rotating the tube around its long axis. When you have mastered the rotation procedure, bring the tube into the Bunsen flame. The blue inner cone must be in contact with a small region of the glass tube approximately 3 cm from one end. Rotate the tube in the flame until the glass becomes soft and begins to sag. The tube must be evenly heated. Even heating is assured by steady rotation of the tube. Keep up the heating and the steady rotation. You should push the two cold ends with slight pressure toward the soft portion to prevent thinning of the soft part. Steady rotation is very important at this stage. If the glass tube is not

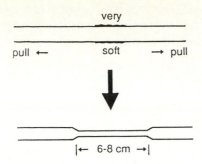

rotated properly, the soft part might become twisted. When the heated part of the tube is very soft, remove the tube from the flame and pull the two ends away from each other to draw the soft part of the tube into a capillary of 6 to 8 cm length and approximately 1 mm inside diameter.

**WARNING:** *Do not touch the hot parts of the tube. Hold the tube for several minutes to let the hot parts cool and become touchable.*

*If you pull too rapidly, too strongly, or too far, the capillary will have an inside diameter that is too small. Don't become discouraged should your first attempt lead to a less than perfect product. Try again. Practice makes perfect in working with glass.*

The longer pieces of glass can be reused in subsequent tries if they are first allowed to cool.

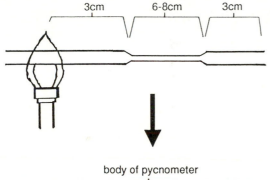

**A-3.** Before proceeding, check carefully that all parts of your glass tube are cool. Then heat the tube approximately 3 cm from the left-end of the capillary as described above and make a section of capillary of 1mm inner diameter and 6 to 8 cm length.

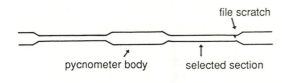

*Do not lay the hot tube on the bench. Hold it for several minutes until it has cooled.*

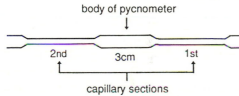

**A-4.** Inspect the two capillary sections and identify the section that comes closest to the desired 1 mm inner diameter and has this diameter for the entire length of the capillary. Make a single scratch at the end of the selected capillary section away from the pycnometer body. Support the tube on the left and right side opposite the scratch with your thumbs and break the capillary at the scratch by bending and pulling apart.

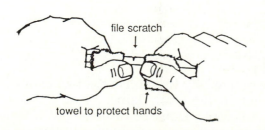

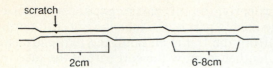

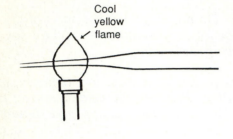

Cool
yellow
flame

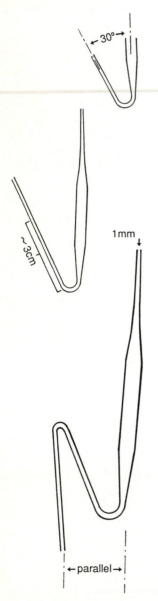

**A-5.** Cut the capillary on the other side of the pycnometer body not more than 2 cm from the point at which the tube begins to thin.

**A-6.** Close the air intake on the burner and reduce the gas flow until you obtain the smallest, cool yellow flame possible. Practice bending the capillary on the piece of glass not used for the construction of the pycnometer. Carefully and cautiously rotate the capillary in the cool flame until it is soft enough to bend into a "V" with a 30 to 40° angle. If the capillary tube becomes too soft, it might close. Should this happen with your pycnometer, you will have to make a new pycnometer.

**A-7.** When you are confident about your skills in capillary bending, take your pycnometer and bend the long capillary of the pycnometer close to the pycnometer body into a "30-40° V". Make a second bend in the capillary approximately 3 cm from the first bend. The last part of the capillary (~3 cm) should be parallel to the long axis of the pycnometer body. The pycnometer should resemble at capital Z. The correct shape and size of the finished pycnometer is shown in the actual-size drawing.

**A-8.** Very carefully fire-polish the two ends of the pycnometer in the cool flame. Don't allow the openings to close. Don't be discouraged if it takes more than one attempt to construct a pycnometer.

## B. DETERMINATION OF THE VOLUME OF THE PYCNOMETER

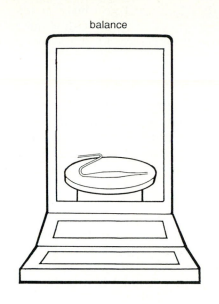

balance

**B-1**. Take a watchglass and your pycnometer to an analytical balance. Zero the balance. Weigh the watchglass and the pycnometer to 0.1 mg. If you determined the standard deviations associated with repeated weighings of the same object in an earlier experiment, you may use that standard deviation as an indicator of the precision of the mass of watchglass + pycnometer. If you do not have this standard deviation, ask your instructor. Should time and the availability of the balance permit repeated weighings of the watchglass + pycnometer, obtain three mass values. After each weighing, arrest the balance, remove the watchglass from the pan, and turn all the mass selection knobs to zero. Calculate the standard deviation of your results.

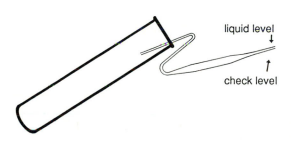

liquid level
↓

↑
check level

**B-2**. Fill a small test tube with distilled water. Place the capillary leg of the Z-shaped pycnometer into the test tube and tip the assembly to bring the pycnometer body below the liquid level in the test tube. The pycnometer will slowly fill with water. Inspect the end of the straight capillary. The liquid level must coincide with the end of the capillary. Carefully wipe off any liquid clinging to the outside of the capillary that was submerged in the water. The filled pycnometer must always be in a horizontal position. If you tilt the pycnometer, the liquid will begin the run out. Place the filled pycnometer on the watchglass. Weigh the watchglass + filled pycnometer to 0.1 mg on the balance you used before.

aspirator

**B-3**. To empty the pycnometer use a suction flask equipped with a rubber seal that has a small hole. Place the straight end of the pycnometer carefully into the hole and turn on the aspirator. Let air pass through the pycnometer for a few minutes. (If the rubber seal has a second hole in it , you will need to cover the second hole with your finger while the pycnometer is being emptied or dried.)

Table 3-1. Density of water, $\rho^T$, g mL$^{-1}$

| °C | g/mL |
|----|---------|
| 15 | 0.99913 |
| 16 | 0.99897 |
| 17 | 0.99880 |
| 18 | 0.99862 |
| 19 | 0.99863 |
| 20 | 0.99823 |
| 21 | 0.99802 |
| 22 | 0.99780 |
| 23 | 0.99751 |
| 24 | 0.99733 |
| 25 | 0.99708 |
| 26 | 0.99682 |
| 27 | 0.99655 |
| 28 | 0.99627 |

**B-4**. Fill the pycnometer again with water and weigh the watchglass + filled pycnometer to 0.1 mg. Repeat this procedure until you have at least three masses for the pycnometer filled with water. Record the laboratory temperature at the time of these weighings.

**B-5**. Calculate the average mass of the watchglass + pycnometer filled with water, the standard deviation and the percent deviation. Express the average with significant figures only. Using this average, calculate the average mass of the water in the pycnometer. Assume that this average water mass has the same standard deviation as the average mass of watchglass + filled pycnometer. Check the temperature of the laboratory or the water you used to fill the pycnometer. With the density for water at this temperature calculate the volume of the pycnometer. Express this volume in significant figures only (Appendix H).

## C. DETERMINATION OF THE DENSITY OF AN UNKNOWN LIQUID

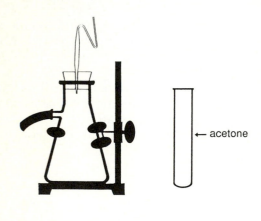

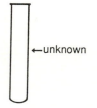

← acetone

←unknown

**C-1**. Obtain a sample of an unknown liquid from your instructor. Record the "unknown" number. Empty the pycnometer with the help of the suction flask. Fill the pycnometer with acetone. Empty the pycnometer. Repeat this procedure twice. Then pass air through the pycnometer for 2 minutes.

**C-2**. Fill the pycnometer with the unknown liquid. Determine the mass of the watchglass + filled pycnometer. Empty the pycnometer. Fill it again with the unknown. Repeat this procedure until you have three values. Use the mass of watchglass + filled pycnometer, mass of the watchglass + empty pycnometer, and the volume of the pycnometer to calculate the density of the unknown liquid. Express your final result with significant figures only and report it with a measure of its precision.

Consult Appendix H or Investigation 2 for details on the calculation of standard deviations and relative standard deviations.

**NOTE:** *Follow your instructor's directions for disposal of your unknown. If you wish to and are allowed to keep your pycnometer, you must empty it, rinse it with acetone, and pass air through the pycnometer for 2 minutes.*

# PRELAB EXERCISES

## INVESTIGATION 3:  DENSITIES OF LIQUIDS

Instructor:_____

Course/Section:_____

Name:_____

ID NO.: _____

Date:_____

1.  Define the following:

Density

Intensive Properties

Extensive Properties

Standard Deviation

Relative Standard Deviation

2.  Give four intensive properties that are frequently measured in the laboratory.

# REPORT FORM

## INVESTIGATION 3:  DENSITIES OF LIQUIDS

Instructor:_____

Section:_____

Name:_____

ID NO.:_____

Date:_____

---

Mass of watchglass + empty pycnometer: _____ g _____ g _____ g

    Average ± standard deviation: _____ ± _____ g

    Relative standard deviation: ± _____ %

Mass of watchglass + pycnometer filled with water: _____ g, _____ g, _____ g

    Average ± standard deviation : _____ ± _____ g

    Relative standard deviation: ± _____ %

        Average mass of water: _____ ± _____ g

Temperature of laboratory/water: _____ °C

Density of water at this temperature: _____ g mL$^{-1}$

Average volume of pycnometer: _____ ± _____ mL

Unknown No.: _____

Mass of watchglass + filled pycnometer: _____ g, _____ g, _____ g

    Average ± standard deviation: _____ ± _____ g

    Relative standard deviation: ± _____ %

    Density of unknown: _____ ± _____ g mL$^{-1}$
(Show calculations on the back of this sheet.)

Are your results accurate?  Are they precise?  Why or why not?  (Hint: Read about precision and accuracy in Appendix H.)  Answer on the back.

        Date _____ Signature _____

# Investigation 4

## CHEMICAL PRODUCTS AND ECONOMICS

### INTRODUCTION

Have you ever wondered how the price you pay for a product in the store is determined? Have you considered prices of products to be outrageously high in comparison to the cost of the materials from which the products were made? Why does a gallon of gasoline cost over one dollar when the 42 gallons in a barrel of crude oil, from which gasoline is made, sells for approximately 24 dollars? Have you recently bought aspirin to alleviate aches and pains for approximately six cents for a 325 milligram tablet, when you can buy two kilograms for $44.70 (a fraction of a cent per tablet) from a chemical supply house? Have you thought about the reason why an ice chest made from expanded polystyrene costs $4.95 when the petroleum from which the polystyrene is derived is worth only a few pennies?

Consumers tend to compare the retail cost of a product with the cost of the raw materials, but do not consider the many costs associated with the conversion of the raw materials to products (labor costs, costs for energy, machinery, buildings, local taxes, state taxes, federal taxes, costs of environmentally acceptable waste disposal, costs for transporting the products to the markets, costs for advertising, and fair profits). Industry depends upon mass production and efficient machinery to keep the cost per item low. Most chemical industries are either very high volume operations or spread the costs of production among many products.

This investigation will provide you with an appreciation for the many "hidden" costs of a product. Because you will not be able to take advantage of the economy of large scale production or diversification, don't be surprised if the cost of your product turns out to be outrageously high.

### CONCEPTS OF THE EXPERIMENT

In this investigation you will become the operator of a chemical plant (on a laboratory scale) that converts Epsom Salt, $MgSO_4 \cdot 7H_2O$, to magnesium sulfate monohydrate, $MgSO_4 \cdot H_2O$, useful as a drying agent. You will buy the Epsom Salt, your raw material, in a grocery store; you will set up the equipment in the laboratory to heat the raw material and drive off the water; you will need expert advice on how to run the process efficiently; you will need to have your operation inspected to make sure your plant is in compliance with local, state, and federal regulations; you will have to package your product and keep track of all your expenses. Finally, you will calculate a price for your product that allows for a profit.

Magnesium sulfate heptahydrate loses its water of hydration in steps. The heptahydrate, forming transparent crystals, may have lost one molecule of water already at room temperature when the air has low humidity. The hexahydrate has the appearance of dull, white, fractured crystals. The hexahydrate will be converted to the monohydrate at approximately 150°C. At 200° the monohydrate loses its water forming anhydrous magnesium sulfate. At much higher temperatures decomposition to magnesium oxide and sulfur trioxide occurs. Therefore, the conversion of the heptahydrate to the monohydrate must be carried out at temperatures below 200° but above 150°C.

## ACTIVITIES

- Purchase Epsom Salt.

- Convert Epsom Salt to $MgSO_4 \cdot H_2O$.

- Keep records of all costs associated with your operation.

- Calculate a retail price for your product.

## SAFETY

*Wear approved eye protection and be careful with chemicals and hot equipment. Follow the details provided by your instructor and by the engineer's plant design suggested below.*

## PROCEDURES

1. Prior to the laboratory period, someone in your lab section must go to a store and purchase a one pound carton of Epsom Salt ($MgSO_4 \cdot 7H_2O$). You will need only 20 g of this salt. The appearance of the crystals must be checked. They must be transparent. If the crystals are opaque and white, you will be starting with $MgSO_4 \cdot 6H_2O$. A sales receipt showing this purchase must be kept as a record and brought to the laboratory.

2. You have made a capital investment in plant and equipment (your laboratory fee) that gives you access to the laboratory and a standard collection of equipment.

3. You are the "personnel" (chemist, engineer, shop foreman, worker, executive, marketing expert, salesman). You must keep a record of the time spent on your trips to the store and your laboratory working time to the nearest one-quarter hour. Pay yourself $5.00 per hour.

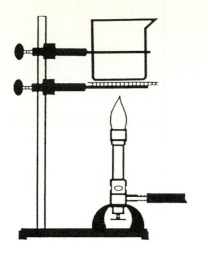

**4**.   You are now the engineer.  Design your experiment.  The process of heating $MgSO_4 \cdot 7H_2O$ should be carried out in a 250-mL beaker.  The beaker is of appropriate size, is less expensive than other reaction vessels, and can withstand the required elevated temperatures.  Because the beaker is made from glass, extra precautions must be taken to prevent breakage and injury to the operator.  A ringstand and two iron rings are needed to secure the beaker above the burner.  The beaker should never be in direct contact with the hot iron ring.  A wire gauze will have to be placed between the beaker and the lower iron ring.  When the material in the beaker is stirred during the heating process, the beaker might work its way off the gauze, fall, and break.  To prevent such a destructive event and the associated loss of time and material, the second iron ring must be placed around the beaker without touching it.

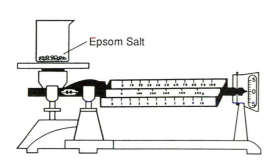

Epsom Salt

**5**.  You are now the chemist and operator.  Charge your beaker with approximately 20 g of Epsom Salt (permissible range 19–20 g).  Place the 250-mL beaker on the triple-beam balance, determine the mass of the beaker, advance the rider by 20 g and carefully add Epsom Salt.  Determine the mass of your sample to 0.1 g.

**6**.  Place the beaker containing Epsom Salt on the ringstand and ascertain that the beaker is secure and all the equipment is ready.  Call your instructor, who acts as the government safety inspector, to inspect your plant for compliance with all regulations.  If you are found to be without eye protection, to have left the balances without having cleaned up any spillages, or caught practicing other unsafe laboratory practices, your instructor may levy fines (point deductions) or in case of drastic violations "close down your plant."

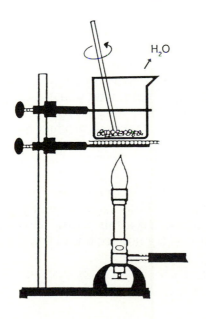

$H_2O$

**7**.  After your plant has passed inspection, begin to gently and cautiously heat the beaker with the crystals.  Record the time you lit the burner.  Stir the crystals to allow the water to escape.  *If you heat too strongly and do not stir the crystals sufficiently, too much water is released too quickly.  This water might dissolve some of the crystals and certainly will cause your material to cake.  Your final product will not be an attractive white powdery substance.  In addition, the equipment will be difficult to clean.*

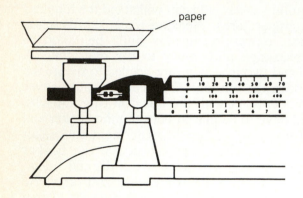

paper

*The yield will be low because some of the material will stick to the walls and will not be recoverable.* Strongly intensify the heating but make sure that your product does not begin to melt. Too much heat will cause the product to take on an unattractive gray coloration and expel sulfur trioxide, an air pollutant, from your reaction mixture. Your plant will be shut down if it pollutes the air.

**8 .** After the dehydration is complete, allow the product to cool. Note the time you turned off the burner. On the triple-beam balance weigh a piece of paper of sufficient size to allow you to package your product in an eye-catching manner. Carefully transfer your product onto the paper and weigh the paper + product. Calculate the mass of your product and the percent yield you achieved.

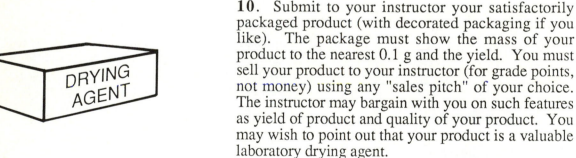

# CLEAN

# UP !

**9** . Your operation will generate by-products and "pollute" the environment. In order for the next generation (the students using the laboratory after you) to enjoy an unpolluted environment, clean all the glassware you used and clean your working area and around the balances. Any pollution incidence will draw a fine in terms of point deductions from your grade.

DRYING
AGENT

**10** . Submit to your instructor your satisfactorily packaged product (with decorated packaging if you like). The package must show the mass of your product to the nearest 0.1 g and the yield. You must sell your product to your instructor (for grade points, not money) using any "sales pitch" of your choice. The instructor may bargain with you on such features as yield of product and quality of your product. You may wish to point out that your product is a valuable laboratory drying agent.

**11** . Calculate a price for one gram of your product based on raw material costs, equipment costs, labor costs, energy costs, and a fair profit. Use the list in the Report Form as a guide for this calculation.

# PRELAB EXERCISES

## INVESTIGATION 4:  CHEMICAL PRODUCTS AND ECONOMICS

Name:_____

Instructor:_____    ID No.:_____

Course/Section:_____    Date:_____

1.  Give the chemical formula for Epsom Salt. _____

2.  Balance the following equation and calculate the molecular masses for all substances in the equation.

$$MgSO_4 \cdot 7H_2O \rightarrow MgSO_4 \cdot H_2O + H_2O$$

3.  Calculate the theoretical amount of magnesium sulfate monohydyrate that can be obtained by dehydration of 19.5 g of magnesium sulfate <u>hexa</u>hydrate.

4.  Describe the correct procedure for weighing a beaker on the triple-beam balance.

(over)

5. (Optional)  Describe the correct way to light a Bunsen burner and to adjust the flame (cool flame, hot  flame).

# REPORT FORM

## INVESTIGATION 4:  CHEMICAL PRODUCTS AND ECONOMICS

Name: _____     ID No.: _____

Instructor: _____     Course/Section: _____

Partner's Name (if applicable): _____     Date:_____

---

A. Chemical Data

    Mass of beaker + Epsom Salt: _____ g    Mass of paper + product: _____ g

    Mass of beaker: _____ g    Mass of paper: _____ g

    Mass of Epsom Salt: _____ g    Mass of product: _____ g

    Percent yield: _____ %

B.  Costs

    1. Raw material

        Cost of package of Epsom Salt:   $ _____ (attach receipt)

        Mass of Epsom Salt package:   _____

        Cost of Epsom Salt used:   $ _____

    2. Energy

        Time burner lit: _____ min.    Time burner turned off: _____ min.

        Cost of gas (min x $0.01/min):  $ _____

    3. Capital investment:  $ _____ (10% of laboratory fee paid)

    4. Personnel (round time to nearest 1/4 hour; minimum 1/4 hour)

| | | |
|---|---|---|
| Trucker (trip to store): | _____ hr x $5.00/hr | = $ _____ |
| Engineer (building of set up): | _____ hr x $5.00/hr | = $ _____ |
| Chemist (development of process*): | _____ hr x $5.00/hr | = $ _____ |
| Worker (performing reaction, clean up): | _____ hr x $5.00/hr | = $ _____ |
| Consultant (help by instructor): | _____ hr x $5.00/hr | = $ _____ |
| Salesman (bargaining with instructor): | _____ hr x $5.00/hr | = $ _____ |
| | Total Personnel Expenses: | $ _____ |

\* Use the time you spent for preparation before coming to the laboratory.

Raw material:        $_____        Total Costs:                                          $_____

Energy:               $_____        Profit before taxes (50% of total costs):  $_____

Capital Investment:  $_____        Sales Price of product:                        $_____

Personnel:            $_____        Price of 1.0 g of product:                     $_____

Percent profit after taxes = 100 x $\dfrac{\text{Profit before taxes - 28\% of profit before taxes}}{\text{Sales price of product}}$ = _____ %

C.  Purity of product (instructor's evaluation):

⬜ Reagent grade    ⬜ Tech grade    ⬜ Poor    ⬜ Unsatisfactory

D.  Government Safety Inspection:

⬜ In compliance with all rules and regulations

⬜ Penalty for infringement (number of grade points to be deducted)

E.  Grade for laboratory work and results (to be negotiated with instructor):

_____ Grade points awarded for product  –  _____ Penalty points  =  _____ Grade

F.  Questions (attach answers on a separate page)

1.  Which of your costs influence the sales price of your product most?  If your competition sells a similar product at a lower price, where would you attempt to economize?  Where would cost-cutting not be successful?

2.  Identify several other costs that would accrue in an industrial setting but were not considered in this experiment.

3.  How could one check that all of the heptahydrate has been converted to the monohydrate?

4.  The yield of magnesium sulfate monohydrate in this experiment sometimes exceeds 100 percent of the calculated yield.  How could such a yield be obtained?

Date _____ Signature _____

# Investigation 5

## ALUM FROM WASTE ALUMINUM CANS

## INTRODUCTION

Discarded beverage cans along the roadside attest to two common American pastimes, thirst-quenching and littering. Aluminum is one of the most indestructible materials used in metal containers. The average "life" of an aluminum can is about one hundred years. The litter from discarded cans is a serious and costly problem. A more serious problem is the depletion of raw materials. Although aluminum is the third most abundant element in the earth's crust, the supply of its useful ores is not inexhaustible. The processes for manufacturing aluminum from its ores require large amounts of electrical energy. Man must, therefore, find ways to "recycle" the materials he uses. A program for "recycling" aluminum cans is in operation in many communities.

In this experiment a chemical process will be used that transforms scrap aluminum into a useful chemical compound. "Alum," potassium aluminum sulfate dodecahydrate, $KAl(SO_4)_2 \cdot 12H_2O$, will be prepared from waste aluminum. Alum is widely used in the dyeing of fabrics, in the manufacture of pickles, as a coagulant in water purification and waste-water treatment plants, and in the paper industry.

Alums are ionic compounds that crystallize from solutions containing sulfate anion, a trivalent cation (e.g., $Al^{3+}$, $Cr^{3+}$, $Fe^{3+}$) and a monovalent cation (e.g., $K^+$, $Na^+$, $NH_4^+$). Most alums crystallize readily as octahedra or cubes which, under the appropriate conditions, may grow to considerable size. Six of the 12 water molecules per formula unit are bound tightly to the trivalent cation. The remaining six are loosely bound to the sulfate anion and monovalent cation.

## CONCEPTS OF THE EXPERIMENT

Although aluminum is a "reactive" metal, it reacts only slowly with dilute acids because its surface is normally protected by a very thin, impenetrable coating of aluminum oxide. On the other hand, alkaline solutions (containing $OH^-$) dissolve the oxide layer and then attack the metal. Aluminum is oxidized (equation 1) in aqueous alkaline medium to the tetrahydroxoaluminate(III) anion which is stable only in basic solution.

$$2Al(s) + 6H_2O(\ell) + 2KOH(aq) \rightarrow 2K[Al(OH)_4](aq) + 3H_2(g) \qquad (1)$$

When sulfuric acid is slowly added to an alkaline solution of this complex anion, one hydroxide ion is removed from each tetrahydroxoaluminate anion causing the precipitation of white, gelatinous aluminum hydroxide, $Al(OH)_3$, (equation 2a or 2b). The excess potassium hydroxide is converted to potassium sulfate (equation 2c).

$$2K[Al(OH)_4](aq) + H_2SO_4(aq) \rightarrow 2Al(OH)_3(s) + K_2SO_4(aq) + 2H_2O(\ell) \qquad (2a)$$

$$[Al(OH)_4]^-(aq) + H^+(aq) \rightarrow Al(OH)_3(s) + H_2O(\ell) \qquad (2b)$$

$$2KOH(aq) + H_2SO_4(aq) \rightarrow K_2SO_4(aq) + 2H_2O(\ell) \qquad (2c)$$

41

On addition of more sulfuric acid, the aluminum hydroxide dissolves forming the hydrated aluminum cation (equation 3a or 3b).

$$2Al(OH)_3(s) + 3H_2SO_4(aq) \rightarrow Al_2(SO_4)_3(aq) + 6H_2O(\ell) \tag{3a}$$

$$Al(OH)_3(s) + 3H^+(aq) \rightarrow Al^{3+}(aq) + 3H_2O(\ell) \tag{3b}$$

Addition of alkali to the $Al(OH)_3$ precipitate will also bring about dissolution by reforming $[Al(OH)_4]^-$. A hydroxide, such as aluminum hydroxide, that can be dissolved by either acid or base is said to be *amphoteric*.

When the acidified aluminum sulfate solution is cooled, potassium aluminum sulfate dodecahydrate ("Alum") precipitates (equation 4a or 4b).

$$Al_2(SO_4)_3(aq) + K_2SO_4(aq) + 24H_2O(\ell) \rightarrow 2K[Al(SO_4)_2]\cdot 12H_2O(s) \tag{4a}$$

$$Al^{3+}(aq) + K^+(aq) + 2SO_4{}^{2-}(aq) + 12H_2O(\ell) \rightarrow K[Al(SO_4)_2]\cdot 12H_2O(s) \tag{4b}$$

The overall reaction (equation 5) is the sum of the reactions described by equations 1, 2a, 2c, 3a, and 4a.

$$2Al(s) + 2KOH(aq) + 4H_2SO_4(aq) + 22H_2O(\ell) \rightarrow 2KAl(SO_4)_2\cdot 12H_2O(s) + 3H_2(g) \tag{5}$$

Equations 1, 2a, 2c, 3a, 4a and 5 are written in molecular form, whereas equations 2b, 3b, and 4b are net ionic equations showing only species that take part in the reactions.


## ACTIVITIES

- Complete the PreLab Exercises before coming to the laboratory.

- Prepare a sample of aluminum from a beverage can for use in the production of alum.

- Measure out reagents.

- Perform the synthesis of alum from aluminum.

- Isolate and purify your product.

- Perform stoichiometric calculations and report your obtained yield.


## SAFETY

*In this experiment you will be working with strong acids and bases. You must wear complete eye protection at all times. Alkaline solutions are particularly hazardous to the eyes. In case any of the reagents used in this Investigation come in contact with your skin or eyes, wash the affected area immediately with lots of water. Notify your instructor.*

*Exercise special care while handling ethanol, a flammable liquid, and while hydrogen, a flammable gas, is being formed. Both of these substances may catch fire when handled improperly.*

## PROCEDURES

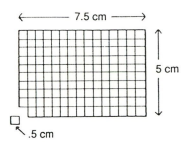

← 7.5 cm →

5 cm

.5 cm

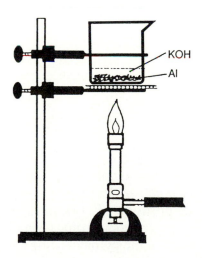

KOH

Al

**1**. Bring to the laboratory an empty, thin-walled aluminum beverage can. If you cannot bring one, ask your instructor. Several students can obtain their needed aluminum from a single can.

**2**. Using scissors provided, cut from your can a piece of aluminum approximately 5 cm x 7.5 cm and scrape off any paint as completely as possible. **Note:** *A considerable amount of time will be saved if you scrape off the paint before you bring the can to the laboratory.* Weigh the cleaned piece on a triple-beam balance. You need between 0.9 and 1.2 grams of aluminum. If your sample is too heavy, cut off any excess. If it is too light, cut and clean some more small pieces from a can. Weigh the final sample to the nearest 0.01 g and record this mass in your laboratory notebook. *If the mass of your sample ends in 00, you may have either spent too long adjusting the amount or you mis-weighed it.* Then cut your aluminum sample into small squares of about 0.5 cm edge length. The squares must be placed into a 250-mL beaker with the least amount of handling.

**3**. *If possible, work at a fume hood.* To the aluminum squares in the 250-mL beaker add 50 mL of 1.4 M potassium hydroxide. Bubbles of hydrogen should soon form from the reaction between aluminum and aqueous hydroxide. Heat the beaker very gently to speed up the reaction, preferably using a hot plate. *If a burner is used, be very careful. Hydrogen-air mixtures may ignite explosively.* When the liquid level in the beaker drops to less than half of its original volume, add distilled water to maintain the volume at approximately 25 mL. No more than 30 minutes should be required to complete the reaction. The final volume of the liquid should be about 25 mL. The reaction is complete when the hydrogen evolution ceases. *During the reactions the initially colorless mixture will turn dark gray or black. The dark material probably comes from the* *decomposition of residual paint or plastic lining. Note the periodic rise and fall of aluminum fragments during the reaction. Suggest an explanation. You may notice pieces of plastic lining remaining after the reaction has stopped.*

**4a**. Filter the hot solution using an aspirator vacuum to remove any solid residue. *Consult Appendix F if you are unfamiliar with the technique of vacuum filtration.* Be sure that the filter flask is securely clamped and the filter paper is moistened before you begin. Filter the solution. When all of the liquid has passed through the filter paper, break the vacuum by disconnecting the rubber tubing from the filter flask. Turn off the aspirator only after the vacuum has been broken.

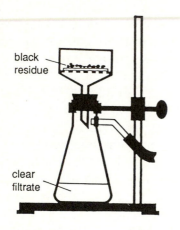

black residue

clear filtrate

**4b**. The filtrate should be clear with any dark residue left on the filter paper. Rinse the beaker twice with 5-mL portions of distilled water, pouring each rinse through the filter residue. Each time turn on the aspirator, filter the solution, break the vacuum, and then turn off the aspirator.

It is okay if your solution is slightly colored at this point.

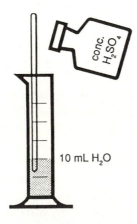

conc. H$_2$SO$_4$

10 mL H$_2$O

**5a**. To prepare 20 mL of 9.0 M H$_2$SO$_4$, if it is not provided, place crushed ice and then distilled water into your 25-mL graduated cylinder to bring the level of the ice/water mixture to the 10-mL mark. *Be sure your cylinder is made of* Pyrex *or similar glass.* Then pour small portions of concentrated (18 M) sulfuric acid *very carefully* from a small reagent bottle into the water in the cylinder. Stir with a glass rod after each small addition. Continue until the liquid level reaches the 20 mL mark on the cylinder. Stir the solution thoroughly, but carefully, and allow it to stand a few minutes. *The diluted acid will be quite hot! When diluting concentrated acids, one*

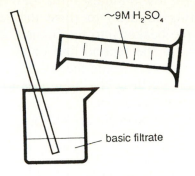

basic filtrate

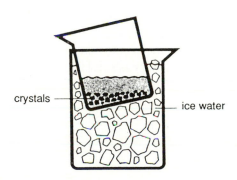

crystals — ——— ice water

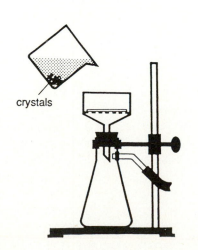

crystals

*should always add the measured, concentrated acid to the water while stirring. Considerable heat will be generated.*

**5b**. Transfer the clear filtrate into a clean 250-mL beaker. Rinse the filter flask with 10 mL of distilled water and pour the rinse water into the beaker. If the filtrate is not yet cool, place the beaker in a cooling bath. *Slowly and carefully, while stirring, add 20 mL of 9.0 M H₂SO₄ to the cooled solution. You will soon notice the appearance of a white precipitate of aluminum hydroxide. You will also notice heat generated by the neutralization reaction. Addition of the last few milliliters of the sulfuric acid will dissolve the Al(OH)₃.* If necessary, warm the solution gently, while stirring, to completely dissolve any Al(OH)₃ that might have formed. The final solution will contain potassium ions (from the KOH used), aluminum ions, and sulfate ions. If, after a few minutes of heating, any solid residue remains, filter the mixture as in step 4 and work with the clear filtrate.

**6**. Prepare an ice-water bath by filling a 600-mL plastic beaker half way with crushed ice. Add water to just cover the ice. Set the reaction beaker (from step 5) into the ice-water bath to chill. Crystals of the alum should begin to form. Allow the mixture to chill thoroughly for about 15 minutes. If crystals do not form, you may have to reduce the volume of solution by boiling away some of the water or you may only have to induce crystallization (consult your instructor). To induce crystallization, you may add one or two very minute seed crystals or you may scratch the inside bottom of the beaker containing the solution. Seed crystals (if desired) can be obtained by placing a drop of solution on the end of a stirring rod and blowing on it until it is dry.

**7**. Clean and reassemble the vacuum filtration equipment. Filter your alum crystals from the chilled solution, transferring as much of the crystalline product as possible to the funnel.

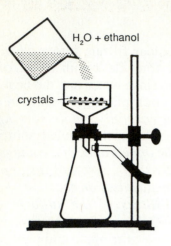

H₂O + ethanol

crystals

**8**. Mix 12 mL ethanol with 12 mL water and chill the mixture for a few minutes in the ice-water bath. Use half of this solution to rinse the remaining crystals from the beaker into the funnel. Rinse the beaker again with the second half of the solution. Use your spatula to distribute the crystals evenly on the filter paper. Allow the aspirator to pull air through the crystals for about 10 minutes. *Ethanol in the wash solution reduces the solubility of the alum.*

While the crystals are drying, weigh a clean, dry 250-mL beaker to the nearest 0.01 g. Record this mass. Use your spatula to transfer all of the air-dried crystals from the filter paper into the beaker. Reweigh the beaker. Record the mass.

**9**. Show the beaker containing your alum to your instructor, report the mass of alum obtained, and request a "product inspection." If your instructor considers your alum to be satisfactory, transfer the alum into the "Alum Storage Bottle" for later use. Otherwise, dispose of the alum as directed by your instructor.

# PRELAB EXERCISES

## INVESTIGATION 5:  ALUM FROM WASTE ALUMINUM CANS

Name:_____

Instructor:_____     ID No.:_____

Course/Section:_____     Date:_____

_____

1.  Give the chemical formula of the alum produced in this investigation.


2.  Write molecular and net equations for the sum (combination) of the reactions described by the equations in the "CONCEPTS OF THE EXPERIMENT" section.

   (a)  (1) + (2a)


   (b)  (3a) + (4a)


   (c)  (1) + (2a) + (3a) + (4a)


3.  Using the definition that molarity (M) is equal to the moles of solute in a given volume of solution divided by this volume of the solution expressed in liters, perform the following calculations:
   (a)  Moles and grams of $H_2SO_4$ in 10 mL of 18.0 M $H_2SO_4$ solution.

       moles = _____     grams = _____

   (b)  Moles and grams of $H_2SO_4$ in 20 mL of 9.0 M $H_2SO_4$ solution.

       moles = _____     grams = _____

   (c)  Moles and grams of KOH in 50 mL of 1.4 M KOH solution.

       moles = _____     grams = _____

4.  How many moles are present in

    (a)  1.15 g of Al?

    (b)  50 mL of 1.4 M KOH solution?

    (c)  20 mL of 9.0 M $H_2SO_4$ solution?

    (d)  30 g (approximately 30 mL) of liquid water?

5.  (a)  Using equation 5 in the CONCEPTS section and the amounts in question 4 (above), which reactant will be the limiting reactant?

    (b)  How many moles of each of the nonlimiting reactants will react?

    (c)  How many moles of alum and hydrogen can theoretically (assuming 100% yield) be formed?

    (d)  How many grams of alum can theoretically be formed?

    (e)  If one mole of hydrogen occupies approximately 24.2 L at room temperature and 1 atmosphere pressure, what volume will the hydrogen calculated in 5c (above) occupy?

# REPORT FORM

## INVESTIGATION 5: ALUM FROM WASTE ALUMINUM CANS

Name: _____ ID No.: _____

Instructor: _____ Course/Section:_____

Partner's Name (if applicable): _____ Date:_____

---

Mass of aluminum used: _____ g          Technical grade ☐

Volume of 1 M KOH used: _____ mL          Reagent grade ☐

Volume of 9.0 M $H_2SO_4$ used: _____ mL

Mass of alum obtained: _____ g          Instructor's initials _____

Yield: _____ %

Using the overall equation (5) and the calculations from the PreLab Exercises that indicate aluminum to be the limiting reactant, show the calculation of your yield by giving the numerical set up and the results. Show all units associated with the quantities you used in your calculation.

(over)

Answer all assigned problems. Attach a separate sheet if necessary.

1. If your product had not been dry, would the yield be higher or lower than the yield based on a dry product? Explain your choice.

2. How many grams of alum can be obtained from 20.0 g of aluminum when the reaction proceeds with 100% yield? How many grams of alum would be obtained if the reaction were to proceed with the yield achieved in your experiment?

3. How many grams of hydrogen are released by the reaction of 1.00 g of aluminum with excess potassium hydroxide solution assuming 100% yield?

4. How many moles of sulfuric acid are required to make 4.74 g of $KAl(SO_4)_2 \cdot 12H_2O$ assuming 100% yield?

5. How is it possible to obtain a yield greater than 100%?

Date _____ Signature _____

# Investigation 6

## GROWING ALUM CRYSTALS

### INTRODUCTION

Crystals are very important. Crystals of carbon – better known as diamonds – are very hard and, therefore, find applications as abrasives to cut and polish hard materials. Diamonds are valued as gemstones because of their optical properties, their hardness, and their rarity. Quartz ($SiO_2$) crystals serve as frequency generators; calcite ($CaCO_3$) crystals are used as polarizers for light waves in microscopes; and crystals of silicon (Si) are the starting material for computer chips.

In nature crystals are formed from molten rocks (magmas), from hot aqueous solutions, or from hot gases. Well-formed crystals, frequently encountered in hydrothermal veins, are spectacular in form and color. Such crystals have always held a fascination for man. Man has collected these crystals, treasured them, ascribed magical powers to them, and studied them. Mineralogy, the description and study of crystals, is one of the oldest scientific disciplines. Good places to see crystals displayed are the mineralogical sections of natural history museums and display cases in the hallways of geology departments.

For a piece of matter to be a crystal, the smooth faces so noticeable on museum specimens are not essential. A crystal is characterized by the orderly, repetitive arrangement in three dimensions of the ions, atoms, or molecules that make up the crystal. The structures of crystals can be determined by a technique known as single-crystal X-ray diffraction. One of the first crystals to be investigated by this technique approximately 70 years ago was sodium chloride (table salt). Sodium chloride crystallizes in the form of cubes. Each sodium ion is surrounded octahedrally by six chloride anions and each chloride by six sodium cations (Figure 6-1).

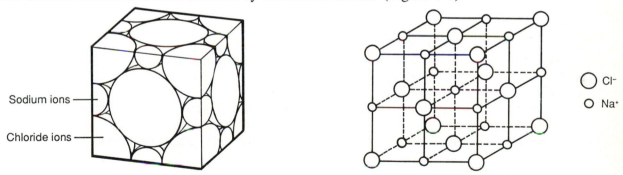

Sodium ions

Chloride ions

Cl⁻

Na⁺

Figure 6-1  Sketch of a cubic sodium chloride crystal and the arrangement of sodium cations and chloride anions in the sodium chloride lattice.

Sodium chloride crystals can be seen by looking at the salt grains from a salt shaker with a magnifying glass. A variety of crystals can be found in nature, especially in mountainous regions with exposed rocks. On your next hike keep your eyes open.

Man's need for crystals now exceeds what nature can supply. The natural supply is either insufficient (quartz crystals) or nature does not form the needed crystals at all (silicon crystals). Therefore, crystals are now grown "artificially". Many gemstones, industrial grade diamonds, and crystals with special optical, electric, and magnetic properties are produced commercially. In many laboratories the search is on for superconducting crystals. Solid state chemistry, the branch of chemistry dealing with the preparation of crystalline materials and the relationships between composition, structure, and properties, is practiced by an ever-increasing number of chemists and physicists. Chemists use the formation of crystals also for an entirely different purpose: for the purification of substances by a process called recrystallization. To recrystallize a solid the solid is

51

Framingham State College
Framingham, Massachusetts

generally dissolved in a suitable solvent at elevated temperature. After removal of undissolved material by filtration, the filtrate is cooled slowly to allow crystals to form. The slower the crystals grow and the larger the crystals become, the purer the substance will be.

This experiment demonstrates crystal growth from an aqueous solution using potassium aluminum sulfate and potassium chromium sulfate dodecahydrate known as alums. These substances crystallize easily in the form of octahedra.

## CONCEPTS OF THE EXPERIMENT

Many substances will dissolve in water forming aqueous solutions. The mass of a substance, also termed solute, that will dissolve in, for instance, 100 grams of water is not unlimited. A solution that contains as much of a substance as will dissolve is called saturated. The mass of substance present in a saturated solution prepared with 100 g of water is known as the solubility of the substance. Solubilities can also be expressed in other units. The solubilities of many substances are compiled in handbooks of chemistry. For instance, the solubility of sodium chloride is 35.6 grams NaCl in 100 g of water at 20°C. The solubility changes with temperature. Most substances are more soluble at higher than at lower temperatures. When a solution saturated at higher temperature is cooled, the cool solution will be supersaturated containing more solute than allowed by its solubility. Supersaturated solutions are unstable and will – in most cases – deposit the excess solute in crystalline form.

Another way to force a solute to crytallize is evaporation of the solvent. When a saturated aqueous solution of a substance is allowed to stand in an open beaker at room temperature, the water will slowly evaporate forcing the solute in excess of the solubility to crystallize.

In an aqueous solution of $KAl(SO_4)_2$ the $K^+$, $Al^{3+}$, and $SO_4^{2-}$ are surrounded by shells of water molecules (they are hydrated). These ions do not have an orderly arrangement in solution. When the compound is forced to crystallize, the ions must begin to join each other in their characteristic order. This process of nucleation may occur spontaneously when the ions of alum collide with appropriate orientation and with sufficiently low kinetic energy to permit them to "stick" to each other and prevent them from rebounding. More frequently, some "foreign" solids (irregularities on the wall of the container, dust particles) will serve as nuclei for the formation of crystals. Once a tiny crystal has formed, ions in their random motion through the solution will hit the faces of the crystal, join the orderly array of ions, and make the crystal grow. To keep the crystal growing, the solution must be cooled to even lower temperatures or solvent must be evaporated continuously. To obtain large cyrstals, small seed crystals are first prepared. A well-formed seed crystal is then suspended in a saturated growing solution and the solvent slowly evaporated. By replenishing the growing solution a huge, perfectly octahedral crystal of alum can be obtained.

## ACTIVITIES

- Complete the PreLab Exercises before coming to the laboratory.
- Prepare a saturated aqueous solution of $KAl(SO_4)_2 \cdot 12H_2O$.
- Grow alum seed crystals.
- Prepare an alum growing solution and seed this solution.
- Grow a large alum crystal.

## SAFETY

*Wear eye protection. Alum solutions strongly irritate eyes.  Store your growing solutions in a place in which they will not be disturbed.*

## PROCEDURES

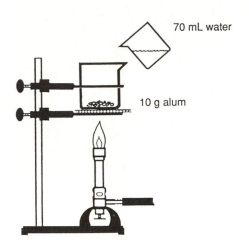

70 mL water

10 g alum

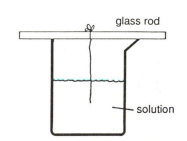

glass rod

solution

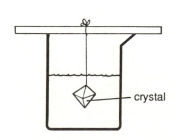

crystal

**1**. Weigh out approximately 10 g of alum [$KAl(SO_4)_2 \cdot 12H_2O$] on a triple-beam balance. Place the alum into a clean 250-mL beaker. For each gram of alum in the beaker add 7 mL of water. Heat the mixture to approximately 60°C.  Stir until all the alum has dissolved.  Should the mixture remain cloudy, let it stand for a few minutes to allow the suspended matter to settle.  Carefully decant the clear solution into another clean 250-mL beaker.

**2**. Tie a piece of thread to a glass rod.  Adjust the length of the thread so that not more than 1/2 inch is submerged.  Smear grease or oil on the part of the thread above the solution to prevent the solution from creeping up the thread.  Place the rod on top of the beaker, cover the beaker with a towel held in place with a rubber band, and store the beaker in a safe place.   Crystals should have formed on the submerged thread, at the bottom of the beaker or in both places by the time you return for your next laboratory session.

**3**. Remove the thread from the solution. Remove all the crystals except the best formed one from the string.  If a suitable crystal has not formed on the string, decant the solution from the crystals at the bottom of the beaker. Inspect the crystals and select one that has a regular octahedral shape and smooth faces. Tie a fine thread to the selected crystal. You might want to keep several crystals as reserve, should something destructive happen to your first-choice seed crystals.

**4**. Prepare a fresh solution of 10 g alum in 70 mL water as described in step 1. Allow this solution to cool to room temperature. Suspend your choice seed

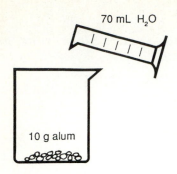

crystal in the solution, cover the beaker with a towel, and keep the beaker undisturbed for a week.

*If your solution is not saturated at the time you add your seed crystal, the crystal will begin to dissolve and may be lost. To prevent such an undesirable occurrence, observe the solution in the vicinity of the seed crystal after its placement into the solution. Should the solution be unsaturated causing the crystal to begin to dissolve, the part of the solution in contact with the crystal will become more concentrated and denser than the solution more remote from the crystal. The denser solution will flow toward the bottom of the beaker. Should you see such a density current, remove the seed crystal, cool the solution further, or dissolve more alum.*

**5**. For several weeks remove the crystal weekly and suspend it in a fresh, saturated alum solution. To keep large crystals completely submerged you might have to prepare larger volumes of alum solutions always maintaining the ratio of 1 g of alum to 7 mL of water. Weigh your crystal each week and describe its color, size and shape. Your instructor will grade your effort at the end of the term. You may keep the crystal after your work has been graded. To preserve the crystal and prevent it from crumbling to a white powder through loss of water, cover it with a clear plastic spray available from your instructor.

**6**. A layered, colored octahedron can be grown by substituting a solution of "chrome alum," $KCr(SO_4)_2$ $\cdot 12H_2O$, for the second or third alum growing solution. Dissolve 8 g $KAl(SO_4)_2 \cdot 12H_2O$ in 56 mL distilled water and 15 g $KCr(SO_4)_2 \cdot 12H_2O$ in 25 mL of water as described in step 1. Hold the beaker with the colorless alum solution toward a light. Slowly pour the chrome alum solution into the colorless alum solution. Mix well and check whether you can still see through the mixed solution. Stop adding the chrome alum solution when it has become difficult to see through the mixed solution. Suspend your crystal in this mixed solution. For subsequent growing stages, use solutions of $KAl(SO_4)_2 \cdot 12H_2O$.

*Your instructor will have a collection container for all waste solutions containing chromium.*

**7**. In your notebook maintain records of the appearance and mass of the crystal at each growing stage. Record how much alum and water you used at each stage. Complete the Report Form using your notebook records.

# PRELAB EXERCISES

## INVESTIGATION 6: GROWING ALUM CRYSTALS

Instructor:_____

Course/Section:_____

Name:_____

ID No.:_____

Date:_____

1. Define:
   Crystal


   Solution


   Saturated solution


   Supersaturated solution


   Decantation


   Solute


2. The solubility of $KAl(SO_4)_2 \cdot 12H_2O$ is approximately 1.0 gram per 7 mL of water at room temperature. How many grams of alum can be dissolved in 100 mL of water?


3. As water evaporates from the solution, the crystal grows. Does the solution in the vicinity of the crystal remain saturated, unsaturated, or supersaturated? Explain.


(over)

4. Why is oil or grease applied to a portion of the thread above the solution?

5. Why is the beaker in which the crystal grows covered but not with an air-tight cover?

6. Calculate the formula mass of $KAl(SO_4)_2 \cdot 12H_2O$.

7. Based on information given in this Investigation, calculate the solubility of alum in moles per 1.0 liter of water at room temperature.

# REPORT FORM

## INVESTIGATION 6: GROWING ALUM CRYSTALS

Name: _____     ID No.: _____

Instructor:_____     Course/Section: _____

Date: _____

Submit your crystal for evaluation and grading on _____.

### Record of Crystal Growth

| Date | Solution | | Mass of Crystal |
|------|----------|---|-----------------|
|      | g Alum* | mL water |  |
|      |         |          |  |

\* Identify with an asterisk the chrome alum solution.

Describe the various stages of the crystal growth. Sketch your final crystal on the back of this page. Measure the angles between the faces on the crystal. Are these angles characteristic of a regular geometric solid?

How many moles of alum were in your final crystal? (Assume that your crystal contained only $KAl(SO_4)_2 \cdot 12H_2O$.)

Date _____ Signature _____

# Investigation 7

## THE SCIENCE OF SOAP

### INTRODUCTION

The Gauls, a Germanic tribe, lived in the area of present-day France at the time the Roman Republic changed to the Roman Empire. The Gauls are said to have discovered the process of making soap by boiling animal fat with a mixture of water and plant ashes. Perhaps Gaius Julius Caesar learned about the beneficial uses of soap during the wars he waged against the Gauls and brought knowledge about the soap-making process to Rome. The Roman empire vanished, but the use of soap survived and spread through Europe. Mankind appeared to have had a yearning for cleanliness. By 800 A.D. Italy and Spain were the leading soap-making countries and soap was common in Europe. Until half a century ago the time-honored process of converting animal fat to soap was the only way to produce a substance that would clean hands, clothes, floors, and dishes and destroy germs. The rapidly increasing population during the recent past brought a quickly escalating demand for soap that could not be met by the limited quantities of animal fat. Synthetic cleaning agents (detergents) made from petroleum-derived hydrocarbons began to take the place of soap from animal fats after World War II. However, soap is still used for personal hygienic and many other applications. Approximately 600,000 metric tons of soap per year help to keep the United States and U.S. citizens clean.

Soaps are sodium or potassium salts of long-chain carboxylic acids. A soap molecule consists of a hydrocarbon chain and a carboxylate group. The formula for the sodium salt of stearic acid, $C_{17}H_{35}COONa$, clearly shows these two parts. The hydrocarbon part of the molecule is

$$CH_3\text{-}CH_2\text{-}CH_2\text{-}CH_2\text{-}CH_2\text{-}CH_2\text{-}CH_2\text{-}CH_2\text{-}CH_2\text{-}CH_2\text{-}CH_2\text{-}CH_2\text{-}CH_2\text{-}CH_2\text{-}CH_2\text{-}CH_2\text{-}CH_2\text{-}C\overset{\displaystyle O}{\underset{\displaystyle O^-\,Na^+}{<}}$$

hydrophobic or lipophilic hydrocarbon chain

hydrophilic or lipophobic carboxylate group

hydro = water, lipo = fat, phobic = fearing or hating, philic = loving

compatible with dirt often consisting of oily and greasy materials that are made up of hydrocarbon-like molecules similar to the carbon chains in soap molecules. When soap molecules encounter an oily dirt particle, the hydrocarbon chains of the soap molecules turn toward the oily particle and penetrate (dissolve in) the oily material. The hydrophilic carboxyl groups cover the surface of the

skin

clean skin

suspended oily dirt

oily particle, interact with the water molecules of the surrounding aqueous phase, force the oily dirt to become suspended in the water and finally to be "washed" away. The suspended dirt particles are kept from agglomerating into films by the electrostatic repulsion between the negatively charged carboxylate groups on the surface of these particles.

The cleaning action of soap is impaired in "hard" water. Hard water contains dissolved calcium and magnesium salts. These salts react with soap to form insoluble precipitates (for instance, calcium stearate) that are observed as "bath-tub rings." After all the calcium and magnesium ions have been precipitated, the soap will lather and clean again. This process, called water softening, of removing calcium and magnesium ions from hard water wastes lots of soap.

The art of making soap from fats and oils, still practiced in many rural households half a century ago, is the subject of this experiment. Because of the ready availability and large variety of commercial cleaning agents, soap making at home is almost a forgotten art.

## CONCEPTS OF THE EXPERIMENT

Animal fats and vegetable oils are rich in compounds called glyceryl esters, such as glyceryl tristearate. Most glyceryl esters found in nature contain more than one kind of carboxylate (fatty acid) moiety. When these esters are heated with a solution of a base such as sodium hydroxide in water, the esters are hydrolyzed into their components, glycerol (the alcohol component) and salts of fatty acids (the carboxylic acid component). This process known as saponification is described by the following equation. Glycerol is soluble in water. The long-chain sodium carboxylates

$$
\begin{array}{c}
\text{glyceryl tristearate} + 3\text{NaOH (aq)} \rightarrow 3\text{NaO-}\overset{\text{O}}{\overset{\|}{\text{C}}}\text{-(CH}_2)_{16}\text{CH}_3 + \text{glycerol} \quad (1) \\
\text{soap}
\end{array}
$$

glyceryl tristearate:
$$\text{H-}\overset{\text{H}}{\underset{}{\text{C}}}\text{-O-}\overset{\text{O}}{\overset{\|}{\text{C}}}\text{-(CH}_2)_{16}\text{CH}_3$$
$$\text{H-}\overset{}{\underset{}{\text{C}}}\text{-O-}\overset{\text{O}}{\overset{\|}{\text{C}}}\text{-(CH}_2)_{16}\text{CH}_3 \ (\ell \text{ or s})$$
$$\text{H-}\overset{}{\underset{\text{H}}{\text{C}}}\text{-O-}\overset{\text{O}}{\overset{\|}{\text{C}}}\text{-(CH}_2)_{16}\text{CH}_3$$

glycerol:
$$\text{H-}\overset{\text{H}}{\underset{}{\text{C}}}\text{-O-H}$$
$$\text{H-}\overset{}{\underset{}{\text{C}}}\text{-O-H}$$
$$\text{H-}\overset{}{\underset{\text{H}}{\text{C}}}\text{-O-H}$$

(soap) are much less soluble in water than glycerol and can easily be precipitated by adding a concentrated sodium chloride solution to the reaction mixture. Filtration separates the soap from the solution. The soap on the filter is purified by washing with cold water.

Examples of carboxylic acids that are combined with glycerol in fats and oils and form soaps are:

| | |
|---|---|
| lauric acid | $CH_3(CH_2)_{10}CO_2H$ |
| palmitic acid | $CH_3(CH_2)_{14}CO_2H$ |
| stearic acid | $CH_3(CH_2)_{16}CO_2H$ |
| oleic acid | $CH_3(CH_2)_7CH=CH(CH_2)_7CO_2H$ |
| linoleic acid | $CH_3(CH_2)_4CH=CHCH_2CH=CH(CH_2)_7CO_2H$ |

# ACTIVITIES

- Complete the PreLab Exercises before coming to the laboratory.

- Before coming to the laboratory become familiar with the terms "hard" and "soft water", the general characteristics and formulas of soaps, and the formulas for carboxylic acids. Know how to perform stoichiometric calculations. Check on the hardness of the local water.

- Measure the required quantities of oil, sodium hydroxide solution, and ethanol.

- Saponify the oil or fat.

- Isolate and purify the soap.

- Test the chemical behavior of soap.

- Make stoichiometric calculations.

# SAFETY

*In this experiment you will be working with concentrated alkaline solutions. Wear safety goggles at all times. One drop of the hot alkali solution in the eye could cause blindness.*

*In case of contact of any chemical with the skin or clothing, wash immediately with lots of water. Notify your instructor. Be particularly careful to keep the ethanol away from open flames.*

# PROCEDURES

## A. Preparation of Soap

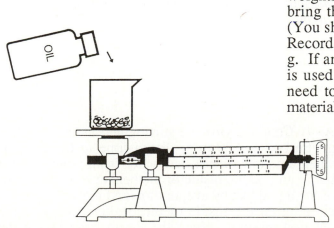

**A-1.** Weigh a clean, dry 400-mL beaker on the triple-beam balance to the nearest 0.1 g. Advance the weights by 10 g and carefully transfer enough oil to bring the balance beam to its equilibrium position. (You should have 9.5 to 10.5 g of oil in the beaker.) Record the actual mass transferred to the nearest 0.1 g. If animal fat or a semisolid vegetable shortening is used in place of a liquid vegetable oil, you will need to use a spatula to assist in transferring the material.

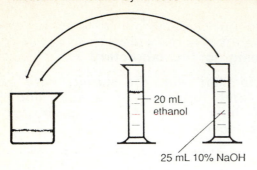

**A-2.** Pour 95% ethanol from a storage bottle into a graduated cylinder until the liquid level rests at the 20-mL mark. Add the ethanol to the oil and swirl the mixture until it has become homogeneous. Pour 25 mL of the 10-percent aqueous sodium hydroxide solution into the beaker containing the oil and ethanol. Mark the liquid level with a grease pencil on the outside of the beaker.

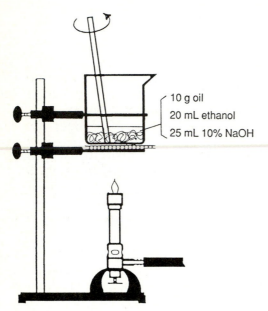

**A-3.** Before continuing work, place a wet towel, large enough to cover your beaker, in a readily accessible location. Then place the beaker on a wire gauze supported on a ring clamp above a laboratory burner. Light and adjust the burner for a low cool flame. Heat and stir the mixture *gently*. The mixture must be kept boiling gently and not vigorously. If foam rises in the beaker, the mixture is being heated too strongly. Continue heating the mixture. The mixture <u>must be heated and stirred</u> at the same time. Stirring will mix the reagents, allow the reaction to proceed more quickly, and prevent dangerous overheating and bumping.

**Warning:** *Do not permit the flame to come near the upper part of the beaker where ethanol vapors might easily ignite. Should your mixture ignite, do not panic. Turn off your burner at the bench gas valve. Place the wet towel on top of the burning beaker. The beaker must be kept covered until the flames are extinguished. Notify your instructor. In most cases you will be able to continue after replacing part of the ethanol lost during the fire.*

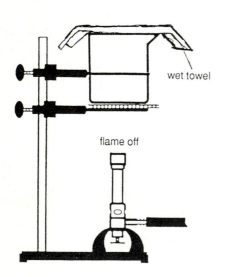

**A-4.** Continue with the stirring and the heating to a gentle boil for 20 min. From time to time, squirt small amounts of distilled water from your squeeze bottle around the inside of the beaker to replace part of the liquid lost by evaporation and to rinse particles clinging on the inside of the beaker back into the reaction mixture. Keep the liquid level inside the beaker between 25 mL and the original 50 mL. Foaming commonly occurs when the saponification

is complete.  After 20 min your beaker should contain an off-white to yellowish viscous mass (a mixture of soap, glycerol, and excess sodium hydroxide solution).  At this point the volume should be approximately 25 mL.  Turn off the burner.  Using a folded towel, remove the beaker carefully from the ring support and set it into a 600-mL beaker half-filled with cold tap water.  Let the mixture cool to below room temperature.  Change the water in the 600-mL beaker when the water becomes warm.

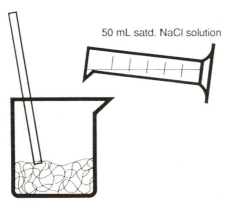

50 mL satd. NaCl solution

**A-5.** Weigh 16 g of sodium chloride into a 250-mL Erlenmeyer flask using the triple-beam balance.  Add 50 mL of distilled water to the salt, stopper the flask, and shake the stoppered flask until most of the sodium chloride has dissolved.  This solution must be a saturated solution of sodium chloride before it can be used in the next procedure.  Should the solution be cloudy, filter it by vacuum filtration (see Appendix F).

**A-6.** Pour the saturated sodium chloride solution into the cool reaction mixture.  Stir the mixture thoroughly.  The excess sodium hydroxide and the glycerol will remain dissolved; the soap will precipitate.  To obtain pure soap, the reaction mixture must be mixed thoroughly with the sodium chloride solution.  Press the soap against the beaker wall with your spatula to break up chunks of soap floating in the mixture.  *If this step is not properly executed, sodium hydroxide and glycerol will be trapped in the soap and will interfere with subsequent experiments to be performed with the soap.*

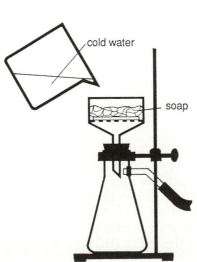

cold water

soap

**A-7.** Set up the aspirator (vacuum) filtration equipment (Appendix F).  Place a filter paper in the funnel.  Moisten the paper with distilled water.  Turn on the water aspirator and *decant* the supernatant liquid through the funnel.  Then transfer the soap into the funnel proceeding as described in Appendix F for vacuum filtration.

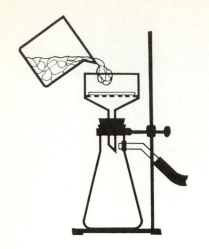

**A-8.** Add a few chunks of ice to 50 mL of distilled water in a 100-mL beaker and allow the water to chill. Use half of the cold water (no ice) to rinse any remaining soap from the beaker into the funnel (while the aspirator is shut off). Use a spatula to scrape soap from the interior of the beaker. With your spatula stir the soap/ice-water mixture in the funnel into a slurry. Be careful not to tear the filter paper. Then turn on the aspirator to filter the mixture. Allow the aspirator to pull air through the soap for about 10 min to help dry it. When the soap is dry, remove a small sample for testing.

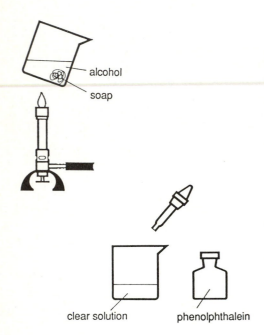

alcohol

soap

clear solution

phenolphthalein

**A-9.** To check for residual sodium hydroxide in your soap, place half a spatula-tip full of the soap into a small beaker and add 10 mL of 95-percent ethanol. Heat the mixture *carefully* over a low flame with gentle swirling until the soap has dissolved. Then cool the beaker to room temperature and add three to five drops of phenolphthalein indicator. If the solution turns bright pink, wash the soap again with cold water. Repeat this procedure until the solution produces only a pale pink color or no color at all upon addition of phenolphthalein.

*Soap made this way should be allowed to age a few days before it is used. During the aging process any NaOH present will react with carbon dioxide from the air and produce sodium bicarbonate ($NaHCO_3$), which is much less caustic.*

**A-10.** When your soap is free of sodium hydroxide, keep it on the filter in the vacuum apparatus and let air pass through the soap for 10 minutes. In the meantime, weigh a clean piece of notebook paper on the triple-beam balance. Then place your dry soap in the center of this sheet of paper. Discard the filter paper. Fold the notebook paper around the soap to make a neat package. Weigh this package.

**A-11.** Calculate the mass of soap obtained; the amount of soap that can theoretically be prepared from the mass of oil (fat) that you used under the assumption that your starting material was glyceryl tristearate; and the yield (%) you achieved in your experiment.

## B. Some Properties of Soaps

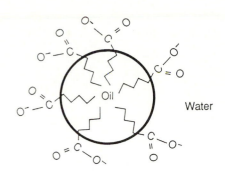

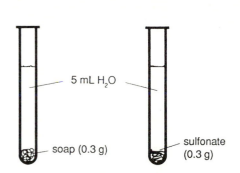

**B-1.** Use some of the soap you just made to wash your hands. If the soap was prepared correctly and washed properly with salt water and cold water, it should not feel greasy (caused by excess oil) and it should not make your hands feel slick (caused by the presence of sodium hydroxide). Rinse your hands free of lather and soap. Then place a small dab of oil on your hands and thoroughly "grease" your hands. Try to wash off this fat with tap water alone. Then use your soap. Give an explanation for the observed results.

**B-2.** Rub some chalk dust into your hands. Use soap and water to rinse your chalky hands. Observe carefully. Does the soap produce lather? What does it take to get your hands clean?

**B-3.** Prepare a clear dispersion of your soap by placing 0.3 g of the soap weighed on a triple-beam balance into a small test tube and adding 5 mL of distilled water. Warm and shake the mixture until you have an almost clear dispersion. Cool the dispersion by setting the test tube into a small beaker filled with cold tap water. Proceed similarly using 0.3 g of sodium dodecylbenzenesulfonate to prepare a detergent suspension.

**B-4.** Take a clean small test tube to your instructor to obtain 10 mL of a calcium chloride solution that contains 40 mg $Ca^{2+}$ per liter. By definition "hard" water is water with a $Ca^{2+}$ concentration equal to or larger than 40 mg $L^{-1}$.

**B-5.** Add 3 mL of the $Ca^{2+}$ solution to each of two test tubes. Add to one a few drops of your soap solution and to the other a few drops of the sulfonate solution. Observe and record what happens.

**B-6.** Place approximately 1 mL of your soap solution in a small test tube. Then add one drop of the stock $Ca^{2+}$ solution to this test tube. Mix well and describe your observations (cloudiness, precipitates, appearance of precipitates). Add another drop of $Ca^{2+}$ solution and continue the addition until no changes can be observed. Perform a similar experiment with your sulfonate solution.

# PRELAB EXERCISES

## INVESTIGATION 7: THE SCIENCE OF SOAP

Name:_____

Instructor:_____     ID No.:_____

Course/Section:_____     Date:_____

1. Write the balanced equation that must be used for stoichiometric calculations associated with the saponification of glyceryl tristearate.

2. Calculate the molecular mass of glyceryl tristearate.

3. How much soap can be obtained from 100 g of oil and excess sodium hydroxide assuming 100% yield?

4. Define saponification.

(over)

5. Define "hard water".

6. Regions with abundant limestone strata generally have hard water.  Explain the connection between water hardness and limestone.

# REPORT FORM

## INVESTIGATION 7:  THE SCIENCE OF SOAP

Name: _____     ID No.: _____

Instructor: _____     Course/Section: _____

Partner's Name (if applicable): _____     Date: _____

---

Mass of oil/fat used: _____ g

Mass of soap obtained: _____ g

   Yield: _____ %

Show yield calculations below

**Washing with Soap**
Observations and explanations

**Observations:**  (Answer on an attached sheet if more space is needed.)

   Soap to hard water

   Sulfonate to hard water

   $Ca^{2+}$ solution to soap

   $Ca^{2+}$ solution to sulfonate

(over)

Answer all assigned problems.  Attach a separate sheet if necessary.

1.  A dirty person takes a bath every day.  He fills the bathtub with 100 liters of water that contains 162 mg calcium hydrogen carbonate [$Ca(HCO_3)_2$] per liter.  How many kilograms of soap will be used per month (30 days) to remove all the calcium ions as calcium stearate from the bath water?  Assume that calcium stearate is completely insoluble in bath water.  The formula of calcium stearate is [$CH_3(CH_2)_{16}CO_2$]$_2$Ca.

2.  Which similarities exist between soap and sodium dodecylbenzenesulfonate?

3.  Why is dodecylbenzenesulfonate a better cleaning agent in hard water than soap?  Will it also work in extremely hard water containing, for instance, 0.06 moles of $Ca^{2+}$ liter$^{-1}$?  Use your experimental results in answering this question.

Date _____ Signature _____

# Investigation 8

## ISOLATION OF A FATTY ACID

## INTRODUCTION

Historically, the term "fatty acid" is applied to long-chain carboxylic acids obtained by hydrolysis of animal fats. The salts of long-chain carboxylic acids are called soaps. The preparation of "soaps" by alkaline hydrolysis of fats or oils is discussed in Investigation 7.

$$CH_3CH_2CH_2CH_2CH_2CH_2CH_2CH_2CH_2CH_2CH_2CH_2CH_2CH_2CH_2CH_2CH_2 \overset{\displaystyle O}{\underset{OH}{C}}$$

stearic acid
an example of a long-chain, saturated (no C=C bonds) carboxylic acid

$$CH_3CH_2CH=CHCH_2CH=CHCH_2CH=CHCH_2CH_2CH_2CH_2CH_2CH_2CH_2 \overset{\displaystyle O}{\underset{OH}{C}}$$

linolenic acid
an example of a long-chain, polyunsaturated carboxylic acid

$$CH_3CH_2CH_2CH_2CH_2CH_2CH_2CH_2CH_2CH_2CH_2CH_2CH_2CH_2CH_2 \overset{\displaystyle O}{\underset{O^-Na^+}{C}}$$

sodium palmitate
an example of soap

Fatty acids are "natural" products initially synthesized by plants from carbon dioxide and water with sunlight as the source of energy. The fatty acids and glycerol are converted to triglycerides which are called fats when they are solid at room temperature and oils when they are liquid. Foods that contain much fat or oil are rich sources of energy. In organisms these fatty acids are broken down stepwise from the carboxyl-end generating one molecule of acetic acid, $CH_3COOH$, and a fatty acid shortened by two carbon atoms in each step. Ultimately, all the carbon atoms in a fatty acid are converted to carbon dioxide and all hydrogen atoms to water. The energy released in the exothermic oxidation of one mole (284 g) of stearic acid to carbon dioxide and water is 2696.1 kcal. This energy is used to drive cellular processes. If too much fat is consumed, some of it is deposited making a person overweight (fat).

Fats and oils are convenient sources of fatty acids. Alkaline hydrolysis of fats and oils produces soaps and acidic hydrolysis "free" fatty acids. The soaps and fatty acids obtained from fats and oils are not pure compounds but mixtures of saturated and unsaturated carboxylic acids of various chain-lengths. Although methods are now available for the separation of such mixtures of carboxylic acids, pure compounds are often not required. These fatty acids find many uses, for instance in the production of detergents and cosmetic formulations, in finishing processes for textiles, and as additives to lubricants, plastics, and dyes and paints.

In this experiment the sodium salt of a fatty acid (a soap) is converted to the free fatty acid.

## CONCEPTS OF THE EXPERIMENT

Acids are substances that dissociate upon dissolution in water into acid anions and hydronium ions. Strong acids (e.g., hydrochloric acid) are completely dissociated, whereas weak acids (e.g., carboxylic acids) remain largely undissociated (equations 1a and 1b).

$$\underset{\text{strong acid}}{HCl} \xrightarrow{H_2O} H_3O^+ + Cl^- \quad (1a) \qquad \underset{\text{weak acid}}{C_{17}H_{35}COOH} \underset{H_2O}{\rightleftharpoons} C_{17}H_{35}COO^- + H_3O^+ \quad (1b)$$

Sodium salts of weak acids (e.g., sodium stearate) and of strong acids (e.g., sodium chloride) are completely dissociated in solution into sodium cation and acid anions. The dissociation of a weak acid is characterized by a dissociation constant K (equation 2).

$$K = \frac{[C_{17}H_{35}COO^-][H_3O^+]}{[C_{17}H_{35}COOH]} \approx 10^{-5} \qquad (2)$$

In an aqueous solution of sodium stearate or any long-chain carboxylate ion, the concentration of the acid anion will be high and the concentrations of the hydronium ion and undissociated acid low. When the strong hydrochloric acid is added to a sodium stearate solution, the hydronium ion concentration increases by many orders of magnitude. To bring the value of the fraction in equation 2 (the reaction quotient) back to the value of K, the stearate anions must combine with $H^+$ to form undissociated stearic acid. An excess of hydrochloric acid will shift the equilibrium (equation 1b) even further toward undissociated stearic acid which is much less soluble in water than sodium stearate and precipitates. The precipitated stearic acid is filtered, washed with water to remove impurities, and purified by recrystallization. Because stearic acid is insoluble in water but soluble in hot ethanol, the crude product is dissolved in hot ethanol, the mixture filtered to remove undissolved material, and the hot filtrate treated with ice. This treatment reduces the solubility of the stearic acid by increasing the concentration of water in the solution and lowering the temperature. The purity of the product is then checked by determining its melting point.

## ACTIVITIES

- Complete the PreLab Exercises before coming to the laboratory.
- Weigh out a sample of soap.
- Convert the soap to a fatty acid.
- Isolate and purify the fatty acid.
- Obtain the melting point of the isolated product.

## SAFETY

*Wear approved eye protection at all times. Use caution in working with acids and be particularly careful in heating ethanol. Ethanol is highly flammable.*

## PROCEDURES

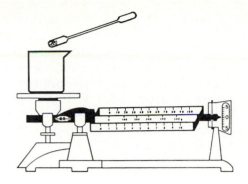

**1.** Weigh a clean 400-mL beaker on a triple-beam balance to ±0.1 g. Transfer approximately 10 g of the soap into the beaker. Weigh the beaker with the soap to ±0.1 g. Record the mass of soap.

**2.** Add 100 mL distilled water to the soap in the beaker. Heat and stir the mixture until the soap is dispersed. Should the mixture foam excessively, remove the burner temporarily and stir the mixture until the foam subsides. Then continue heating.

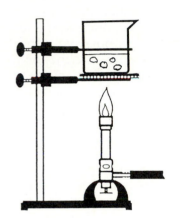

**3.** Calculate the volume of 3 M hydrochloric acid that contains twice the amount of HCl needed to react with the mass of your soap in the beaker. Use in this calculation the result from your PreLab Exercise in which you found the volume of 3 M hydrochloric acid (in mL) needed to react stoichiometrically with 10 g of sodium stearate.

**4.** Take your clean graduated cylinder to the hood and from the "3 M HCl" storage bottle pour the required volume of hydrochloric acid into the graduated cylinder. Pour the hydrochloric acid into the stirred, hot soap dispersion. Add pieces of ice to the mixture until its temperature has dropped to 0°C. *When this temperature has been reached, the ice will melt very slowly.*

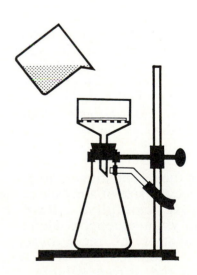

**5.** Isolate the precipitated carboxylic acid by vacuum filtration (Appendix F). To prevent the filter from becoming clogged, keep liquid in the funnel during filtration. Allow the filter to become dry only after the last part of the mixture has been poured into the funnel.

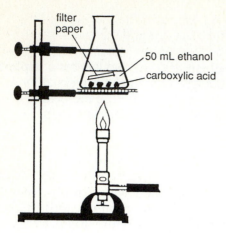

filter paper

50 mL ethanol

carboxylic acid

**6.** Add ice to 200 mL distilled water to bring the temperature to 0°C. Pour 50 mL of this cold water (without ice) into the reaction beaker, swirl to loosen any carboxylic acid clinging to the walls, and quickly pour the liquid into the funnel. With your spatula break up the filter cake and crush any clumps of carboxylic acid. *Do not tear the filter paper during this operation.* By suction separate the wash-water from your product. Repeat this procedure two more times. Then press the filter cake with your spatula while the aspirator vacuum is on to remove as much of the water as possible. Allow air to pass through the filter cake for several minutes.

**7.** Weigh a clean and dry 250-mL Erlenmeyer flask on the triple-beam balance to ± 0.1 g. Scrape the carboxylic acid from the filter paper and transfer the acid as quantitatively as possible into the weighed Erlenmeyer flask. *Do not discard the filter paper.* Reweigh the flask, calculate the mass of the product, and record this mass in your notebook as the mass of crude product.

filter paper

**8.** Add the filter paper and then 50 mL of 95% ethanol to the Erlenmeyer flask containing the crude product. Heat the mixture very carefully (*fire hazard*) until most of the acid has dissolved.

**9.** Fold a piece of filter paper of the appropriate diameter accordion-like to produce a fluted filter. Open the folded filter and place it into a powder funnel (a funnel with a short but wide stem). Place the funnel with the filter on top of a 250-mL beaker. Quickly pour the hot solution into the funnel. If the funnel can hold all the solution without overflowing, pour all the solution into it at once. Wait until the solution has passed through the filter.

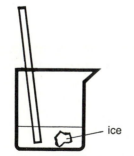

ice

**10.** Remove the funnel and discard the filter. Add a piece of ice twice the size of a marble to the clear filtrate. Observe carefully. Record your observations in your notebook along with a suggested explanation of these observations. Remove the beaker from the ringstand. Add small pieces of ice to the stirred solution until its temperature is 0°C.

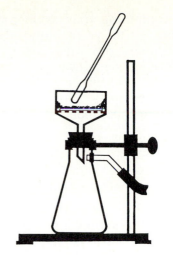

**11.** Collect the precipitated fatty acid by vacuum filtration. After all the liquid has passed through the filter, press the carboxylic acid with your spatula. Draw air through the funnel and continue to press the product with your spatula for 5 minutes.

**12.** Disconnect the aspirator from the suction flask. Add 2 mL of acetone *(not more)* to the filter cake. Mix the acetone and the carboxylic acid with your spatula until a uniform paste is obtained. Spread the paste evenly over the entire surface of the filter paper and reconnect the aspirator. Draw air through the filter cake until a white powder is formed and you can no longer detect the odor of acetone. Break up the dry product with your spatula.

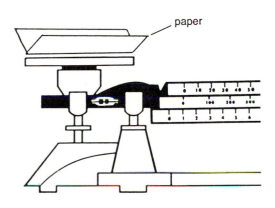

paper

**13.** Use the triple-beam balance and a preweighed piece of paper (half-page size) to determine the mass of the obtained product. Calculate the yield of your reaction assuming that the carboxylic acid is $CH_3(CH_2)_{16}CO_2H$.

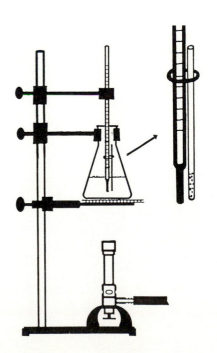

**14.** Fill a 125-mL Erlenmeyer flask three-quarters full with water. Clamp the flask to a ringstand. Obtain from your instructor a melting-point capillary and fill it to a depth of 2 mm with your product. Compact the sample by dropping the capillary through a long glass tube with its lower end against the bench top. Attach the capillary to the thermometer by means of a rubber band. Position the capillary so that the product is next to the bulb of the thermometer. Clamp the thermometer to the ringstand so that the bulb is immersed in the water. Heat the water at a rate of $5-10$ deg $min^{-1}$ and determine the approximate melting point of the carboxylic acid you have obtained. Repeat this determination heating the water rapidly to about 10 degrees below the approximate melting point of the acid. Then reduce the heating rate to 2 degrees per minute and determine the melting point.

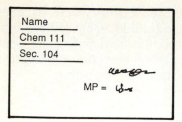

**15.** Fold the paper containing your product into an envelope shape and write your name, course number, section number, the mass of product obtained, and the product's melting point on the outside. Your instructor will direct you on how to store or dispose of your samples. You may use this carboxylic acid in a future experiment.

**16.** Be sure that all observations and data are recorded in your permanent notebook along with the calculations required for various stages of this experiment.

Use this record to complete the Report Form for submission to your instructor.

# PRELAB EXERCISES

## INVESTIGATION 8:  ISOLATION OF A FATTY ACID

Name:_____

Instructor:_____     ID No.:_____

Course/Section:_____     Date:_____

---

1.  Write a balanced equation for the production of stearic acid from sodium stearate.

2.  Calculate the maximal amount (the theoretical amount) of stearic acid that can be obtained from 1.00 g of sodium stearate and excess acid.

3.  Calculate the moles of HCl required to react completely with 1.0 g of sodium stearate.

4.  Calculate the volume of 3 M HCl (3 moles of HCl per liter) that contains the number of moles of HCl required for the reaction with 10.0 g of sodium stearate.

# REPORT FORM

## INVESTIGATION 8:  ISOLATION OF A FATTY ACID

Name: _____     ID No.: _____

Instructor: _____     Course/Section: _____

Partner's Name (if applicable): _____     Date: _____

---

**Experimental data**          (Brand of soap used: _____)

Mass of soap used

    Mass of soap + beaker: _____ g

    Mass of beaker:     _____ g

    Mass of sample:     _____ g

Volume of 3 M HCl used:   _____ mL

Show all calculations.

| Crude product | | Final product | |
|---|---|---|---|
| Mass of flask + acid: | _____ g | Mass of paper + acid: | _____ g |
| Mass of flask: | _____ g | Mass of paper: | _____ g |
| Mass of acid: | _____ g | Mass of acid: | _____ g |

Melting point of the acid obtained: _____ °C

**Calculated values**

Theoretical amount of carboxylic acid: _____ g
Show all calculations on back of this page.

Percent yield of crude product: _____ %          Percent yield of purified product: _____ %

Show all calculations.

Description of the crude product and the purified product.

Answer all assigned problems.  Attach a separate sheet if necessary.

1.  Based upon your obtained yield, how many tons of each of the following would be required for the production of 100 lb of fatty acid?

    a)  soap


    b)  3 M HCl    (Its density is 1.01 g/mL)


    c)  acetone    (Its density is 0.79 g/mL)


2.  What happened to the sodium ion during the conversion of soap to a carboxylic acid?  Where does the sodium ion end up?  In the filter cake?  In the filtrate?  Which anion will be associated with $Na^+$?


3.  Compare the melting point of stearic acid listed in a handbook with the melting point of your product.  Comment on any differences.


Date _____ Signature _____

# Investigation 9

## DETECTING IRON, COPPER, AND LEAD IN ORES

### INTRODUCTION

Iron, copper, and lead are among the most important metals of our technological society. Iron is the fourth most abundant element in the earth's crust. The top three are oxygen, silicon, and aluminum, respectively. Rarely found in nature as the free metal, iron is produced commercially mainly from its oxide ores, *hematite* ($Fe_2O_3$) and *magnetite* ($Fe_3O_4$). *Pyrite* ($FeS_2$) is an important industrial mineral, but its sulfur content makes it unsatisfactory for most types of smelting operations. More than 100 million tons of iron ores are processed annually in the United States, mostly for the production of steel.

Copper, of critical importance for the manufacture of electrical devices, wires, and corrosion-resistant pipes and equipment, is much less abundant than iron constituting only about $10^{-4}$ percent by weight of the earth's crust. Although "native copper" (the free metal) is found in nature, such deposits are relatively rare and most copper is prepared from its sulfide ores such as *chalcocite* ($Cu_2S$) and *chalcopyrite* ($CuFeS_2$). Other common ores include *cuprite* ($Cu_2O$), *azurite* [$Cu_3(CO_3)_2(OH)_2$], and *malachite* [$Cu_3CO_3(OH)_2$]. Malachite specimens are often polished for use in jewelry. The depletion of high-grade ores in the United States, coupled with problems of obtaining guaranteed imports from such copper-rich countries as Chile, has necessitated the use of low-grade ores and recovery of copper from scrap metals. Nearly half of our current copper production is from salvaged metal waste.

Lead, even less abundant in nature than copper, occurs mainly as *galena* ($PbS$) from which nearly 2 million tons of lead are produced annually, about one-fourth in the United States. Some of the other common ores are *cerussite* ($PbCO_3$) and *wulfenite* ($PbMoO_4$). Lead is used mainly in the manufacture of "storage" batteries, cable coverings, pigments, and various alloys. Like copper, nearly half of the current lead production is from reclaimed metal scrap because high-grade lead ores have also been seriously depleted.

It has not been too many years since the days when the prospector and the assayer, lured by rich finds of gold and silver, were common sights in the American West. Much of the history of the metal industries in this country can be traced to discoveries of profitable deposits of copper and other "baser" metals made during the search for silver and gold.

When an ore-bearing vein had been located, samples were collected and brought to the assay office for laborious and usually time-consuming analysis. The assayers determined the concentrations of gold, silver (expressed in ounces of metal per ton of ore), and other metals in the ore. Such an analysis is still one of the first steps in evaluating an ore body. Today's laboratories have an array of analytical instruments for the rapid, precise, and accurate assay of ore samples. The simple inexpensive wet-chemical tests that were the state-of-the-art approximately seventy years ago have, however, not become obsolete. The field geologist may need to do a quick preliminary assay of a promising mineral find and the individual prospector and small mining companies will usually employ the slower, but less expensive, wet-chemical methods.

This experiment provides the opportunity to discover, with the help of qualitative tests you develop with a limited number of reagents for iron, copper, and lead, which of these metals are present in unknown ore samples.

## CONCEPTS OF THE EXPERIMENT

When an unknown ore sample has to be tested for metals contained in it, a small portion of the powdered sample is dissolved. One of the simplest ways to dissolve the sample is treatment with a strong acid, for instance, nitric acid. Warming the ore/nitric acid mixture will speed up the dissolution. The resulting solution will contain the metallic components of the ore as metal nitrates. Nitric acid will not dissolve the host rock (the gangue) that consists largely of silicates. Filtration yields a homogeneous solution that is used for the wet-chemical test.

How does one establish that a certain metal ion is present in solution? Metal ions have the property of reacting with certain reagents that are added as aqueous solutions and forming solids, for instance, by combination of the metal cation with the anion from the added reagent. These solids, called precipitates, often have a characteristic color and are easily observable by the analyst. If under well defined conditions a certain reagent is known to form a precipitate with only one metal cation, then the appearance of a precipitate after the addition of this reagent to the solution to be tested is unequivocal proof that this particular metal is present. However, this ideal situation is encountered only rarely. Usually two or more tests are needed to confirm the presence of a particular metal.

Another possibility of ascertaining the presence of a metal ion is the addition of a reagent that forms a highly colored compound with a particular metal ion. Such a color change is again easily noticed and indicative of the presence of the metal ion.

A solution containing manganese sulfate, obtainable by treatment of a manganese oxide ore with sulfuric acid, can be boiled with a suspension of lead peroxide ($PbO_2$) in concentrated nitric acid. Under these conditions, the pale-pink $[Mn(H_2O)_n]^{2+}$ cation is oxidized to the deep purple permanganate ion, $MnO_4^-$. The appearance of a purple color in the solution is characteristic of manganese. Alternately, solutions containing $Mn^{2+}$ can be made alkaline with sodium hydroxide and then treated with hydrogen peroxide to precipitate brown $Mn_2O_3$.

To detect a metal ion in a solution when no other metal ion is present is easy. The task becomes more complex when several metal ions are present. Under these circumstances, the metal ions must be separated from each other. Such a separation is possible when a reagent can be found that precipitates a particular metal ion but not the others. To achieve clean separations, the precipitations must be carried out under the appropriate conditions. Conditions that are important in this context relate to the acidity (or basicity) of the solution to which reagents are added. Qualitative tests are performed in acidic solutions (acid was added), basic solution (a base was added), or neutral solutions (pH of solution approximately 7). The appropriate reagents for all the metal ions and the pertinent conditions for their detection are known and are described in textbooks of qualitative analytical chemistry. To let you taste the thrill of discovery, to allow you to hone your skills of observation, and to provide the opportunity to experience a research-like experiment, you are asked to develop qualitative tests for $Fe^{3+}$, $Cu^{2+}$, and $Pb^{2+}$. The following seven aqueous reagent solutions are at your disposal:

1. NaOH   sodium hydroxide, 6 M (strong base)
2. $NH_3$   ammonia, 6 M (weak base)
3. HCl   hydrochloric acid, 6 M (strong acid)
4. $H_2SO_4$ sulfuric acid, 3 M (strong acid)

5. $K_4Fe(CN)_6$   potassium hexacyanoferrate(II) (a salt)
6. $NH_4CNS$   ammonium thiocyanate (a salt)
7. KI   potassium iodide (a salt)

After you have explored the reactions of these three metal ions with these reagents, you have the task of developing an analysis sequence (a flow chart) that will allow you to detect each of these metal ions in the presence of the other two. You will have the opportunity to apply your analytical scheme to an unknown ore sample.

## ACTIVITIES

- Complete the PreLab Exercises before coming to the laboratory.

- Explore the reactions of $Cu^{2+}$ solutions, $Pb^{2+}$ solutions, and $Fe^{3+}$ solutions with the reagents under acidic, basic, and neutral conditions.

- Record all observations.

- Develop a flow chart for the identification of each metal ion in the presence of the other two metal ions.

- Dissolve an unknown ore sample.

- Test the "ore" solution for $Fe^{3+}$, $Cu^{2+}$, and $Pb^{2+}$ using your flow chart.

- Analyze an unknown liquid for the possible presence of $Fe^{3+}$, $Cu^{2+}$, and/or $Pb^{2+}$

## SAFETY

*You will be working with some dangerous reagents. Wear safety goggles at all times. In case of accidental contact with chemicals, wash immediately and thoroughly with water. Notify your instructor. Exercise special precautions when heating chemicals to avoid splattering.*

## PROCEDURES

### A. DEVELOPMENT OF TESTS WITH SOLUTIONS KNOWN TO CONTAIN $Fe^{3+}$, $Cu^{2+}$, or $Pb^{2+}$

**A-1.** Copy the Tables from the Report Form into your notebook. Use these Tables to organize your record keeping. Do not fill out the Tables in the Report Form until after you have completed the entire experiment.

**A-2.** Place five drops of the solution containing $Fe^{3+}$ into each of seven test tubes.

**A-3.** Add approximately 0.5 mL (10 drops) of distilled water from the squeeze bottle to each test tube and agitate the mixture.

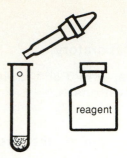

**A-4.** To each of the seven test tubes add, drop by drop, one of the seven reagents listed (i.e., a different reagent to each tube). Agitate after each addition. Keep adding reagent until on further addition of one drop no change is observed or until you have added a maximum of six drops.

*After the dropwise additions of the reagents, you should make one of the following observations:*

- *No change in the appearance of the solution.*
  *If this is the case, make a note of it in your record book. No further tests are necessary for this solution.*
- *Formation of a precipitate.*
  *Continue as described in A-5, A-6 or A-7, as appropriate.*
- *Formation of a colored solution.*
  *Continue as described in A-5, A-6 or A-7, as appropriate.*

*The reagents at your disposal are either acids, bases, or salts. If a precipitate or a colored solution was formed:*

- *Continue with A-5 if the reagent added was an acid.*
- *Continue with A-6 if the reagent added was a base.*
- *Continue with A-7 if the reagent added was a salt.*

**A-5.** If in step A-4 you have added an acid to one of your solutions and observed a change, do the following. Before proceeding further, describe the observation in your notebook. Then add six drops of 6 M NaOH to this test tube and mix thoroughly. Record any observed changes. Add three more drops of the NaOH solution and mix again. Now check your mixture by dipping a clean stirring rod into the mixture and transferring a drop onto a strip of *red* litmus paper. If the mixture is basic, the litmus will turn blue. Should the solution not yet be basic, add three more drops of NaOH solution, stir, and repeat the litmus paper test. Record all observations.

**A-6.** If in step A-4 you have added a base to one of your solutions and observed a change, do the following. Before proceeding further, describe the observation in your notebook. Then add six drops of 6 M $HNO_3$ to this test tube and mix thoroughly. Add three more drops of the acid and mix again. Check your mixture by transferring a drop onto a strip of *blue* litmus paper. If the mixture is acidic, the litmus will turn red. Should the solution not yet be acidic, add three more drops of $HNO_3$ solution, stir, and repeat the litmus paper test. Record all observations.

**A-7**. If in step A-4 you have added a salt and observed a change, do the following. Mix the contents of the test tube thoroughly and record your observations in your notebook. Pour half of the mixture into a clean test tube. Add nine drops of 6 M NaOH to one of these test tubes and nine drops of 6 M $HNO_3$ to the other. Stir and test these mixtures for basicity (as in step A-5) or acidity (as in step A-6) using red or blue litmus, respectively. Record all observations.

**A-8**. When you have finished the test with the $Fe^{3+}$ solution, discuss your results with your instructor. Your instructor might suggest further experiments or ask you to work with the next ion.

**A-9**. Wash the test tube with tap water and loosen any materials clinging to the walls with a test tube brush. Rinse the test tubes with distilled water from your squeeze bottle. Repeat procedures A-1 through A-8 with the $Cu^{2+}$ and $Pb^{2+}$ solutions.

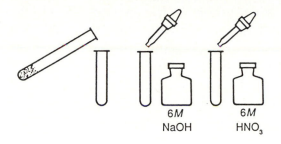

6M
NaOH

6M
$HNO_3$

# B. FLOW CHART FOR THE IDENTIFICATION OF $Fe^{3+}$, $Cu^{2+}$, $Pb^{2+}$

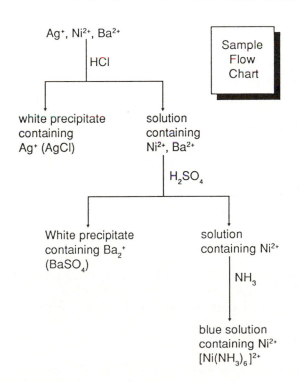

$Ag^+$, $Ni^{2+}$, $Ba^{2+}$

HCl

white precipitate containing $Ag^+$ (AgCl)

solution containing $Ni^{2+}$, $Ba^{2+}$

$H_2SO_4$

White precipitate containing $Ba_2^+$ (BaSO$_4$)

solution containing $Ni^{2+}$

$NH_3$

blue solution containing $Ni^{2+}$ $[Ni(NH_3)_6]^{2+}$

Sample Flow Chart

**B-1**. Using the results obtained from procedure A, work out an analysis scheme that will allow the identification of each one of these three ions in a solution that might contain all three ions.

Your scheme should be presented in the form of a flowchart as illustrated by the following example for the analysis of a mixture of silver ion ($Ag^+$), nickel ion ($Ni^{2+}$), and barium ion ($Ba^{2+}$). Chemical formulas may be omitted at the discretion of your instructor. Note that there are many different analysis schemes possible, but some are more efficient than others.

**B-2**. Have your scheme checked by your instructor. If it is considered satisfactory, test your scheme on a known solution containing $Cu^{2+}$, $Pb^{2+}$ and $Fe^{3+}$.

FeCl$_3$    Cu(NO$_3$)$_2$    Pb(NO$_3$)$_2$

**B-3**. To prepare a known solution of $Fe^{3+}$, $Cu^{2+}$, and $Pb^{2+}$, add to a clean test tube five drops each of your $Fe^{3+}$, $Cu^{2+}$, and $Pb^{2+}$ solutions. With this solution perform the tests outlined in your analysis scheme. When you have to separate a precipitate from a solution, use a centrifuge as described in Appendix E.

## C.  DISSOLUTION OF THE ORE SAMPLE

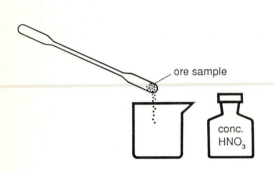

ore sample

conc. HNO$_3$

**UNDER HOOD**

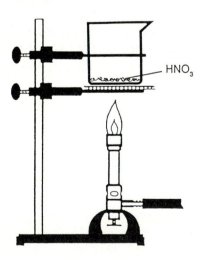

HNO$_3$

**C-1**. Secure an ore sample from your instructor. These samples will generally be in powder form to insure faster reaction with the acid, but chunks of natural sample may be on display for you to look at.

**C-2**. Using a spatula transfer an amount of powdered sample about the size of a medium bean into a 100-mL beaker. Take the beaker into the hood. *Carefully* add concentrated nitric acid until the powder is just covered with the acid.

**C-3**. Attach two rings to a ringstand at the proper height to allow room for a Bunsen burner and place a wire gauze on the bottom ring. Carry this assembly into the hood and put your beaker on the wire gauze. *Position the second ring as shown in the drawing to help prevent the beaker from being knocked from the support.* After the powder is just covered with the nitric acid, brown fumes of NO$_2$ should soon be generated. (Some ores may not produce the brown fumes.) Heat *carefully* with the Bunsen burner if the reaction does not start spontaneously. *As soon as brown fumes appear* (or the reaction begins), remove the flame. When the reaction slows down, heat gently with the Bunsen burner. When the volume has dropped to one-half, add a small amount of acid and some water. When the reaction has subsided, most of the ore should be dissolved. There will always be some insoluble residue (gangue). Sometimes a yellow solid (which is sulfur) will be produced. Dilute carefully with distilled water to three times the original volume and *boil* the solution carefully for five minutes.

**CAUTION:**  *Do not inhale the brown fumes. Do not spill any nitric acid on your skin or clothes. In case of contact of nitric acid with your skin or* *clothes, rinse the affected area immediately with lots of water. Ask one of your fellow students to notify your instructor immediately.*

**C-4**. Decant about 1 mL (20 drops) of the clear liquid from the solids into a test tube.

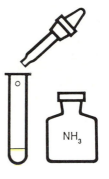

**C-5**. *Carefully* add ammonia, drop by drop, to the solution until a precipitate appears. Add 10 more drops of ammonia.

**C-6**. Into this mixture add 3-5 drops of acetic acid and carefully swirl until the solution becomes clear. It might take a minute or two for the solution to clear. Vigorous stirring will help. Should the solution still be cloudy, add one drop of 6 M nitric acid. Note any observations. This clear solution will be analyzed for $Cu^{2+}$, $Fe^{3+}$, and $Pb^{2+}$.

## D.  TESTING THE DISSOLVED ORE SAMPLE AND AN UNKNOWN FOR $Fe^{3+}$, $Pb^{2+}$, AND $Cu^{2+}$

**D-1**. Place five drops of the clear solution obtained in procedure C into a clean test tube and perform those reactions that will allow you to identify $Fe^{3+}$, $Pb^{2+}$, and $Cu^{2+}$. (See procedure B for flow chart.)

**D-2**. Your instructor may provide a liquid sample for you to analyze in addition to the dissolved ore sample.

**D-3** Using the observations made during this experiment, as recorded in your laboratory notebook, complete the Report Form and submit it to your instructor for grading. Do not use the Report Form as a laboratory notebook. Describe your observations in Tables 1, 2, and 3 with brief notes, or chemical equations, or both.

# PRELAB EXERCISES

## INVESTIGATION 9: DETECTING IRON, COPPER, AND LEAD IN ORES

Instructor:_____

Course/Section:_____

Name:_____

ID No.:_____

Date:_____

---

1. List eight acids, bases or salts that you may use in this Investigation.

2. During this Investigation you will use four common laboratory acids. Give the name and formula of each.

3. Define the term "ore" and give examples for iron, copper, and lead ores.

4. Why is a glass stirring rod and not a spatula used when testing whether a solution is acidic or basic?

(over)

5. Describe the expected state and appearance of $NO_2$.

6. Write a balanced equation that shows the formation of a precipitate when lead ion and sulfate are combined.

# REPORT FORM

## INVESTIGATION 9: DETECTING IRON, COPPER, AND LEAD IN ORES

Name: _____     ID No.: _____

Instructor: _____     Course/Section: _____

Partner's Name (if applicable): _____     Date: _____

---

## Approved Analysis Scheme

$$Fe^{3+}, Cu^{2+}, Pb^{2+}$$

## Analysis Results

Unknown identification code (ore sample)_____

Check ions found       ❑ $Fe^{3+}$       ❑ $Cu^{2+}$       ❑ $Pb^{2+}$

Unknown identification code (liquid unknown)_____

Check ions found       ❑ $Fe^{3+}$       ❑ $Cu^{2+}$       ❑ $Pb^{2+}$

Describe how you applied the approved analysis scheme as shown above to identify the cations in your unknown(s).

Date _____     Signature _____

TABLE 1. Results of qualitative tests carried out on the $Fe^{3+}$ solution.

| Reagent | Addition of reagent | Addition of 6 M HNO$_3$ | Addition of 6 M NaOH |
|---|---|---|---|
| NaOH | | | |
| NH$_3$ | | | |
| HCl | | | |
| H$_2$SO$_4$ | | | |
| K$_4$Fe(CN)$_6$ | | | |
| NH$_4$CNS | | | |
| KI | | | |

TABLE 2. Results of qualitative tests carried out on the $Cu^{2+}$ solution.

| Reagent | Addition of reagent | Addition of 6 M HNO$_3$ | Addition of 6 M NaOH |
|---|---|---|---|
| NaOH | | | |
| NH$_3$ | | | |
| HCl | | | |
| H$_2$SO$_4$ | | | |
| K$_4$Fe(CN)$_6$ | | | |
| NH$_4$CNS | | | |
| KI | | | |

TABLE 3. Results of qualitative tests carried out on the $Pb^{2+}$ solution.

| Reagent | Addition of reagent | Addition of 6 M HNO$_3$ | Addition of 6 M NaOH |
|---|---|---|---|
| NaOH | | | |
| NH$_3$ | | | |
| HCl | | | |
| H$_2$SO$_4$ | | | |
| K$_4$Fe(CN)$_6$ | | | |
| NH$_4$CNS | | | |
| KI | | | |

# Investigation 10

## INTRODUCTION

Analytical chemists are frequently called upon to determine the concentration of inorganic and organic substances in the air, in drinking water, in soils, and in food. The goal of these determinations is the identification of substances harmful to life and the quantification of materials needed for life processes to proceed optimally. Examples of substances that should not be in food are pesticides and herbicides. Drinking water should not contain excessive concentrations of nitrate and toxic heavy metals. However, food items must contain compounds of elements (known as essential elements) at concentrations that provide at least the minimal required daily doses of these elements. Examples of essential elements are chromium, iron, silicon, selenium, nickel, cobalt, and manganese. Because people naturally do not want to be exposed to toxic substances that might cause illness ranging from skin rashes to cancer, water, food, soil and air are tested for health-threatening substances. These tests will identify "hot spots" of pollution and should lead to quick remedial action. This reasonable and health-preserving attitude toward toxic substances often fails to prevail where self-inflicted exposure to harmful substances through smoking, consumption of alcohol, or sniffing drugs is the issue. The understandable desire to keep the concentrations of toxic substances in air, water, soil, and food as low as possible, leads to the idea that a concentration of zero should be the goal.

Although the concept of zero concentration has philosophical merit, its practical meaning is questionable. Every test has what is called its detection limit. Qualitatively, the detection limit is the lowest concentration of a substance that just produces a positive test. The detection limit for a particular molecule or ion will vary with the type of test that is performed. For instance, color tests can be used to detect the presence of $Cu^{2+}$ in aqueous solution. The "blueness" of the aquated $Cu^{2+}$ ion is a test with a very high detection limit; $Cu^{2+}$ must be present at a concentration of 0.01M or higher. However, when ammonia is added to a dilute $Cu^{2+}$ solution, the very intense color of $[Cu(NH_3)_4]^{2+}$ allows much lower concentrations of $Cu^{2+}$ to be detected. The ammonia test for $Cu^{2+}$ has a low detection limit. The same reagent will have different detection limits for different molecules or ions. For instance, ammonia cannot be used at all as a reagent to detect $Zn^{2+}$ (zinc ion), whereas $Cu^{2+}$ can easily be detected.

No test that is useful for environmental samples has a detection limit that even comes close to a zero concentration. Graphite furnace atomic absorption spectrometry, for instance, has a detection limit of $1 \times 10^{-9}$ moles/L for arsenic. Any concentration smaller than $1 \times 10^{-9}$ moles/L cannot be "seen" by this test. To say that no arsenic is present is not stating the truth because arsenic could be present at a concentration of $1 \times 10^{-11}$ moles/L. In terms of molecules, this undetectable concentration is far from zero. A solution containing $1 \times 10^{-11}$ moles arsenic per liter would still have 6,000 billion arsenic-containing molecules in one liter of solution. Therefore, the detection limit of the test must always be given whenever a substance is reported "not to be present." It is better to state such a "negative" analytical result in the following manner: Test XYZ indicated that substance UVW cannot be present at a concentration higher than xyz moles per liter, the detection limit of the test. Knowledge of the detection limits is very important for scientific, political, and legal reasons.

This experiment will introduce you to the concept of detection limits through the semiquantitative determination of detection limits for tests for $Fe^{3+}$, $Ba^{2+}$, and $Pb^{2+}$.

## CONCEPTS OF THE EXPERIMENT

The detection limit of a test for a certain substance is determined by preparing a series of solutions with decreasing concentrations of the substance. The reagent is then added under carefully controlled conditions (temperature, concentration of the reagent, volume of the solution to be tested, volume of the reagent added, time between mixing and evaluation). The concentration of the substance is then identified that produces a just barely observable response. This qualitative definition of detection limit is replaced in quantitative work by the following statement: The detection limit is the concentration of analyte that produces a detector response equal to twice (or three-times) the standard deviation of the detector noise. For this experiment the qualitative definition will suffice.

The solutions are prepared by dilution from stock solutions available in the laboratory. When a volume of the more concentrated solution, $V_C$, of concentration $M_C$ is mixed with a volume of distilled water, $V_W$, the total volume of the dilute solution, $V_D$, will be $V_C + V_W$. The concentration of the dilute solution is $M_D$. During the dilution process the number of moles of substance originally present in the volume $V_C$ remains constant. The number of moles of substance in a solution of volume $V_C$ and concentration $M_C$ is obtained as the product $V_C M_C$ provided the volume is expressed in liters. The concentration of the dilute solution can then be

$$V_C M_C = V_D M_D \tag{1}$$

calculated from equation 1. This relationship can also be used to calculate the volume $V_C$ of a more dilute solution of a certain molarity $M_D$.

When the tests are performed, a strict protocol must be followed. The same volume of the solution to be tested must always be placed in the test tubes; the same number of drops of the reagent must be added. Any change of these conditions will influence the value for the detection limit.

When a substance is to be detected, a property of this substance that is proportional to its concentration must be measured. Examples of such properties are color, electrical conductivity, ability to form precipitates, and absorption of infrared radiation. These properties are measured by detectors that may be as complex as a Fourier-Transform Infrared Spectrometer or as simple as a filter colorimeter. In this experiment your eyes will serve as the detector responding to the intensity of the red color of the $Fe^{3+}-SCN^-$ complex and judging whether or not a precipitate of barium sulfate or lead iodide is present in your test solution.

The addition of ammonium thiocyanate solution to an acidic solution of $Fe^{3+}$ produces a red

$$Fe^{3+} + SCN^- \rightarrow [Fe(SCN)]^{2+} \tag{2}$$
$$\text{yellow} \quad \text{colorless} \quad \quad \text{red}$$

$$K = \frac{\left[Fe(SCN)^{2+}\right]}{\left[Fe^{3+}\right]\left[SCN^-\right]} = 250 \tag{3}$$

complex (equation 2). The equilibrium constant for this reaction defined by equation 3 has the value of 250 at 25°. In the experiment the thiocyanate concentration is practically constant in the test solution at low initial $Fe^{3+}$ concentration. The $Fe^{3+}$ concentration is decreased. As the test is carried out with lower and lower $Fe^{3+}$ concentrations, the concentration of the complex at equilibrium will also decrease, the red color of the solution will become fainter and fainter, and finally will not be discernible any more. The concentration of the complex can be calculated from a rearranged version of equation 3 (equation 4). In these equations the concentrations are the

$$[Fe(SCN)^{2+}] = K \times [SCN^-] \times [Fe^{3+}] = K'[Fe^{3+}] \qquad (4)$$

concentrations of the species at equilibrium and not the concentrations of $Fe^{3+}$ and $SCN^-$ in solution before the reaction has taken place. Solving this equation with the initial concentration of $Fe^{3+}$ equal to the detection limit will show that most of the iron is present in form of $\{Fe(SCN)\}^{2+}$. The total iron concentration is far from being equal to zero, although the thiocyanate test does not indicate the presence of $Fe^{3+}$ any more.

Addition of a solution of sodium sulfate to an aqueous solution of a soluble barium salt precipitates barium sulfate, a fine white solid (equation 5) that can be easily seen against a black

$$Ba^{2+}(aq) + SO_4^{2-}(aq) \rightleftarrows BaSO_4 \downarrow \qquad (5)$$

background. The precipitation of barium sulfate is governed by the solubility product constant $K_{sp}$ (equation 6). The concentrations in equation 6 are again concentrations at equilibrium. Barium

$$K_{sp} = [Ba^{2+}][SO_4^{2-}] = 1 \times 10^{-10} \qquad (6)$$

sulfate will not precipitate from a solution containing barium cations and sulfate anions until the product $[Ba^{2+}][SO_4^{2-}]$ is larger than $1 \times 10^{-10}$. Under the conditions for this experiment the sulfate ion concentration is constant. From equation 6 the barium concentration, at which barium sulfate will just begin to precipitate at the constant sulfate concentration, can be calculated. This concentration is much smaller than the detection limit you will find in the experiment. The reason for this discrepancy can easily be found. Assume that the solubility product has been exceeded to allow one $BaSO_4$ "molecule" to precipitate. This molecule has a diameter of a few Ångstroms ($1 \text{Å} = 1 \times 10^{-10}$ m). It is floating in the test solution but is much too small to be seen with the naked eye or a light microscope. The light microscope cannot make visible particles with diameters smaller than approximately 3,000 Å. The unaided eye could probably see a barium sulfate particle of 0.01 mm diameter. Under the simplifying assumption that this particle is a cube with a volume of $1 \times 10^{-9}$ cm$^3$ and a mass of $4.5 \times 10^{-9}$g (density of $BaSO_4$ = 4.5 g cm$^3$), the particle will consist of approximately $2 \times 10^{-11}$ moles of barium sulfate. Because this particle would be observed to precipitate from 0.5 mL of solution, the concentration of $Ba^{2+}$ before precipitation must have been $4 \times 10^{-8}$ M, the concentration that would be reported as the detection limit. However, according to equation 6 the $Ba^{2+}$ concentration at the beginning of precipitation from a solution 1 M with respect to sulfate is $1 \times 10^{-10}$ M. The several 100-fold difference in these concentrations will become even larger when the fact is considered that many tiny, not visible particles form, all of which need to grow to visible size.

Addition of a 0.1 M solution of potassium iodide to a lead nitrate solution causes yellow platelets of lead iodide to precipitate (equation 7). The solubility product constant for lead iodide is

$$\underset{\text{colorless}}{Pb^{2+}} + \underset{\text{colorless}}{2I^-} \rightleftarrows \underset{\text{yellow}}{PbI_2 \downarrow} \qquad (7)$$

$1 \times 10^{-9}$. Considerations similar to the ones outlined for barium sulfate apply to lead iodide. However, lead iodide has the advantage of being yellow. The precipitate, therefore, is easier to see. Even when the particles of lead iodide are not discernable, the yellow color of the test system indicates their presence.

## ACTIVITIES

- Complete the PreLab Exercises before coming to the laboratory.

- Dilute solutions by properly measuring small volumes of salt solutions and mixing them with the required volumes of distilled water.

- Perform calculations related to dilutions and solubility product expressions.

- Perform tests with the metal ion solutions ($Fe^{3+}$ – $SCN^-$, $Ba^{2+}$ – $SO_4^{2-}$, and $Pb^{2+}$ – $I^-$).

- Estimate the detection limits for these tests.

## SAFETY

*Wear approved eye protection in the laboratory. Soluble salts of barium and lead are toxic. Do not bring these solutions in contact with your skin or ingest them. If you spill solutions on your hands, don't panic, but rinse the affected area with copious amounts of water. Handle all chemicals in the proper way with due respect.*

## PROCEDURES

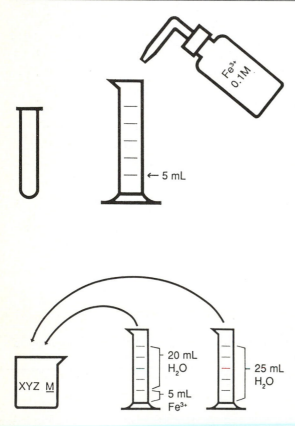

**1.** Clean six beakers (250 mL or smaller volume) and wipe them dry inside. Similarly clean a 25-mL graduated cylinder. Clean at least 10 test tubes and arrange them in a row in a test tube rack.

**2.** Take a clean test tube and the 25-mL graduated cylinder to the squeeze bottle labeled "0.1 M $Fe^{3+}$". Squirt approximately 0.5 mL (10 drops) of this solution into the test tube. Fill the dry graduated cylinder to the 5.0-mL mark with the $Fe^{3+}$ solution. Return to your station and fill the graduated cylinder with distilled water from a squeeze bottle to the 25-mL mark. Pour this solution into one of the beakers. Fill the cylinder again to the 25-mL mark with distilled water. Pour the 25 mL of water into the beaker that contains the diluted $Fe^{3+}$ solution. Mix the solution by swirling the liquid. Rinse the graduated cylinder with distilled water and clamp the cylinder upside down to a ringstand to allow the residual rinsewater to drain from the cylinder. Calculate and record the concentration of $Fe^{3+}$ in the diluted solution. Label the beaker containing the solution with the molarity.

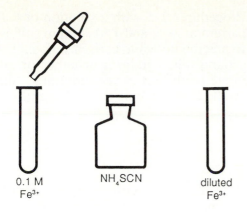

0.1 M Fe³⁺        NH₄SCN        diluted Fe³⁺

**3**. Transfer approximately 0.5 mL (10 drops) of the more dilute $Fe^{3+}$ solution into a test tube. Add to the test tube containing the more dilute $Fe^{3+}$ solution and the one containing the 0.5 mL of 0.1 M $Fe^{3+}$ (from procedure 2) two drops (~0.1 mL) of the ammonium thiocyanate solution. Shake, observe, and record your observations. Also record the concentration given on the label of the ammonium thiocyanate solution. Discard excess solutions in a manner described by your instructor.

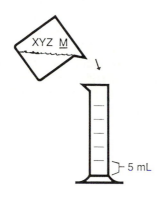

XYZ M

— 5 mL

**4**. Pour 5.0 mL of your most dilute $Fe^{3+}$ solution into the clean graduated cylinder. Fill with distilled water to the 25-mL mark. Pour this solution into a dry beaker. Fill the cylinder again with distilled water to the 25-mL mark. Pour this water into the same beaker. Swirl the liquid. Clean the graduated cylinder. Place 0.5 mL (10 drops) of the resulting solution into a test tube and mix with 2 drops of ammonium thiocyanate solution. Record your observation. Calculate the molarity of this solution and label the beaker with this molarity. Arrange the test tubes in a row in order of decreasing concentration of $Fe^{3+}$.

Decreasing [Fe³⁺]

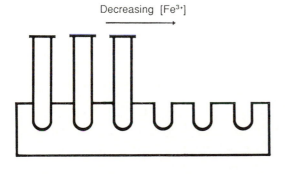

**5**. Repeat procedure 4 until the solution remains colorless upon addition of ammonium thiocyanate. Then make solutions intermediate in concentration between the solution that did not produce any color and the next more concentrated solution that did give a color reaction. Identify the concentration of the solution that gave a color that can just barely be seen. Report this concentration as the detection limit of the ammonium thiocyanate test for $Fe^{3+}$.

Detection limit

Decreasing color        Colorless

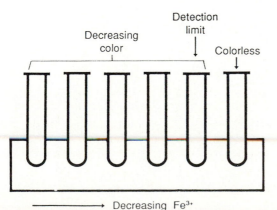

Decreasing Fe³⁺

0.1 M
Ba(NO$_3$)$_2$

**6**. Repeat Procedures 1-5 with a solution of 0.1 M Ba(NO$_3$)$_2$ (barium nitrate) and 1 M sodium sulfate as the reagent to precipitate white barium sulfate. Use a black background when judging whether or not a white precipitate is present in your mixture. Shake the test tube well before making this judgment.

0.1 M
Pb(NO$_3$)$_2$

**7**. Repeat Procedures 1-5 with a solution of 0.1 M Pb(NO$_3$)$_2$ (lead nitrate) and 1 M potassium iodide as the reagent to precipitate yellow lead iodide. Use a white background when judging whether or not a yellow lead iodide precipitate is present in solution. Lead iodide might take a while to precipitate from very dilute solutions. Give these solutions five to ten minutes before evaluating them.

# PRELAB EXERCISES

## INVESTIGATION 10:  DETECTION LIMITS

Name:_____

Instructor:_____     ID No.:_____

Course/Section:_____     Date:_____

1.  Define:
    Concentration

    Detection limit

    Solubility product

    Equilibrium constant

2.  A solution of iron(II) sulfate (10 mL, 0.2 M) is mixed with distilled water (70 mL).  Calculate the $Fe^{2+}$ concentration in the dilute solution under the assumption that the volumes are additive.

3.  A dilute solution (150 mL, 0.05 M) must be prepared from a 1.2 M solution of copper sulfate. How many milliliters of the concentrated solution and of distilled water are to be mixed?

4.  The solubility product of lead sulfate is $1.6 \times 10^{-8}$.  A solution has a sulfate concentration of 2 moles/L.  Calculate the concentration of $Pb^{2+}$ that can be present in this solution without precipitation occurring.

# REPORT FORM

## INVESTIGATION 10: DETECTION LIMITS

Name: _____  ID No.: _____

Instructor: _____  Course/Section: _____

Partner's Name (if applicable): _____  Date: _____

Equations for the reactions between

$Fe^{3+}$ and $SCN^-$:

$Ba^{2+}$ and $SO_4^{2-}$:

$Pb^{2+}$ and $I^-$:

| Concentration of Metal Ion | Observations | | |
|---|---|---|---|
| | $Fe^{3+}$ | $Ba^{2+}$ | $Pb^{2+}$ |
| | | | |
| | | | |
| | | | |
| | | | |
| | | | |
| | | | |
| | | | |
| | | | |

Estimated concentration in test tubes of $SCN^-$: _____ M

$SO_4^{2-}$: _____ M

$I^-$: _____ M

Detection Limits for $Fe^{3+}/SCN^-$ reaction: _____

$Ba^{2+}/SO_4^{2-}$ reaction: _____

$Pb^{2+}/I^-$ reaction: _____

Concentration of metal ion (calculated from solubility product constant) at which precipitation would just begin at the estimated concentration of the reagent:

$Ba^{2+}$: _____ M

$Pb^{2+}$: _____ M

## OPTIONAL

The ratio (D/C) of the detection limit found in the experiment (D) to calculated concentration at which the metal ion just begins to precipitate or first gives a detectable color (C) is:

$Fe^{3+}$: _____

$Ba^{2+}$: _____

$Pb^{2+}$: _____

Why is the ratio D/C not one?

How many moles of $Fe^{3+}$, $Ba^{2+}$, or $Pb^{2+}$ were present in 0.5 mL of the 0.1 M solutions of the ions?

Date _____ Signature _____

# Investigation 11

## INTRODUCTION

All living systems require traces of certain metal ions such as $Cr^{3+}$, $Mn^{2+}$, $Fe^{3+}$, $Co^{2+}$, and $Cu^{2+}$. Known as essential elements, these ions play critical roles in a number of biochemical reactions. Other metal ions and their compounds are toxic to most organisms. Poisonings by mercury and lead are familiar examples. Much is still unknown about the interactions between metal ions and biological systems. The emerging field of bioinorganic chemistry explores these interactions on a molecular level. Essentiality and toxicity are two characteristics that do not exclude each other. To support life functions, organisms including man must maintain in the body certain concentrations (generally very low) of the essential elements. These elements are provided in the form of various compounds by the water we drink and the food we eat. If the body receives insufficient daily doses of the essential elements, biochemical reactions are impaired and sickness or even death may result. If food provides excesses of these trace elements, different illnesses and death may occur. Therefore, the diet of any living organism must contain not only the proper elements, but also the proper range of their concentrations.

We take advantage of the concentration-dependent action of these elements in a number of familiar systems. For example, zinc oxide ointments are employed as topical agents for the treatment of certain fungal infections. Zinc oxide is not very soluble in water. The concentration of $Zn^{2+}$ ions present in the fluids at the treated site, therefore, is rather low and absorption of these ions through the skin does not raise the zinc concentration in the body to dangerous levels. (In the average man approximately two grams of zinc are present serving as an essential constituent of many enzymes.) However, the $Zn^{2+}$ concentration from the zinc oxide is high enough to be toxic to the fungus causing the disease. Exposure to this zinc concentration causes the death of the fungus and cures the fungal infection.

Copper salts, usually copper sulfate pentahydrate, $CuSO_4 \cdot 5H_2O$, is added to swimming pools and decorative ponds to prevent the growth of algae. Copper is an essential element for algae. Copper must be present in the water at a concentration of approximately 1 mole of copper in 10 million liters of water. However, at higher concentrations of copper (approximately 1 mole of copper in 100,000 liters of water) the algae cannot live. To keep a swimming pool free of algae and attractive to swimmers, the concentration of $Cu^{2+}$ in the pool water must be maintained at the level lethal to the algae. Concentrations too low will promote algae growth, and concentrations too high will stain swim wear and may become toxic to the swimming pool users.

In this experiment the "algae-swimming pool-copper concentration" system is used as a pretext to make you familiar with concentration units and the preparation of solutions by dilution and by dissolving weighed amounts of copper sulfate in water.

## CONCEPTS OF THE EXPERIMENT

A solution consists of a substance, the solute, that is dissolved in a solvent. A very important solvent is water. When a solute is dissolved in the solvent water, an aqueous solution is formed.

One of the most important characteristics of a solution is its concentration. The concentration of a solution can be expressed in various ways.

- moles of solute per liter of solution: molarity, mol $L^{-1}$
- moles of solute per kilogram of solvent: molality, mol $kg^{-1}$
- milligrams of solute per 1 million milligrams of solvent: mg $kg^{-1}$ (ppm)
- micrograms of solute per 1 million milligrams of solvent: $\mu g$ $kg^{-1}$ (ppb)

In this experiment only the concentration units mol $L^{-1}$ solution and mg $kg^{-1}$ solvent will be used.

A solution of a certain molarity is prepared by calculating the grams of solute needed and then dissolving this amount of solute in distilled water to a certain volume. This operation is performed with volumetric flasks, long-necked flasks with a mark on the neck of the flask indicating the volume. The weighed solute is quantitatively transferred into the volumetric flask. Water is then added to the mark. The molarity of the solution can then be calculated from the mass of solute (m), the molecular mass of the solute (MM), and the volume of the solution (V) given on the flask according to equation 1.

$$M\left(mol\ L^{-1}\right) = \frac{m(g)}{MM\left(g\ mol^{-1}\right)V(L)} \tag{1}$$

When solutions of very low concentrations are to be prepared, the weighing-dissolution procedure cannot be used. Suppose a solution with 0.1 mg of solute per liter of solution must be prepared. Although the analytical balance can weigh to 0.1 mg, the error in such a weighing can easily be as large as the sample to be weighed. One could increase the volume of the solution to be prepared to 10 liters. The mass of solute now required is 1.0 mg, a mass that is still too small to be determined precisely on the analytical balance. In addition, 10-L volumetric flasks are hard to find and cumbersome to handle. The "solution" to this problem is dilution.

A more concentrated solution is prepared by weighing a large amount of the solute precisely. Then a certain volume ($V_C$) of this concentrated solution is transferred into a volumetric flask and diluted with distilled water to the mark indicating a volume $V_D$. If the concentration of the more concentrated solution was $M_C$, the concentration of the diluted solution, $M_D$, can be obtained from equation 2. In such a dilution process the number of moles in the volume $V_C$ is the same as the number of moles in the volume of $V_D$ after dilution. The number of moles of solute in a volume V

$$M_D = \frac{M_C V_C}{V_D} \tag{2}$$

of a solution of molarity M is given by M x V when the volume V is expressed in liters. Equation 2 is obtained when the expressions for moles of solute in the volume $V_C$ and the volume $V_D$ are set equal. Because a volume appears in the numerator and the denominator in equation 2, any volume unit (liter, milliliter, gallons, ...) can be used for calculations based on this equation, provided $V_C$ and $V_D$ are expressed in the same units. This dilution process can be used repeatedly until the desired solution of low concentration is obtained.

Equation 2 is also useful when the volume $V_C$ of a more concentrated solution required to prepare a certain volume of a dilute solution of molarity $M_D$ is to be calculated. In this case, equation 2 is rearranged to isolate the unknown $V_C$ and bring it to the left side of the equation.

When a concentration of a very dilute solution is expressed in ppm (mg of solute per kg of solvent), the density of this solution and the formula for the solute are needed to convert the ppm concentration to mol $L^{-1}$ (molarity) concentration. A solution of "N" ppm of solute with molecular mass MM and a density $\rho$, will have N mg of solute in 1000 g of solvent. The total mass of the solution is $(1000 + N \times 10^{-3})$ g.  Density is defined as mass per unit volume.  The volume of the solution will then be $(1000 + N \times 10^{-3})$g$/\rho$(g mL$^{-1}$).  Volume V contains N x 10$^{-3}$ (g)/MM(g mol$^{-1}$) moles of solute.  Molarity is defined as the "number of moles of solute per liter of solution." Because the solution has N x 10$^{-3}$/MM moles in a volume V that must be expressed in liters [(1000 + N x 10$^{-3}$)/($\rho$ x 10$^{3}$)], the molarity of the solution is equal to the quotient of (N x 10$^{-3}$/MM) divided by V.  A general expression for this conversion is given in equation 3.  Because

$$M\left(\text{mol L}^{-1}\right) = \frac{N \times 10^{-3}(g) \times \rho\left(g\,mL^{-1}\right) \times 10^{3}\left(mL\,L^{-1}\right)}{MM\left(g\,mol^{-1}\right)\left[1000 + N \times 10^{-3}\right](g)} = \frac{N(ppm)\,\rho\left(g\,mL^{-1}\right)}{MM\left(g\,mol^{-1}\right)\left[1000 + N \times 10^{-3}\right](g)} \quad (3)$$

$[1000 + N \times 10^{-3}]$ is approximately equal to 1000 and the density of very dilute aqueous solutions is about equal to the density of pure water (which is approximately 1 g mL$^{-1}$ at room temperature), equation 3 can be simplified (equation 4).

$$M\left(\text{mol L}^{-1}\right) = \frac{N(ppm)}{MM\left(g\,mol^{-1}\right)\,1000}$$

$$(4)$$

Note that in equation 4 only the units needed for the numbers entering the calculation are given.

## ACTIVITIES

- Complete the PreLab Exercises before coming to the laboratory.
- Weigh the required amount of copper sulfate pentahydrate within ± 2 mg on the analytical balance without spilling any of the salt.
- Prepare a solution of copper sulfate.
- Perform calculations needed for the dilution of copper sulfate solutions.
- Dilute copper sulfate solutions to predetermined concentrations.
- Determine the density of the 1 ppm $Cu^{2+}$ solution.
- Complete the Report Form and solve the assigned calculations (problems).

## SAFETY

*Wear approved eye protection.  Exercise standard precautions when handling reagents.  Avoid spilling solid copper sulfate and copper sulfate solutions.*

## PROCEDURES

**1.** Place some tap water into a 100-mL volumetric flask. Stopper the flask and invert it. Hold the stopper with your fingers to prevent it from falling out. No water should leak out of the inverted flask. If you notice leakage, consult your instructor.

**2.** Inspect the flask for cleanliness. No foreign material should be in the flask. If you detect solid material in the flask, use water, detergent, and a test tube brush to clean the flask. Rinse the flask thoroughly with tap water whether or not you had to clean it with a detergent. Then rinse the flask with distilled water. Let all the water drain from the flask. The flask does not have to be dry inside.

**3.** Obtain from the $CuSO_4 \cdot 5H_2O$ storage bottle several tenths of a gram of the blue salt. Place the salt into a mortar and grind it with a pestle into a fine powder.

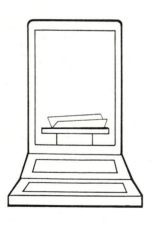

**4.** Weigh a clean, creased piece of *smooth* paper (~5 x ~8 cm) on the analytical balance to the nearest 0.1 mg. Arrest the balance, but do not change any of the mass dials. Leave your paper on the pan of the balance. Check your PreLab Exercise for the amount of $CuSO_4 \cdot 5H_2O$ required for the preparation of 100 mL of a 0.004 M solution of copper sulfate. Add the masses of your creased paper and the required copper sulfate pentahydrate. Appropriately adjust the mass dials on the balance to this new mass. Scoop some of the powdered copper sulfate from the mortar into your scoopula. *Perform the next step without spilling any of the copper sulfate.*

**5.** Release the balance slowly and carefully. Hold the scoopula with the salt in one hand and bring the end of it over the creased paper on the balance pan. Hold the scoopula steady with its front end approximately 1 cm above the center of the paper. With your other hand tap the hand holding the scoopula lightly to force copper sulfate to fall from the scoopula onto the paper. Watch the light window on the balance. When the desired mass is approached, add very small amounts of the salt and let the balance come to equilibrium before you add more copper sulfate. *Keep in mind that addition of salt to the paper is easier than removing salt from the paper.* **Note:** *Should you need practice with the method of addition needed here, load your scoopula, hold it over the mortar and tap your hand holding the scoopula to let the salt fall into the mortar. Practice "tapping off" very small amounts of salt.* When you have added sufficient copper sulfate and you are within ± 2 mg of the required mass, return the salt still on your scoopula to the mortar. Weigh the paper with the copper sulfate to the nearest 0.1 mg. Arrest the balance. Record the result of your weighing.

**6.** Place a funnel on your 100-mL volumetric flask. Carefully (don't lose any salt) remove the paper with the copper sulfate from the balance. Let the salt slide from the paper into the funnel. Tap the paper with your scoopula to transfer any salt crystals adhering to the paper into the funnel. Rinse the funnel with distilled water until all crystals have been washed into the flask. Fill the flask half full with distilled water. Remove the funnel. Swirl the solution until all the crystals have dissolved. Should the solution remain cloudy, add one or two drops of dilute hydrochloric acid.

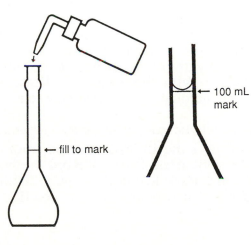

← 100 mL mark

← fill to mark

**7.** Fill the flask to the mark with distilled water. Add the last volume increments drop by drop until the meniscus of the solution touches the 100-mL mark. Stopper the flask. Mix the solution thoroughly by inverting the flask 10 to 15 times and gently swirling the flask. Pour the solution into a clean, dry 250-mL beaker.

**8.** Calculate to three significant figures the molarity of the solution you have prepared (solution I). Label the beaker with this molarity. Then calculate the volume of this solution required for the preparation of 100 mL of a $1.3 \times 10^{-4}$ M solution of copper sulfate. Have your instructor check your calculation before proceeding.

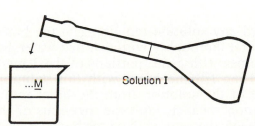

...M        Solution I

**9.** Rinse a clean 50-mL buret twice with 10-mL portions of the copper sulfate solution. Then pour approximately 10 mL of the solution into the buret. Record the volume reading to the nearest 0.01 mL. Save the solution remaining in the beaker.

**10**. Rinse your 100-mL volumetric flask thoroughly with distilled water. Drain the volume of 0.004 M copper sulfate solution required to prepare 100 mL of a $1.3 \times 10^{-4}$ M solution as accurately as possible from the buret into the volumetric flask. Record the final buret reading. Fill the volumetric flask with distilled water, stopper, and mix thoroughly.

Calculate the molarity of the solution you just prepared to three significant figures (solution II).

**11**. Your next assignment is the preparation of a solution containing 1 mg of $Cu^{2+}$ per liter of solution by dilution of the copper sulfate solution obtained in procedure 10. First calculate the molarity of a solution containing 1 mg $Cu^{2+}$ per liter of solution. Then find the volume of the solution prepared in procedure 10 required to make 100 mL of a solution containing 1.00 mg $Cu^{2+}$ per liter of solution. Check with your instructor before proceeding.

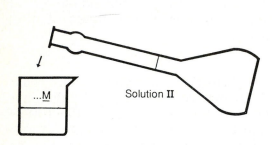

Solution II

**12**. Empty the buret and rinse it three times with 10-mL portions of the solution in the volumetric flask (solution II). Fill the buret with the solution. Drain approximately 5 mL of solution from the buret into a "waste solution" beaker. Record the buret reading. Empty the solution remaining in the volumetric flask into a clean, dry 250-mL beaker. Label the beaker with the molarity of the solution. Save this beaker with the solution. Rinse the volumetric flask thoroughly with distilled water.

**13**. Drain the volume calculated for the preparation of 100 mL of the 1 mg $Cu^{2+}$/L solution from the buret into the volumetric flask. Record the buret reading. Fill the flask to the mark with distilled water, stopper, and mix thoroughly (solution III).

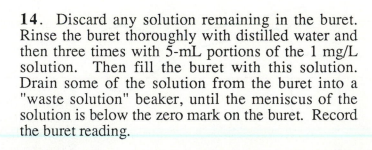

**14**. Discard any solution remaining in the buret. Rinse the buret thoroughly with distilled water and then three times with 5-mL portions of the 1 mg/L solution. Then fill the buret with this solution. Drain some of the solution from the buret into a "waste solution" beaker, until the meniscus of the solution is below the zero mark on the buret. Record the buret reading.

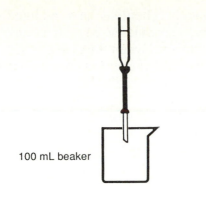

100 mL beaker

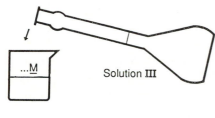

...M

Solution **III**

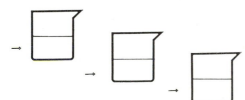

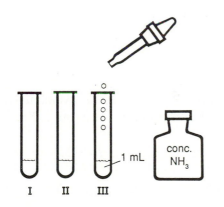

conc.
NH₃

1 mL

I     II     III

**15**. Weigh a clean and *dry* 100-mL beaker on the analytical balance to the nearest 0.1 mg. Make sure that no extraneous materials (water or dirt from your hand or from the lab bench) adhere to the beaker. Drain the 1 mg/L solution from the buret into the weighed beaker until the meniscus is close to the 50-mL mark on the buret. Wait two minutes to read the buret. Record the reading. Reweigh the beaker containing the solution to the nearest 0.1 mg. Record all masses. From the volume and the mass of the solution calculate the density of the 1mg/L solution. Pour the solution remaining in the volumetric flask into a clean, dry 250-mL beaker. Record in your notes the laboratory temperature at this time.

**16**. Copper(II) ions in aqueous solution have a characteristic light-blue color. Inspect your three $Cu^{2+}$ solutions by looking through the solutions from the outside of the beaker. Estimate the concentration of the $Cu^{2+}$ solution that produces sufficient color to be just noticeable. You might want to compare the color to a beaker containing distilled water.

**17**. Copper(II) ions form the intense blue tetraammine complex, $[Cu(NH_3)_4]^{2+}$ with ammonia. This complex can be used to detect $Cu^{2+}$ in quite dilute solutions. Pour 1 mL each of your three solutions into separate small test tubes. To each test tube add 5 drops of concentrated (15 M) ammonia. Observe and record your observations. Estimate the concentrations of a $Cu^{2+}$ solution that would still produce a noticeable blue color.

**18**. Dispose of all solutions as directed by your instructor. Clean all glassware you used.

## CALCULATIONS

1. The concentrations of very dilute solutions are often expressed in terms of mass of solute in 1 million mass units of solvent. This concentration unit is abbreviated as "ppm" (parts per million by mass). For instance, a 1 ppm solution of $Cu^{2+}$ in water has 1 mg $Cu^{2+}$ dissolved

in 1,000,000 mg (= 1000 g = 1 kg) of water. Unfortunately, chemists have adopted the concentration unit moles per liter of solution (molarity, abbreviated M). It is easy to calculate the grams of dissolved substance in any mass of solvent when the concentration is given in ppm units, and in any volume of solution when the concentration is given in terms of molarity. However, the conversion of "ppm" concentration to "molarity" concentration requires that a mass of a solution is converted to a volume. This conversion can be accomplished when the densities of the solutions are known. For your solution III, calculate the following and enter your answers on the Report Form:
  a.  the milligrams of $Cu^{2+}$ in 1,000.0 g of water.
  b.  the milligrams of $Cu^{2+}$ in 1.000 L of solution.
  c.  the milligrams of $Cu^{2+}$ in 1,000.0 g of solution.

2. Based on the answers for problem 1 above, can you recommend the practice of converting ppm concentrations to molarities for very dilute aqueous solutions using the assumption that the densities of such dilute solutions are the same as the density of distilled water?

3. Would this numerical equality between mass of solvent, volume of solvent and volume of solution also hold for liquids with densities different from the density of water?

4. Methanol has a density of 0.792 g $mL^{-1}$. A solution of $CuSO_4 \cdot 5H_2O$ in methanol has a concentration of 1.00 mg $Cu^{2+}$ per 1000.0 g methanol. Calculate the molarity (mol $L^{-1}$) of this solution. Base any assumptions you have to make on the experience you gained in this experiment.

5. Copper(II) ions are toxic to algae only above a certain concentration. At lower concentrations $Cu^{2+}$ is essential for algal growth. For a certain species of algae the optimal concentration for growth is 2.0 x $10^{-8}$ M, the lethal concentration 1.0 ppm $Cu^{2+}$. Calculate for a 5000-gallon swimming pool (1 gallon = 3.985 liters):
  a.  the kilograms of $CuSO_4 \cdot 5H_2O$ needed to achieve a $Cu^{2+}$ concentration in the pool optimal for algal growth resulting in an algal bloom.
  b.  the kilograms of $CuSO_4 \cdot 5H_2O$ needed to prevent algae from infesting the swimming pool.

6. When the ppm concentration unit is used, the compound or ion to which this concentration refers must be given. The ppm unit is defined as mass of compound or ion per 1,000,000 mass units of solvent.
  a.  Calculate the ppm concentration with respect to $Cu^{2+}$ and with respect to $SO_4^{2-}$ for an aqueous solution that is known to be 1.0 ppm with respect to $CuSO_4 \cdot 5H_2O$.
  b.  Calculate the concentration of $Cu^{2+}$ and of $SO_4^{2-}$ in moles per liter of solution in a 1.0 x $10^{-8}$ M aqueous solution of $CuSO_4 \cdot 5H_2O$.

# PRELAB EXERCISES

## INVESTIGATION 11:   COPPER(II) AND ALGAE

Name:_____

Instructor:_____     ID No.:_____

Course/Section:_____     Date:_____

1. Name six metal ions essential to living systems.

2. Name two metal ions that are toxic to living systems.

3. Define "density of solution".

4. Calculate the mass of $CuSO_4 \cdot 5H_2O$ required for the preparation of 100.0 mL of a $4.00 \times 10^{-3}$ M solution.

5. Calculate the volume of $4.00 \times 10^{-3}$ M $CuSO_4 \cdot 5H_2O$ required for the preparation of 100.0 mL of a $1.3 \times 10^{-4}$ M $CuSO_4 \cdot 5H_2O$ solution.

# REPORT FORM

## INVESTIGATION 11:   COPPER(II) AND ALGAE

Name: _____    ID No.: _____

Instructor: _____    Course/Section:_____

Partner's Name (if applicable): _____    Date:_____

---

Show calculations on separate pages.  Attach these pages to the Report Form.

Formula mass of $CuSO_4 \cdot 5H_2O$: _____ g mol$^{-1}$

Mass of $CuSO_4 \cdot 5H_2O$ needed for 100 mL $4.00 \times 10^{-3}$M solution: _____ g

### SOLUTION I

Mass of paper + $CuSO_4 \cdot 5H_2O$: _____ g    Concentration of Solution I calculated

Mass of paper: _____ g    from experimental data:_____ mol L$^{-1}$

Mass of $CuSO_4 \cdot 5H_2O$: _____ g

### SOLUTION II

Volume of Solution I needed to make 100 mL of a $1.3 \times 10^{-4}$ M solution:_____ mL

Solution I in buret:    Final buret reading: _____ mL

Initial buret reading: _____ mL

Volume of Solution I diluted: _____ mL

Concentration of Solution II calculated from experimental data: _____ mol L$^{-1}$

### SOLUTION III

Concentration (mol L$^{-1}$) of a solution containing 1.00 mg $Cu^{2+}$ per liter of solution: _____ mol L$^{-1}$

Volume of Solution II needed to make 100 mL of Solution III:_____ mL

Solution II in buret:    Final buret reading: _____ mL

Initial buret reading: _____ mL

Volume of Solution II diluted: _____ mL

Concentration of Solution III calculated from experimental data: _____ mol L$^{-1}$

## DENSITY OF SOLUTION III (Algicide Solution)

Solution III in buret:   Final buret reading: _____ mL

Initial buret reading: _____ mL

Volume transferred into beaker: _____ mL

Mass of beaker + solution: _____ g

Mass of beaker: _____ g

Mass of solution: _____ g

Density of Solution III: _____ g mL$^{-1}$

Laboratory temperature: _____ °C

Density of water at this temperature: _____ g mL$^{-1}$ (from table in Investigation 3)

## QUALITATIVE TESTS

Estimated concentration of $Cu^{2+}$ that can just be
detected by the light-blue color of hydrated $Cu^{2+}$: _____ mol L$^{-1}$

Estimated concentration of $Cu^{2+}$ that can
just be detected by the color of $[Cu(NH_3)_4]^{2+}$: _____ mol L$^{-1}$

## ANSWERS TO THE CALCULATIONS DESCRIBED IN THE SECTION FOLLOWING PROCEDURE 18. (Attach your calculations for these questions.)

1.  a.  _____ mg $Cu^{2+}$/1,000 g $H_2O$    b.  _____ mg $Cu^{2+}$/1.000 L solution

c.  _____ mg $Cu^{2+}$/1,000 g solution

2.  _____

_____

3.  _____

_____

4.  _____ M

5.  a.  _____ Kg $CuSO_4 \cdot 5H_2O$    b.  _____ Kg $CuSO_4 \cdot 5H_2O$

6.  a.  _____ ppm $Cu^{2+}$    _____ ppm $SO_4^{2-}$

b.  _____ mole L$^{-1}$ $Cu^{2+}$    _____ mole L$^{-1}$ $SO_4^{2-}$

Date _____   Signature _____

# Investigation 12

## ALKA-SELTZER® AND A LITTLE GAS

### INTRODUCTION

The "gastric juice" in the stomach contains hydrochloric acid. This acid aids the digestion of foods. Its secretion increases after a meal. An excess of hydrochloric acid (hyperacidity) may result from any of a number of causes, among them "overeating." The results are generally unpleasant and the remedy most commonly sought is some way of quickly neutralizing excess acid.

A popular home remedy used to be a tablespoon of baking soda ($NaHCO_3$) that consumed excess acid (equation 1). Various commercial preparations are now more commonly employed, among them such alkaline compounds as magnesium oxide ("milk of magnesia") (equation 2).

$$HCO_3^-(aq) + H^+(aq) \longrightarrow H_2O(\ell) + CO_2(g) \tag{1}$$

$$MgO(s) + 2H^+(aq) \longrightarrow Mg^{2+}(aq) + H_2O(\ell) \tag{2}$$

With any of these compounds, dosages must be carefully controlled to avoid reducing the stomach acidity too much, a condition producing undesirable physiological effects. Alka-Seltzer® uses a combination of sodium hydrogen carbonate, citric acid, and calcium hydrogen phosphate to neutralize only excess stomach acid.

A very large number of these acid-neutralizing tablets are produced annually. To assure that the production process works properly and the tablets have the composition claimed on the label, randomly selected samples must be analyzed. Because the hydrogen carbonate ion is the reagent that "consumes" the excess acid, the content of hydrogen carbonate in a tablet is a good measure of the quality of the process and the effectiveness of the remedy.

In this experiment the determination of the percent by mass of hydrogen carbonate in Alka-Seltzer® or a similar product will illustrate the application of gas laws to analytical problems.

### CONCEPTS OF THE EXPERIMENT

An Alka-Seltzer® tablet "fizzes" when dropped into water because the dissolved sodium hydrogen carbonate, also known as sodium bicarbonate or baking soda, reacts with dissolved citric acid to liberate carbon dioxide (equation 1), a gas at room temperature. This reaction does not occur at room temperature when the reagents are mixed as solids. The volume of gaseous carbon dioxide liberated from a weighed piece of a tablet will be used to calculate the percent hydrogen carbonate ion in the tablet. To make sure that all the hydrogen carbonate in the sample has reacted, the tablet is added to 6 M hydrochloric acid containing a large excess of acid. The carbon dioxide formed (equation 3) is collected and its volume measured. To perform this task in a quantitative

$$NaHCO_3(aq) + HCl(aq) \longrightarrow NaCl + H_2O + CO_2 \uparrow \tag{3}$$

way you will use an apparatus consisting of an Erlenmeyer flask in which the reaction between sodium hydrogen carbonate and hydrochloric acid will occur, a buret which will allow the volume of generated carbon dioxide to be measured, and a leveling bulb, the function of which will be explained shortly. To prevent the carbon dioxide from escaping into the atmosphere, the buret and the leveling bulb are partially filled with water. The part of the buret not filled with water, the rubber tube connecting the buret to the Erlenmeyer flask, and the Erlenmeyer flask are filled with air occupying a volume V. When carbon dioxide is formed, the number of moles of gas in the volume occupied by air increases. If we insist that the volume remain constant, then the total pressure P in the constant volume V must increase. That the pressure must increase under these

$$PV = nRT \qquad (4)$$

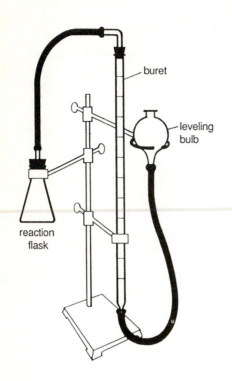

buret

leveling bulb

reaction flask

conditions is clearly shown by applying the ideal gas law (equation 4). When only air ($n_a$ moles of air) occupies

$$P_a = \frac{n_a RT}{V} \qquad (5)$$

the volume V, the pressure can be calculated (equation 5). When $n_c$ moles of carbon dioxide are added to the volume V, the total number of moles will be $n_c + n_a$, and the total pressure P will be given by equation 6.

$$P = P_c + P_a = \frac{(n_c + n_a)RT}{V} \qquad (6)$$

The goal of the experiment is the determination of the number of moles of carbon dioxide, $n_c$, liberated by a piece of the tablet. Equation 6 would allow $n_c$ to be calculated provided all other quantities are known. The gas constant R is known, the temperature can be easily measured, the volume and $n_a$ could be determined, but the accurate and precise measurement of the total pressure cannot be carried out with equipment available in the laboratory of this course.

Because it is much easier to measure a volume change than pressure, the experiment is conducted to have the gas (air and carbon dioxide) in the apparatus at the barometric pressure when the carbon dioxide evolution has ceased. This condition is easily achieved with your apparatus. The volume of a gas sample will change with the temperature, the pressure, and the moles of gas (see equation 4). In your experiment the temperature (the laboratory temperature) will remain constant. When the pressure is kept constant too, the volume will only depend on the number of

$$V = (n_c + n_a)\frac{RT}{P} = K(n_c + n_a) \qquad (7)$$

moles of gas (equation 7). It is convenient to keep the pressure of the gas inside the apparatus at the barometric pressure. Equation 7 can now be rewritten for the sample of gas in the apparatus before any carbon dioxide has formed (equation 8a) and after all the carbon dioxide has evolved (equation 8b). The volume change under these conditions is proportional to the moles of carbon

$$V_{before} = n_a\frac{RT}{P_b} \qquad (8a)$$

$$V_{after} = (n_a + n_c)\frac{RT}{P_b} \qquad (8b)$$

dioxide, $n_c$, liberated (equation 9). When the volume change has been measured, $n_c$ is the only unknown in equation 9.

$$\Delta V = V_{after} - V_{before} = (n_a + n_c)\frac{RT}{P_b} - n_a\frac{RT}{P_b} = n_c\frac{RT}{P_b} \qquad (9)$$

How does one make sure that the volumes are measured at barometric pressure? This task can be accomplished with the help of the leveling bulb. When the buret and the leveling bulb are partially filled with water and the bulb is resting in the ring, the water level in the buret will be higher than the level in the bulb. Let us analyze this situation with respect to pressure. Pressure equal to the barometric pressure is exerted by the atmosphere on the surface of the liquid in the leveling bulb. Because the liquid levels are stationary, the barometric pressure must be equal to the sum of the pressure exerted by the air in the apparatus ($P_a$) on the surface of the liquid in the buret and the pressure exerted by the liquid water column in the buret ($P\ell$) corresponding to the difference in height between the liquid levels in the buret and the leveling bulb (equation 10). When the liquid levels in the buret and the bulb are at the same height, $\Delta h$ is zero and $P\ell$ is zero,

$$P_b = P_a + P_\ell \qquad (10)$$

and the gas pressure inside the apparatus is equal to the barometric pressure (equation 10). The matching of the levels is accomplished by raising or lowering the leveling bulb. In this manner the volume of the gas in the apparatus (represented by the initial buret reading) is found. This buret reading is not the actual volume of the gas in the apparatus at barometric pressure. The actual volume is not known and is not required because it cancels when the volume change is calculated at constant pressure. When carbon dioxide is formed, the water level in the buret drops. While carbon dioxide is evolving, the liquid levels do not have to be the same in the buret and the bulb. The bulb may rest in the ring. However, keeping the levels matched during the reaction and, thus, the pressures inside and outside the apparatus equal will reduce errors caused by a leaky apparatus. When the liquid level in the buret does not change any more, the bulb and buret levels are matched again and the buret read. The difference between the final and initial buret reading is the volume of carbon dioxide produced by the sample.

Before the moles of carbon dioxide evolved can be calculated according to equation 9, two corrections must be applied.

The first correction is needed because the gas in the apparatus is not only air and carbon dioxide but also contains gaseous water (water vapor). The gas is saturated with water vapor. It is known that the vapor pressure of water, the pressure exerted by the gaseous water when the gas sample contains as much water vapor as possible (saturated with water vapor), is only dependent on the temperature. The vapor pressure of water was determined at various temperatures. The values at temperatures close to room temperature are listed as millimeters of mercury (torr) in Table 12.1. To find out how to correct for the presence of water vapor, we confine at barometric pressure the air in the apparatus to the volume $V_a$ and the carbon dioxide to the volume $\Delta V$ (the difference in the buret readings). The gas in the volume $\Delta V$ consists of carbon dioxide and water

**Table 12.1** Vapor pressure of water

| C° | torr | C° | torr |
|----|------|----|------|
| 15 | 12.8 | 26 | 25.2 |
| 16 | 13.6 | 27 | 26.7 |
| 17 | 14.5 | 28 | 28.4 |
| 18 | 15.5 | 29 | 30.0 |
| 19 | 16.5 | 30 | 31.8 |
| 20 | 17.5 | 31 | 33.7 |
| 21 | 18.6 | 32 | 35.7 |
| 22 | 19.8 | 33 | 37.7 |
| 23 | 21.1 | 34 | 39.9 |
| 24 | 22.4 | 35 | 42.2 |
| 25 | 23.8 | 36 | 44.6 |
|    |      | 37 | 47.1 |

vapor. The total pressure of these two gases is equal to the barometric pressure and equal to the sum of the pressures (partial pressures) of the carbon dioxide, $P_c$, and water vapor. $P_{H_2O}$ (equation 11). Because $P_b$ is known and $P_{H_2O}$ can be found in Table 12.1, the partial pressure of carbon dioxide can be calculated. This procedure

$$P_b = P_c + P_{H_2O} \qquad (11)$$

is known as "correcting for vapor pressure of water".

The carbon dioxide sample liberated in the reaction is now characterized by its volume (difference in buret readings), by its partial pressure, $P_c$, and its temperature (laboratory temperature). Application of equation 9 could now be used to calculate the moles of carbon dioxide generated. However, a second correction to the volume of collected carbon dioxide must be applied because carbon dioxide is somewhat soluble in water as you know from carbonated beverages.

The carbon dioxide generated in the reaction in the Erlenmeyer flask is in contact with the aqueous hydrochloric acid and with the water in the buret. To prevent carbon dioxide from dissolving in the water in the buret this water was saturated with carbon dioxide before being placed in the buret (see Procedure 4). Therefore, no carbon dioxide will be lost by dissolving in the "buret" water. However, the hydrochloric acid in the Erlenmeyer flask will become saturated with carbon dioxide. The carbon dioxide dissolved in this acid will not be measured. Experiments beyond our laboratory capabilities showed that 0.80 mL of gaseous carbon dioxide at 760 torr and 25°C will dissolve in each mL of 6.0 M hydrochloric acid. (This value varies with temperature and pressure but we will treat it as a constant for this experiment.) With this information and the volume of hydrochloric acid placed into the Erlenmeyer flask, the volume of carbon dioxide lost by dissolution in the solution in the flask can be calculated and added to the measured volume of released carbon dioxide.

Before you begin to add the two volumes of gaseous carbon dioxide, consider that the volumes of gases change with temperature and pressure. Gas volumes should only be added when they correspond to the same temperature and pressure. It is unlikely that the temperature and pressure for the volume of carbon dioxide dissolved in 1 mL 6 M HCl (given for 25°C, 760 torr) will be the same as the temperature and pressure (laboratory temperature, barometric pressure) at which the liberated carbon dioxide was collected. Therefore, one of the gas volumes must be converted from condition $P_1$, $V_1$, $T_1$ to condition $P_2$, $V_2$, $T_2$ (number of moles of gas is constant). This conversion can be accomplished with the help of equation 12 that is easily derived from the ideal

$$\frac{P_1 V_1}{T_1} = \frac{P_2 V_2}{T_2} \qquad (12)$$

gas law. For the sake of uniformity, the volume of collected carbon dioxide should be converted to 25°C and 760 torr. Now you can add the volumes of carbon dioxide and calculate from this total volume the number of moles of carbon dioxide (equation 4), the number of moles of hydrogen carbonate (equation 1), the number of grams of hydrogen carbonate in your piece broken from the tablet and, finally, using the mass of your sample, the percent by mass of hydrogen carbonate in the tablet.

To obtain a more reliable value for the percent hydrogen carbonate, average your experimental results. Express the precision of your average in terms of the standard deviation and the relative standard deviation (Appendix H).

## ACTIVITIES

- Complete the PreLab Exercises before coming to the laboratory.

- Review calculations with the ideal gas law before coming to the laboratory.

- Assemble the apparatus needed for the experiment.

- Quickly weigh a moisture-sensitive solid on an analytical balance to 0.1 mg.

- Determine volumes of carbon dioxide at room temperature and barometric pressure.

- Use the ideal gas law for calculations involving the gaseous carbon dioxide including corrections for water vapor and solubility of carbon dioxide in 6 M hydrochloric acid.

- Complete the Report Form

## SAFETY

*You will be working with hydrochloric acid. Wear safety goggles at all times. Under no circumstances should the reaction of Alka-Seltzer® (or a similar preparation) with acid or water be performed in a closed container. An explosion could result. Be very careful when working with glassware, particularly when tightening stoppers and slipping rubber tubings on glass tubes. Excessive force in these operations might lead to breakage of the glass and painful cuts.*

## PROCEDURES

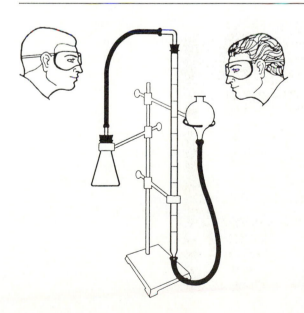

1. This experiment requires more than one pair of hands. Therefore, choose a lab partner or work with the partner assigned to you by your instructor. Although the work will be divided between the two partners, each partner must maintain a complete set of notebook records and submit separate PreLab Exercises and Report Forms.

2. Assemble the apparatus shown in the drawing. Attach a buret clamp to a ringstand about 30 cm above the base. The arm of the clamp must extend over the base of the stand. Attach at approximately a 45 degree angle to the first clamp another buret clamp

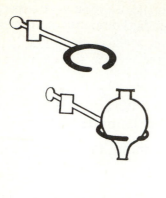

and a clamp with a small ring that has a small opening. These also need to be near the top of the ringstand. Connect a 60-cm long rubber tubing to the lower end of a 50-mL buret. Fasten the buret securely to the center clamp. The lower end of the buret should be 15 cm above the base of the ringstand. Obtain a No. 00 one-hole rubber stopper that holds a short piece of glass tube with a 90-degree bend and a one-hole stopper with a piece of straight glass tube fitting a 125-mL Erlenmeyer flask.

Fasten the 125-mL flask securely to the left clamp. Connect the two stoppers with the 60-cm long rubber tubing. Insert the small stopper into the top of the buret and the larger stopper into the Erlenmeyer flask. Connect the rubber tubing attached to the bottom of the buret to a leveling bulb. Place the bulb into the ring.

3. Obtain from your instructor an Alka-Seltzer® tablet or a tablet of a similar preparation. *The tablet must be kept dry and must be protected from atmospheric moisture. Make sure your hands are dry when handling the tablets. Keep the tablet wrapped in aluminum foil or plastic wrap whenever you are not breaking off a sample.* Weigh a piece of smooth paper on the analytical balance. Record its weight to $\pm 0.1$ mg. Place the whole tablet on the paper and quickly determine the mass of the paper plus tablet. Record the mass of the tablet in your notebook. Place the tablet back in the foil wrapper. From the tablet break off approximately one tenth. Weigh this fragment on the analytical balance by first placing a piece of smooth paper on the balance pan. Lay the fragment of the tablet on the paper. Quickly determine the mass of paper + fragment. Remove the fragment. Weigh the paper. Calculate the mass of the fragment. Use this result to estimate the size of the fragment weighing 0.2 to 0.3 g. Such a fragment will be used in procedure 8.

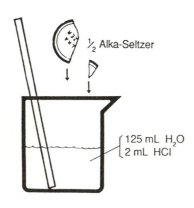

½ Alka-Seltzer

125 mL H₂O
2 mL HCl

4. To prepare water saturated with carbon dioxide, pour 125 mL of distilled water into a 250-mL beaker and carefully add 2.0 mL of concentrated (12 M) hydrochloric acid. Stir the mixture. Drop half of the tablet including all small fragments into the acidic solution. Stir the mixture until bubbling has ceased.

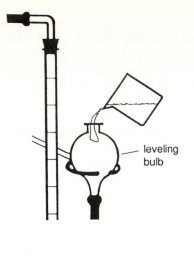

leveling
bulb

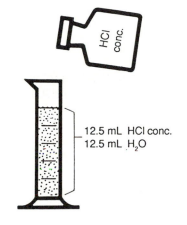

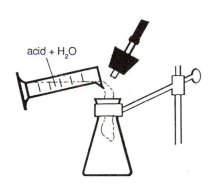

12.5 mL  HCl conc.
12.5 mL  $H_2O$

acid + $H_2O$

**5**. Unstopper the Erlenmeyer flask.  Pour the solution that is saturated with carbon dioxide (prepared in procedure 4) into the leveling bulb.  Lift the bulb and slide the rubber tubing through the opening in the ring.   Raise and lower the leveling bulb several times until all gas bubbles have been expelled.  Place the bulb back into the ring.

**6**. If it is not provided, prepare 25 mL of ~6 M hydrochloric acid by filling a graduated cylinder to the 12.5-mL mark with distilled water.  While stirring, slowly add concentrated (12M) HCl until the solution reaches the 25-mL mark. Stir thoroughly, but carefully; then remove the stirring rod and read (to the nearest 0.5 mL) and record the volume of the liquid in the cylinder.

**7**. Carefully pour about 5 mL of the 6 M hydrochloric acid into the Erlenmeyer flask.  Read (to the nearest 0.5 mL) and record the level of the acid remaining in the cylinder.  From the difference in the two readings, calculate the volume of the acid transferred to the flask.  Wipe dry, if necessary, the inside neck of the flask and the stopper and restopper the flask.

*Plan the next steps carefully and perform them* quickly.  *Alka-Seltzer® deteriorates rapidly on exposure to moist air.  Be sure your hands are dry.*

**8**.  Take your remaining piece of the Alka-Seltzer® tablet that is wrapped in foil to the analytical balance. Place a piece of smooth paper on the balance pan. Quickly unwrap the tablet, break off a piece estimated to weigh 0.2 to 0.3 g.  Wrap the remaining part of the tablet with the foil.  Place the piece on the paper on the pan and weigh the paper + Alka-Seltzer® piece.  If the piece weighs more than 0.3 g, chip off a small piece and reweigh.  Repeat this procedure until the mass of the piece is between 0.2 and 0.3 g.  Weigh the piece + paper to 0.1 mg.  Your partner should remove the piece from the pan and continue with procedure 9.  Reweigh the paper left on the balance pan.  Calculate and record the exact mass of the piece of Alka-Seltzer®.

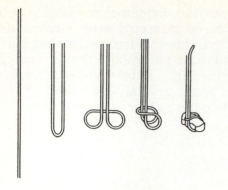

**9.**   Using a thin thread about 20 cm long, quickly form a loop as shown, slip the weighed tablet piece into the loop, and pull it snug. While you are doing this, have your partner continue with procedure 10.

**10.**   Unstopper the Erlenmeyer flask and raise the leveling bulb until the liquid level in the buret is between the 0-mL and 2-mL mark.

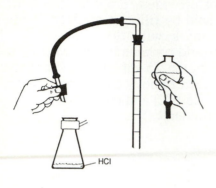

HCl

**11.**   Then carefully suspend the tablet sample about 2 cm above the liquid surface. Hold the thread in place and slowly push the stopper snugly into the neck of the flask. Be sure the stopper fits tightly. *Do not exert too much pressure on the stopper. You might break the neck of the flask and cut yourself.*

**12.**   Lower the leveling bulb to a position approximately 10 cm below the ring clamp. Watch the water level in the buret. The level should drop and come to rest above the level in the bulb. When the liquid level in the buret does not come to rest and slowly descends, the system is leaking. Check both stoppers for leaks. Carefully tighten the two stoppers with a twisting motion and some downward pressure. *Do not push too hard. You might break the buret or flask and injure your hand.* Repeat the test for leaks. If you continue to have a leak, moisten the connections or consult your instructor.

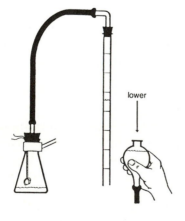

lower

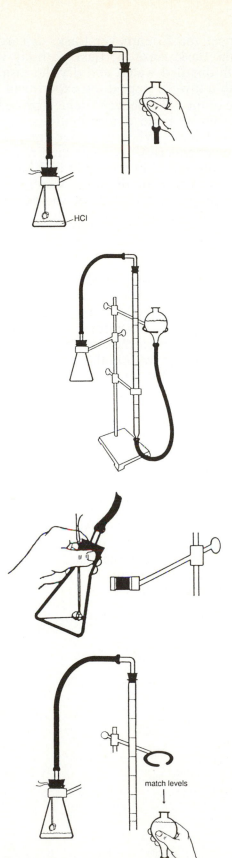

**13**. When your system has passed the leak test, raise the bulb until the liquid level in the buret and the liquid level in the bulb are at the same height. *Under these conditions the liquid level in the buret should be between the 1-mL and 5-mL marks. If the liquid level in the buret is below the 5-mL mark, your system was not gas-tight. You will have to unstopper the buret and repeat procedures 10 and 12.* When the buret reading is in the correct range, record the reading to the nearest 0.01 mL.

**14**. After you have read the liquid level in the buret with the level in the bulb at the same height, replace the bulb into the ring. The critical "fast steps" are now finished.

**15**. Carefully unclamp the Erlenmeyer flask and tilt it until the acid contacts the suspended piece of Alka-Seltzer®. The sample will soon slip out of the thread. Be sure it has dropped into the solution; then reclamp the flask in position and wait until the $CO_2$ evolution has ceased.

*It is advisable to keep the liquid levels in the buret and bulb at the same height during gas evolution by lowering the bulb at the same rate at which the level in the buret drops. This precaution will minimize errors caused by leaks.*

**16**. Loosen the flask clamp just enough to permit you to shake the flask gently. Continue careful checking until the liquid level in the buret has stabilized. Then lower the leveling bulb to match liquid levels and read and record the liquid volume in the buret to 0.01 mL.

To check whether your system was gas-tight during the reactions, place the bulb into the ring. The liquid in the buret should rise, and come to rest, and remain at rest. If the liquid level in the buret keeps creeping up, your system is leaking and the results of the experiment are unreliable. You will have to repeat the experiment after the system has been made gas-tight.

**17.** Unstopper the Erlenmeyer flask, discard the solution in the flask, rinse the flask with tap water and then with distilled water, and dry the inside of the flask with a towel. Repeat the experiment with another piece of the Alka-Seltzer® tablet (procedures 7-16) until four reliable determinations of $CO_2$ volumes are obtained. When you have completed a satisfactory set of experiments, clean the Erlenmeyer flask, discard the liquid in the buret and bulb, and rinse all glassware with distilled water. Your instructor will let you know what you should do with the apparatus.

**18.** In your notebook you should have four sets of data for the mass of a tablet piece and the volume of $CO_2$ collected. Obtain the barometric pressure and the temperature in the laboratory. Calculate the partial pressure of carbon dioxide in your sample. Calculate the volume which the carbon dioxide would occupy at the pressure of 1 atm (760 torr) and 25°C. Then add the volume of carbon dioxide at 25°C and 760 torr that remained dissolved in the hydrochloric acid in the Erlenmeyer flask. From the total volume of $CO_2$ at 25°C and 760 torr calculate the moles of $CO_2$ liberated by each piece of Alka-Seltzer® using the ideal gas law. Write a balanced equation for the reaction of hydrogen carbonate ($HCO_3^-$) with acid to produce $CO_2$. Calculate the grams of $HCO_3^-$ in each piece of Alka-Seltzer® and the percent by mass of $HCO_3^-$. Average the percent-by-mass data and calculate the standard deviation and relative standard deviation of your average. Complete the Report Form.

# PRELAB EXERCISES

## INVESTIGATION 12: ALKA-SELTZER® AND A LITTLE GAS

Name:_____

Instructor:_____     ID No.:_____

Course/Section:_____     Date:_____

1. Define the standard state for gases.

2. What volume will be occupied by one mole of an ideal gas under standard conditions? by 0.1 mole at 25°C and 760 torr? by 10 g of carbon dioxide under standard conditions?

3. Solve the equation $PV = (n_c + n_a)RT$ for $n_c$.

4. A sample of gaseous carbon dioxide was collected over water at 21°C and 753 torr. The volume under these conditions was measured to be 42.05 mL. How many moles of carbon dioxide are in this sample?

5. A piece of a tablet weighing 0.2506 g generated 32.01 mL of carbon dioxide at standard conditions (already corrected for water vapor). Calculate the percent hydrogen carbonate in the sample.

# REPORT FORM

## INVESTIGATION 12: ALKA-SELTZER® AND A LITTLE GAS

Name: _____   ID No.: _____

Instructor: _____   Course/Section: _____

Partner's Name (if applicable): _____   Date: _____

---

## Data

Barometric pressure: _____ torr _____ atm

Laboratory temperature: _____ °C _____ °K

Vapor pressure of water at the temperature of the lab: _____ torr

Partial pressure of $CO_2$ in the buret: _____ torr

| Measured or calculated quantity | Experiment No. | | | |
|---|---|---|---|---|
| | **1** | **2** | **3** | **4** |
| Mass of tablet fragment (g) | | | | |
| Final buret reading (mL) | | | | |
| Initial buret reading (mL) | | | | |
| Volume of $CO_2$ collected (mL) | | | | |
| Pressure of $CO_2$ + $H_2O$ | | | | |
| Partial pressure of $H_2O$ | | | | |
| Partial pressure of $CO_2$ | | | | |
| Volume of $CO_2$ at 760 torr and 25°C | | | | |
| Volume of 6 M HCl (mL) | | | | |
| Volume of $CO_2$ at 760 torr and 25°C dissolved in HCl | | | | |
| Total volume of $CO_2$ liberated by tablet fragment | | | | |
| Moles of $CO_2$ | | | | |
| Mass of $HCO_3^-$ | | | | |
| % $HCO_3^-$ by mass | | | | |

Average percent by mass of $HCO_3^-$: _____ %
Standard Deviation: ± _____
Relative Standard Deviation: ± _____


Mass of one tablet: _____ g
Average Mass of $HCO_3^-$ per tablet: _____ g
Average Mass of $NaHCO_3$ per tablet: _____ g
(if all $HCO_3^-$ were present as $NaHCO_3$)


Sample calculations of Experiment No. _____ are attached.


Date _____    Signature _____

# Investigation 13

## MOLECULAR MASS OF A VOLATILE HYDROCARBON

## INTRODUCTION

When chemists prepare a new compound, they must characterize the compound by determining such properties as melting point, boiling point, color, magnetic and electric behavior, interaction with radiation, and elemental composition. The elemental composition of an organic compound can be found by its combustion to carbon dioxide and water. From the masses of the combustion products and the mass of the compound combusted, the percent composition with respect to carbon and hydrogen can be calculated. Similar methods make it possible to find the percentages of nitrogen, chlorine, bromine, iodine, sulfur, phosphorus, and other elements in a compound. From the percent composition and the atomic masses of the elements, empirical formulas can be calculated.

For instance, a hydrocarbon, with hydrogen and carbon as the only elements, consists of 83.62% carbon and 16.38% hydrogen. This composition, using as atomic masses 12.01 for carbon and 1.008 for hydrogen, leads to an empirical formula of $C_3H_7$ for this hydrocarbon. Compounds with formulas $C_6H_{14}$, $C_9H_{21}$, and $C_{12}H_{28}$ all have the same percent composition as $C_3H_7$. To decide which formula describes the molecule of the hydrocarbon, the molecular mass must to be known.

Elemental analyses, the determinations of the composition of a substance in terms of elements, were very important investigations during the first half of the 19th century, approximately 150 years ago. The atomic mass of an element was defined at that time as the mass that combines with 1.00 gram of hydrogen. Skilled chemists could always find the mass of an element that is associated in a compound with one gram of hydrogen. For instance, the simplest gaseous hydrocarbon was known to have 3.0 grams of carbon for each gram of hydrogen (a combining ratio C/H of 3:1). Depending on the molecular formula one assumes for this hydrocarbon (CH, $CH_2$, $CH_3$, $CH_4$), one obtains 3, 6, 9, or 12 as the atomic masses for carbon. Which atomic mass is the correct one can be easily decided on the basis of the molecular formula. This formula can be deduced from the molecular mass. The determination of molecular masses, therefore, was and still is very important. Before reliable methods became available for the determination of molecular masses, the atomic masses of the elements were not known with any certainty. Old papers contain many incorrect atomic masses and many incorrect molecular formulas. In the meantime - thanks to the many methods developed for the determination of molecular masses - the confusion about atomic masses was cleared up.

In this experiment, you will be introduced to a method for the determination of the molecular mass of a volatile hydrocarbon based on the ideal gas law. The method is applicable to substances that are gases at room temperature or can be easily volatilized by heating.

## CONCEPTS OF THE EXPERIMENT

The ideal gas law (equation 1) relates the pressure and volume of a gas to its temperature and the number of moles. In this equation R is the universal gas constant, the value of which will

$$PV = nRT \qquad (1)$$

depend on the units used to express pressure and volume. The temperature is always expressed in degrees Kelvin. If the volume of a gas sample, its pressure, and its temperature are known, n (the number of moles of gas in the sample) can be calculated. To obtain the molecular mass (MM) of the gas, n must be expressed in terms of the mass of the gas sample and its molecular mass (equation 2). Substitution of equation 2 into equation 1 produces a relation (equation 3) from which the molecular mass can be calculated if the mass, temperature, pressure and volume of the sample is known.

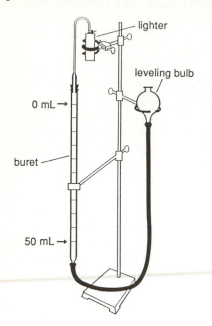

$$n = \frac{m}{MM} \qquad (2)$$

$$PV = \frac{m}{MM} RT \qquad (3)$$

In this Investigation a hydrocarbon, the molecular mass of which is to be determined, is available as a liquid in a disposable lighter that is weighed on an analytical balance. The lighter is connected to the top of the buret, the lower end of which is connected to a leveling bulb. The bulb is filled with water. When the water levels in the bulb and in the buret are at the same height, the pressure of the gas inside the buret is equal to the barometric pressure that is read on the laboratory barometer. When the system is air-tight, the lever on the lighter is depressed, the hydrocarbon in the lighter reservoir vaporizes and the vapor travels through the tube into the buret where it pushes water from the buret into the bulb. When approximately 40 mL of water have been displaced, the lever is released. The water levels in the buret and bulb are matched again (barometric pressure outside the buret) and the volume is read on the buret. The difference between initial and final volume readings gives the volume of volatile hydrocarbon collected. The temperature of the gas sample is equal to the laboratory temperature measured with a thermometer. The lighter is weighed again. The difference in mass (lighter before and after gas release) provides the mass of the gas sample collected in the buret. All quantities are now known (P = barometric pressure, T = laboratory temperature, V = volume of gas, m = mass of gas) to calculate the molecular mass. Before the calculation can be carried out, the gas pressure must be corrected. The gas in the buret is a mixture of the hydrocarbon and water vapor. The total gas pressure in the buret is the sum of the pressure exerted by the hydrocarbon gas and the water vapor (equation 4). When the gas is saturated with water vapor (as is the gas in

$$P_{total} = P_{barometric} = P_{HC} + P_{H2O} \qquad (4)$$

the buret), $P_{H2O}$ (vapor pressure of $H_2O$) depends on the temperature. The temperature is known. Table 13-1 shows the dependence of the vapor pressure of water on temperature. From these data the partial pressure of the hydrocarbon can be obtained from equation 4. This pressure and the other relevant data will allow you to calculate the molecular mass of the hydrocarbon using equation 3. A more detailed discussion of the situation in the gas phase inside the buret, including an explanation why the air initially in the buret does not cause any problems, can be found in the section "Concepts of the Experiment" of Investigation 12.

Table 13-1 Vapor Pressure of Water

| T,°C | torr | T,°C | torr |
|------|------|------|------|
| 15.0 | 12.8 | 19.5 | 17.0 |
| 15.5 | 13.2 | 20.0 | 17.5 |
| 16.0 | 13.6 | 20.5 | 18.1 |
| 16.5 | 14.1 | 21.0 | 18.7 |
| 17.0 | 14.5 | 21.5 | 19.2 |
| 17.5 | 15.0 | 22.0 | 19.8 |
| 18.0 | 15.5 | 22.5 | 20.4 |
| 18.5 | 16.0 | 23.0 | 21.1 |
| 19.0 | 16.5 | 23.5 | 21.7 |

## ACTIVITIES

- Complete the PreLab Exercises before coming to the laboratory.
- Assemble the apparatus for the determination of molecular masses.
- Collect several sets of data for the hydrocarbon in disposable lighters.
- Calculate the partial pressure of the hydrocarbon.
- Calculate the molecular mass of the hydrocarbon, its average, standard deviation, relative standard deviation, and precision.
- Complete the Report Form.

## SAFETY

*You are working with volatile hydrocarbons under pressure. Do not allow open flames to come in contact with the hydrocarbon gas. Be careful when connecting hoses to glass tubes. Too much force could break the glass and cut your hand. Wear approved eye protection while in the laboratory.*

## PROCEDURES

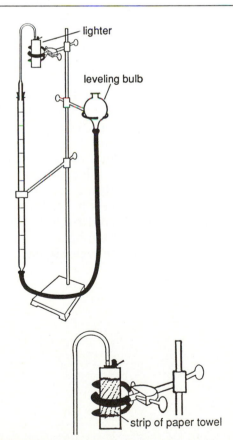

**1**. Assemble the apparatus for the determination of molecular masses consisting of a 50-mL buret, a leveling bulb, and a disposable lighter. Fasten the buret and the lighter with clamps to the ringstand as shown in the drawing. To hold the leveling bulb, attach a ring clamp with an open ring to the ringstand. Connect the lower end of the buret to the leveling bulb with a 60-cm rubber hose. Carefully slide a 50-cm long piece of 1.2 to 1.6 mm (I.D.) polyethylene tubing onto the capillary end of the glass tube protruding from a No. 00 rubber stopper. Remove the metal part and the flint and wheel from the disposable lighter. Weigh the partly dismantled lighter on the analytical balance to the nearest 0.1 mg. Wrap the middle 2 cm of the lighter with a piece of folded paper. Clamp the lighter onto the ringstand with the top clamp. Make sure the paper is between the claws of the clamp and the lighter body. Loosely insert the stopper into the top of the buret. Carefully slip the end of the plastic tubing onto the metal tube on top of the lighter. You may have to hold this tubing with one hand to prevent it from slipping off the connection on the lighter while you or your partner releases the hydrocarbon in Procedure 4. Loosen the stopper on the buret to allow air in the buret to be in contact with the air outside the buret.

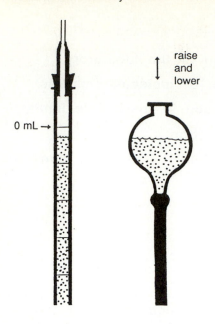

**2**. Pour approximately 200 mL tap water into the leveling bulb. Raise and lower the leveling bulb several times to expel all air bubbles that might be trapped in the rubber hose. Air bubbles left in the rubber hose that come loose during the experiment will produce erratic results. Place the leveling bulb in the ring.

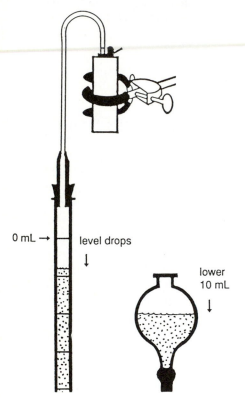

**3**. Place the stopper tightly into the top of the buret. Check that the plastic tubing is securely stuck on the glass tube and on the metal tube portion of the lighter. Lower the leveling bulb. Watch the water level in the buret. If it is free of leaks, the level of the water in the buret will move down and become stationary. If it keeps slowly descending, you have an air leak in your system. Check the stopper and the connection between the stopper and the lighter. Repeat the leak test. When your system is leak free, you are ready to begin the experiment. If you cannot produce a leak free system, check with your instructor. If you have a leak free system, open the system by loosening the rubber stopper at the top of the buret. Raise the glass bulb until the level in the buret is slightly below zero. Place the stopper back into the top of the buret. Run another leak check. Hold the bulb so that the levels in the bulb and buret are at identical heights. Read the buret. Record this volume reading. Place the bulb back on the ring at a level that is below the level of water in the buret.

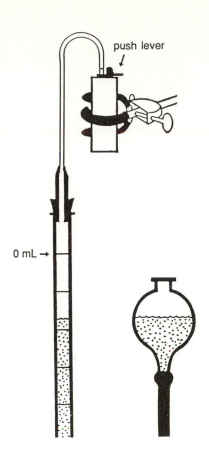

push lever

0 mL →

**4.** While holding the tubing tightly to the lighter, push down on the lever on the lighter to release hydrocarbon from the reservoir. Under atmospheric pressure and room temperature the hydrocarbon in the lighter is a gas. Therefore, the hydrocarbon will vaporize and move into the buret where the water level will begin to sink. When the water level has fallen to approximately the 43-mL mark, release the lever on the lighter. Lower the bulb and the ring clamp until the water levels in the bulb and in the buret match again. Allow your system to come to equilibrium with room temperature and the hydrocarbon gas to become saturated with water vapor. Two minutes should be sufficient time. With the levels matched, read the buret. Record your reading. To minimize errors from any leaks, you might want to lower the leveling bulb with the falling water level in the buret, keeping the two water levels always at the same height, and the pressure inside the buret equal to the barometric pressure. You might want to ask your neighbor to help you.

**5.** Read a thermometer and a barometer in the laboratory. Record the temperature and barometric pressure of the laboratory at the time of the experiment.

**6.** Open the buret to the atmosphere by loosening the stopper. Repeat Procedures 3-5 at least three more times.

**7.** From your data set, calculate values for the molecular mass of the hydrocarbon. Average the values, calculate the standard deviation and the relative standard deviation. Under the assumption that the hydrocarbon in the lighter is a saturated hydrocarbon of the general formula $C_nH_{2n+2}$, use your average molecular mass to identify the hydrocarbon by formula and name. Calculate the theoretical molecular mass and the accuracy of your experimental value as the difference between the theoretical and the experimental value. Express the accuracy also as the percent deviation of your experimental molecular mass from the theoretical molecular mass.

# PRELAB EXERCISES

## INVESTIGATION 13: MOLECULAR MASS OF A VOLATILE HYDROCARBON

Instructor:_____

Course/Section:_____

Name:_____

ID No.:_____

Date:_____

1. Define:    Ideal gas law

    Barometric pressure

    Vapor pressure of water

    Degrees kelvin

2. Rearrange the $PV = \dfrac{m}{MM} RT$ to an equation of the form $MM = f(P_1, T_1, m,$ etc.$)$.

3. A gaseous hydrocarbon collected over water at a temperature of 21°C, and a barometric pressure of 753 torr occupied a volume of 48.1 mL. The hydrocarbon in this volume weighs 0.1133 g. Calculate the molecular mass of the hydrocarbon.

# REPORT FORM

## INVESTIGATION 13: MOLECULAR MASS OF A VOLATILE HYDROCARBON

Name: _____   ID No.: _____

Instructor: _____   Course/Section: _____

Partner's Name (if applicable): _____   Date: _____

---

Temperature in Laboratory: _____ °C  _____ K

Barometric Pressure:  _____ torr  _____ atm

Vapor Pressure of Water at the Laboratory Temperature:  _____ torr

| | Experiment | | | |
|---|---|---|---|---|
| | 1 | 2 | 3 | 4 |
| Initial Mass of Lighter | | | | |
| Final Mass of Lighter | | | | |
| Mass of Hydrocarbon Released | | | | |
| Final Volume Reading | | | | |
| Initial Volume Reading | | | | |
| Volume of Hydrocarbon | | | | |
| Gas Pressure in Buret | | | | |
| Vapor Pressure of Water | | | | |
| Partial Pressure of Hydrocarbon | | | | |
| Molecular Mass of Hydrocarbon | | | | |

Average Molecular Mass: _____

Standard Deviation: _____

Relative Standard Deviation: _____

Formula for saturated hydrocarbon based on molecular mass: _____

Name for hydrocarbon: _____

Molecular Mass Calculated for Formula: _____

Accuracy (theoretical-experimental MM): _____

$\dfrac{100\ (\text{theoretical - experimental MM})}{\text{theoretical MM}}$: _____ %

Show calculations.

Date _____ Signature _____

# Investigation 14

## INTRODUCTION

Without abundant energy in useful form at reasonable cost the productivity of American industry would plummet and the lifestyle of the American people would deteriorate sharply. The materials that do the work for us are coal, petroleum, natural gas, and the uranium isotope with mass number 235. Coal and natural gas are burned in power plants in which the chemical energy in these fossil fuels is converted to electricity. The electricity lights, warms, and cools houses and powers electric motors for a multitude of tasks. Crude petroleum is refined to gasoline, diesel, and aviation fuel that keep the transportation system running. The splitting of uranium-235 in fission reactors produces heat that vaporizes water; the steam in turn drives a turbine connected to an electric generator. Hydroelectric and geothermal power plants provide a small percentage of the total energy used. About 90 percent of the approximately 71 billion British thermal units used annually in the United States comes from petroleum, natural gas, and coal, non-renewable fossil energy resources. Of particular concern are the liquid fuels – gasoline, diesel, and kerosene – that are derived from petroleum. Fifteen million barrels of petroleum products are burned every day in the United States in internal combustion engines in our cars, in furnaces in our homes, and under the boilers of our industrial plants. At the current rate of consumption, a few decades from now insufficient petroleum might be available to satisfy the world demand for this energy source. The United States imports approximately half of the petroleum it consumes.

To reduce the amount of non-renewable petroleum used, alternate energy sources must be tapped. Solar energy, the radiation streaming from the sun to the earth, is such an energy source. In a nuclear fusion reaction, hydrogen is converted to helium in the center of the sun. The energy set free in this reaction leaves the sun in the form of visible, ultraviolet, and other forms of electromagnetic radiation, a small fraction of which the earth intercepts. The sun's energy resource, hydrogen, is also non-renewable; however, with the amount of hydrogen in the sun, our star will shine for another five billion years providing plenty of time to develop yet unthought of energy sources. The problem with solar energy is not one of insufficient availability, but rather one of finding efficient and economical ways of collecting and utilizing this form of energy. Indirectly, some of the solar energy can be used to drive windmills and generate electricity. Uneven heating of the atmosphere by the sun causes air movements. Solar radiation can be directly converted to electricity by appropriate cells. In this way radiation from the sun can be turned into heat by solar collectors and the heat converted to electricity.

To have energy from the sun available also during nights and cloudy days, sufficient solar energy must be collected and stored during sunny day times. Several storage systems for solar energy have been developed. This experiment introduces a solar energy storage system based on a low-melting salt, sodium thiosulfate pentahydrate, $Na_2S_2O_3 \cdot 5H_2O$. The heat of fusion of this salt will be determined. This heat must be known for the design of a solar energy storage system.

## CONCEPTS OF THE EXPERIMENT

When a solid is slowly heated, it will melt at its melting point. As long as molten and solid substances are present during the melting process, the temperature will not change although heat is still being added to the heterogeneous mixture. The added heat is used to convert the solid to the liquid by overcoming the lattice energy of the crystalline solid. This heat is known as the enthalpy of fusion because this process is carried out at constant pressure (barometric pressure) and causes the solid to fuse. When all the solid has melted, additional heat will increase the temperature of the now homogeneous melt. When a homogeneous melt at the temperature of its melting point is allowed to lose heat to the surroundings, the melt will begin to solidify. This temperature, known as the freezing point (FP), is equal to the melting point (MP) and will not change as long as liquid and solid coexist. During the process of the conversion of a melt to a crystalline solid, the heat (enthalpy) of crystallization, numerically equal to the enthalpy of fusion, is liberated. Heat of crystallization is sometimes referred to as heat of solidification. When all the liquid has crystallized, the temperature of the solid will begin to decrease as heat dissipates into the surroundings. Figure 14-1 shows the change of temperature during a melt-freeze cycle. Equation 1 summarizes the phase changes and the energies associated with them for the case of sodium thiosulfate pentahydrate. The enthalpy of fusion, $\Delta H_{fus}$, is positive because the system (the

$$Na_2S_2O_3 \cdot 5H_2O(s) \xrightarrow{\text{MP}} Na_2S_2O_3 \cdot 5H_2O(\ell) \quad \Delta H_{fus} = +$$

$$(1)$$

$$Na_2S_2O_3 \cdot 5H_2O(\ell) \xrightarrow{\text{FP}} Na_2S_2O_3 \cdot 5H_2O(s) \quad \Delta H_{cryst} = -$$

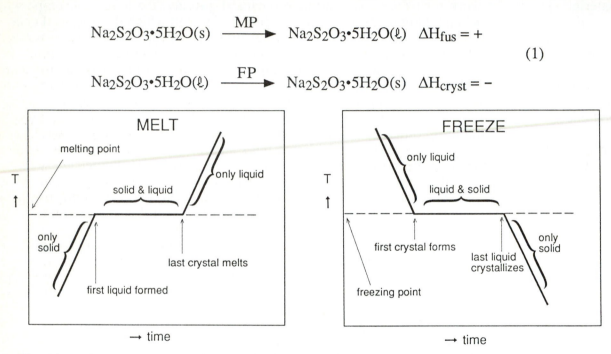

Fig. 14-1. Temperature changes during heating of a solid from below to above its melting point and during cooling of melt to below its freezing point.

sodium thiosulfate) takes up heat. The enthalpy of crystallization or solidification, $\Delta H_{cryst}$, is negative because the system loses energy. This thermodynamic sign convention is easy to remember when you consider your checking account as the system: any addition is positive (it increases your usable balance); any withdrawal is negative. The quanity of heat needed to completely melt a sample of a solid and the quantity of heat released by crystallization depends on the mass of the sample. These heats are extensive properties. To make these heats independent of the size of the samples and, thus, an intensive property and characteristic of the substance, these enthalpies are given for one gram of a substance (specific enthalpies) or one mole of the substance (molar enthalpies of fusion or crystallization).

Sodium thiosulfate pentahydrate has received attention as a storage agent for solar energy because it forms crystals easily, it can be easily transported and filled into heat-storage containers, it has a low melting point (48°C) permitting liquefaction at a temperature easily achieved by solar collectors, and its specific enthalpy of fusion is high enough to collect useful amounts of solar energy per unit mass. How can the specific enthalpy of fusion/crystallization of $Na_2S_2O_3 \cdot 5H_2O$ be determined? In principle, this determination is without difficulty. Weigh a sample of the salt, melt it completely at the melting point, let the melt crystallize, and measure the heat given off by the sample during the period between the appearance of the first crystal and the crystallization of the last droplet of liquid. The heat liberated during the crystallization process is measured in a calorimeter.

A calorimeter is a well-insulated vessel that ideally prevents any heat from escaping to the surroundings. In this experiment a styrofoam cup serves as the calorimeter. The calorimeter contains a substance, for instance water, the mass of which is exactly known ($m_w$). When the vial with the melt at the freezing point is immersed in the water and crystallization begins, the enthalpy of crystallization is released to the water; the temperature of the water will rise.

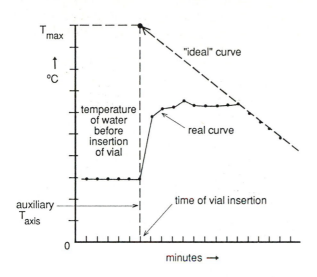

Fig. 14–2. Temperature–time plot for the crystallization of $Na_2S_2O_3 \cdot 5H_2O$ in the calorimeter water.

Under the assumption of instantaneous crystallization, transfer of heat to the water, and mixing of the water, the temperature-time plot shown in Fig. 14-2 would be obtained. The temperature of the cold calorimeter water, $T_{cw}$, rises to $T_{max}$. When the mass of water, the mass of $Na_2S_2O_3 \cdot 5H_2O$, and the temperature of the cold water are chosen appropriately, the temperature readings that will be used to find $T_{max}$ by extrapolation will approximately equal the last measured temperature of the molten thiosulfate. This situation is convenient because correction for heat released by the vial and the now solid salt on cooling to a temperature below the freezing point are either zero or so small that they can be neglected. All the data are now available to calculate the enthalpy of crystallization of the sample according to equation 2.

Equation 2 is based on the first law of thermodynamics: energy cannot be lost. The

$$m_w C_w (T_{cw} - T_{max}) = \Delta H_{cryst} = \Delta H_{cryst}^{spec} \cdot m_{salt} \qquad (2)$$

$C_w$: specific heat capacity of water, 4.18 J g$^{-1}$ degree$^{-1}$

$m_w$: mass of water in the calorimeter

$T_{cw}$: temperature of cold water in the calorimeter

$T_{max}$: temperature of water (vial, salt) after instantaneous heat transfer

$\Delta H_{cryst}$: enthalpy of crystallization

$\Delta H_{cryst}^{spec}$: specific enthalpy of crystallization

$m_{salt}$: mass of $Na_2S_2O_3 \cdot 5H_2O$ in the vial

energy liberated in form of heat by the crystallizing salt must be equal (but of the opposite sign) to the energy taken up by the water when its temperature rises to $T_{max}$. From often repeated and careful experiments the energy needed to raise one gram of water by one degree C is known to be 4.18 J (the specific heat capacity of water). Because $m_w$ grams of water were raised from $T_{cw}$ to

$T_{max}$, the left-hand side of equation 2 is the heat gained by the water. In this equation only $\Delta H_{cryst}$ (or $\Delta H_{cryst}^{spec}$) is unknown and can be calculated from the experimental data.

How can $T_{max}$ be determined? The experimental temperature-time curve labeled in Figure 14-2 as "real curve" does not directly give $T_{max}$. However, the "real curve" coincides with the "ideal curve" at times somewhat longer than the time corresponding to the maximum in the "real curve". The experimentally obtained "real curve" has a straight-line section at longer times that can be extended (extrapolated) to the time the vial with the melt was immersed in the calorimeter. The point at which the extended straight line intersects the auxiliary temperature axis erected at the time of immersion parallel to the temperature axis gives $T_{max}$. For the extrapolation to be useful, the straight-line portion of the curve must be well defined by taking sufficient number of temperature readings after steady cooling is observed.

This discussion thus far assumed that the water in the calorimeter is the only substance whose temperature rises when the $Na_2S_2O_3 \cdot 5H_2O$ crystallizes or when the hot water is added. This assumption is not correct. Heat is also taken up by the thermometer with which the water is stirred and by the styrofoam cup. To account for the heat taken up by the system except the water, a calorimeter constant K defined as the heat absorbed by the system except water per degree (°C) temperature rise is experimentally determined. Without this correction factor, the heat of crystallization of the sodium thiosulfate will be underestimated. A certain mass of water, $m_{rt}$, at room temperature, $T_r$, is placed into the calorimeter. Another mass of water, $m_b$, is brought to boil. The boiling water at temperature $T_b$ is poured into the water in the calorimeter. The temperature is measured every minute and a plot similar to Figure 14-2 prepared. The linear portion of the curve is extrapolated to time zero (the time of mixing) and $T_{w,max}$ is read from the temperature axis. Application of the first law of thermodynamics to this experiment produces

$$m_{rt}(T_r - T_{w,max})C_w + K(T_{w,max} - T_r) = m_b(T_{w,max} - T_b)C_w \qquad (3)$$

equation 3. This expression states that the heat lost by the boiling water (right side of equation 3) is equal to the heat gained by the water at room temperature plus the heat gained by the system parts exclusive of the water. All terms but K are known. Therefore, K (J per degree) can be calculated.

Equation 2 must now be modified to include the calorimeter constant. When the heat liberated by crystallization warms the water in the calorimeter, the other system parts are also warmed. The heat of crystallization of the sample of thiosulfate is, therefore, not only taken up by the water $[m_wC_w(T_{cw} - T_{max})]$ but also by the other parts of the system that are warmed from $T_{cw}$ to $T_{max}$. The heat required to bring about this temperature increase for the system exclusive of the water is $K(T_{max} - T_{cw})$. The sum of these two terms of heat gained will be equal to the enthalpy of crystallization (equation 4). The enthalpy of crystallization and the specific enthalpy of

$$K(T_{cw} - T_{max}) + m_wC_w(T_{cw} - T_{max}) = \Delta H_{cryst} = \Delta H_{cryst}^{spec} \cdot m_{salt} \qquad (4)$$

crystallization can be calculated from this equation.

This experiment was designed to permit the determination of a fairly accurate enthalpy of crystallization. A more accurate determination would take into account the heat capacity of the vial, the temperature of the melt at the time of immersion, and a few additional fine points. In this experiment the temperature of the melt and the vial at the time of immersion are close to the temperatures of the calorimeter water used for extrapolation thus making the amount of heat extracted by the vial and the salt from the calorimeter water very small.

## ACTIVITIES

- Complete the PreLab Exercises and study the section "Concepts of the Experiment" before coming to the laboratory.  Without a good understanding of the concepts, the experiment will be difficult to carry out.

- Construct a calorimeter from two styrofoam cups.

- Determine the calorimeter constant.

- Determine the specific enthalpy of crystallization of $Na_2S_2O_3 \cdot 5H_2O$.

## SAFETY

*Wear eye protection and exercise proper care in heating and in transferring hot equipment and reagents.*

## PROCEDURES

This experiment requires more than one pair of hands and eyes.  Therefore, select a partner or work with the partner assigned by your instructor.  Each of you must keep a complete set of notes, carry out all the calculations, and submit a Report Form.

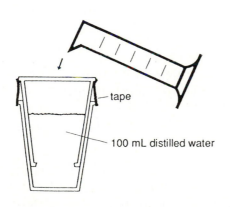

tape

100 mL distilled water

**1.** Obtain two styrofoam cups.  Carefully cut out the bottom of one cup.  Push the bottomless cup firmly into the intact cup.  Use a strip of masking tape to hold the two cups together.  Make a pencil mark 2 cm below the rim of the inner cup.  Weigh the two cups on the triple-beam balance to 0.01 g.  Place 200 mL distilled water into a 250-mL beaker.  Set the beaker aside for later use in Procedure 14.

**2.** With a graduated cylinder measure 100 mL of distilled water into the cups.  Weigh the cups with the water to the nearest 0.01 g on the triple-beam balance.

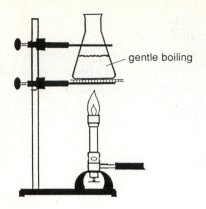

gentle boiling

**3.** Pour 80 mL of distilled water into a 125-mL Erlenmeyer flask. Heat the water to boiling. Adjust the burner to keep the water gently boiling. The thermometer used in this experiment has limited range. The temperature of the boiling water cannot be determined with this thermometer. Find the boiling point of the water at the barometric pressure in the laboratory from the pressure/boiling point graph.

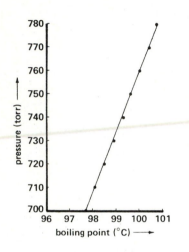

**4.** Use a thermometer with 0.2 degree marks to stir the water in the styrofoam cups in a circular motion at a rate of about one circle per second. Do not let the thermometer drag on the bottom of the cup or rub the sides of the cup. Practice reading the thermometer to 0.05 degrees. Stir the water and practice reading the temperature every minute for four minutes. To read the thermometer stop stirring. Bring your eyes to the height of the liquid in the thermometer capillary when reading the temperature.

**5.** Prepare a table in your notebook that will allow you to record the temperature readings.

| Time (min.) | 1 | 2 | 3 | 4 | 5 | 6 | 7 | 8 | 9 | 10 | 11 | 12 |
|---|---|---|---|---|---|---|---|---|---|---|---|---|
| Temp (°C) | | | | | | | | | | | | |

Read and record all temperatures to the nearest 0.05°. *TIMING IS CRITICAL IN THE NEXT STEPS. Be sure you and your partner have planned your work carefully for efficient operation.*

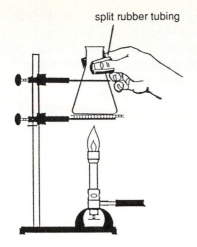

split rubber tubing

**6**. Stir the water in the calorimeter. Read and record the temperature of the calorimeter water every minute for 5 minutes. Between the fifth and the sixth minute have your partner remove the flask of boiling water from the ringstand using two short pieces of split rubber tubing to protect the two fingers holding the flask.

add boiling water to pencil mark

**7**. Exactly at the start of the sixth minute pour the boiling water into the calorimeter until the water level just touches the pencil mark. Continue reading the time and temperature every minute. The temperature will rise quickly to a maximum and then decrease slowly. Keep stirring and recording temperatures every minute until the temperature change per minute is constant for at least five minutes. *It is very important that you have at least five points in the straight-line segment of the temperature time region (Figure 14-2).*

 Remove the thermometer from the calorimeter. Let all the water from the stem and bulb of the thermometer drain into the cups. Dry the thermometer with a towel. Determine the mass of the calorimeter and water to 0.01 g on a triple-beam balance. Don't empty the calorimeter since it will contain the room-temperature water needed in subsequent procedures.

**8**. Pour 450 mL of distilled water into a 600-mL beaker. Bring the water to a gentle boil. Adjust the burner to keep the water boiling gently.

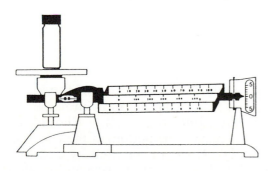

**9**. Weigh a clean, dry screw-cap vial (O.D. 2.5 cm, 7 cm high with cap) and the cap on a triple-beam balance to 0.01 g.

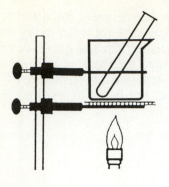

**10**. Fill a 20-cm long test tube half full with crystals of $Na_2S_2O_3 \cdot 5H_2O$. Set the test tube containing the sodium thiosulfate into the boiling water in the 600-mL beaker. The crystals will soon begin to melt. You must not allow all the crystals to melt. Time your work to prevent all the crystals from melting and the melt from being heated to higher temperatures.

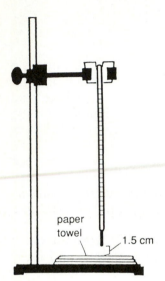

paper towel    1.5 cm

**11**. While the sodium thiosulfate is heated, carefully push the upper end of the thermometer (0.2 degree divisions) into a one-hole No. 6 rubber stopper until the top of the thermometer has almost reached the top of the stopper. This operation will be easier when the end of the thermometer is moistened with water. Check that the thermometer is held tightly in the stopper and cannot slip out. Attach a clamp to a ringstand and open the claws of the clamp sufficiently to allow the stopper to rest in the claws without slipping through. Raise or lower the clamp until the thermometer bulb is suspended approximately 1.5 cm above the base of the ringstand. Place a twice-folded towel on the base of the ringstand to serve as insulation.

*Steps 14, 15, and 16 must be completed within 10 MINUTES. Plan your work. Prepare in your notebook a time/temperature table (see Procedure 5) allowing for entries of up to 40 minutes.*

**12**. When almost all the crystals in the test tube have melted, remove the tube from the beaker and pour the melt into the previously weighed vial until the liquid level is just below the neck of the vial. Use pieces of rubber tubing on your fingers to protect them.

**13**. Lift the thermometer, slide the vial with the thiosulfate melt into place, and immerse the thermometer bulb into the melt. Should the thermometer bulb rest on the bottom of the vial, raise the clamp slightly. Now stir the melt by circularly moving the vial without lifting it. Be careful not to allow the thermometer to hit the walls or the bottom of the vial. Begin to read and record the temperature every minute as soon as the temperature has dropped to 48°C. This temperature is defined as the temperature for "time zero."

**14.** Continue recording the temperature every minute. When the temperature reaches 45°C (supercooled condition), carefully remove the thermometer. Continue recording each elapsed minute. Place the thermometer into the 250-mL beaker of room-temperature water until the thermometer registers room temperature, then wipe it dry and....

**15.** ...begin to stir the calorimeter water with the thermometer, reading and recording the temperature every minute for an additional 5 min. Between the additional fifth and sixth minutes....

one crystal

**16.** ...have your partner drop one tiny crystal of thiosulfate into the vial and close the vial tightly. At the start of the sixth minute....

**17.** ...slide the sealed vial carefully into the calorimeter. The thiosulfate must still be liquid except for the added "seed" crystal. Stir the calorimeter water around the vial as described in Procedure 4. Read and record the temperature every minute to the nearest 0.05°C. The temperature will slowly rise to a maximum, stay near the maximum for a few minutes, and then begin to drop. Continue stirring and recording the temperature until seven consecutive readings show a constant *rate* of cooling. (Be sure you do not unscrew the vial cap during the stirring.)

If the salt began to crystallize before the vial was completely immersed in the calorimeter water, you will have to remove the vial, let all water drain back into the calorimeter, tie a string around the vial's neck, and reheat the vial by submerging it in the hot water contained in the 600-mL beaker, and begin again with Procedure 14.

**18.** When the cooling rate has remained constant for 7 min and the temperature is 2.5 degrees or more below the maximal reading, remove the thermometer and vial, and wipe the vial dry. Carefully unscrew the cap and wipe off any water adhering to the neck of the vial. Reseal the vial and weigh it to the nearest 0.01 g. Record this mass. Reweigh the calorimeter plus water if you suspect it to be different than that obtained at the end of Procedure 7.

When you have finished, tie a piece of string around the neck of the vial and heat the vial in boiling water until the crystals melt. Pour the molten salt into the specified waste container, then wash and dry the vial.

**19**. Plot your temperature/time data and determine the temperatures $T_{w,max}$ and $T_{max}$ by extrapolation. Calculate the calorimeter constant and the specific enthalpy of crystallization of $Na_2S_2O_3 \cdot 5H_2O$. Complete the Report Form.

# PRELAB EXERCISES

## INVESTIGATION 14: HEAT STORAGE FOR SOLAR HEATING

Name: _____

Instructor: _____    ID No.: _____

Course/Section: _____    Date: _____

1. Define:

    Specific heat capacity

    Specific enthalpy of crystallization or solidification

    Calorimeter constant

2. Rearrange equation 3 to express K, the calorimeter constant, as a function of $\Delta T$, m, and $C_w$.

3. Rearrange equation 4 to isolate $\Delta H_{cryst}^{spec}$ on the left-hand side of the equation.

4. Give the units for:

    K (calorimeter constant)

    $\Delta H_{cryst}$

    $\Delta H_{cryst}^{spec}$

    Specific heat capacity

# REPORT FORM

## INVESTIGATION 14: HEAT STORAGE FOR SOLAR HEATING

Name:_____     ID No.: _____

Instructor:_____     Course/Section: _____

Partner's Name (if applicable):_____     Date:_____

---

Barometric pressure _____ torr     Boiling point of water at this pressure _____ °C

### Directly Determined Data

Mass of empty calorimeter ($M_C$) (Procedure 1):                                                          _____ g

Mass of calorimeter ($M_C$) + (~100 mL) cold water ($M_{CW}$) (Procedure 2):        _____ g

Mass of calorimeter ($M_C$) + cold water ($M_{CW}$) + boiling water ($M_{BW}$) (Procedure 7):_____ g

Mass of empty vial and cap ($M_V$) (Procedure 9):                                                        _____ g

Mass of vial and cap ($M_V$) and thiosulfate ($M_{salt}$) (Procedure 18):                  _____ g

### Extrapolated Data

For the boiling water-cold water experiment:

Temperature of cold water (extrapolated from data of Procedure 6):        $T_r$_____ °C

Temperature of mixture (extrapolated from data of Procedure 7):        $T_{w,max}$ _____ °C

For the thiosulfate experiment:

Temperature of cold water (extrapolated from data of Procedure 15):        $T_{cw}$_____ °C

Temperature of water after crystallization of thiosulfate (extrapolated from data of Procedure 17):

$T_{max}$_____ °C

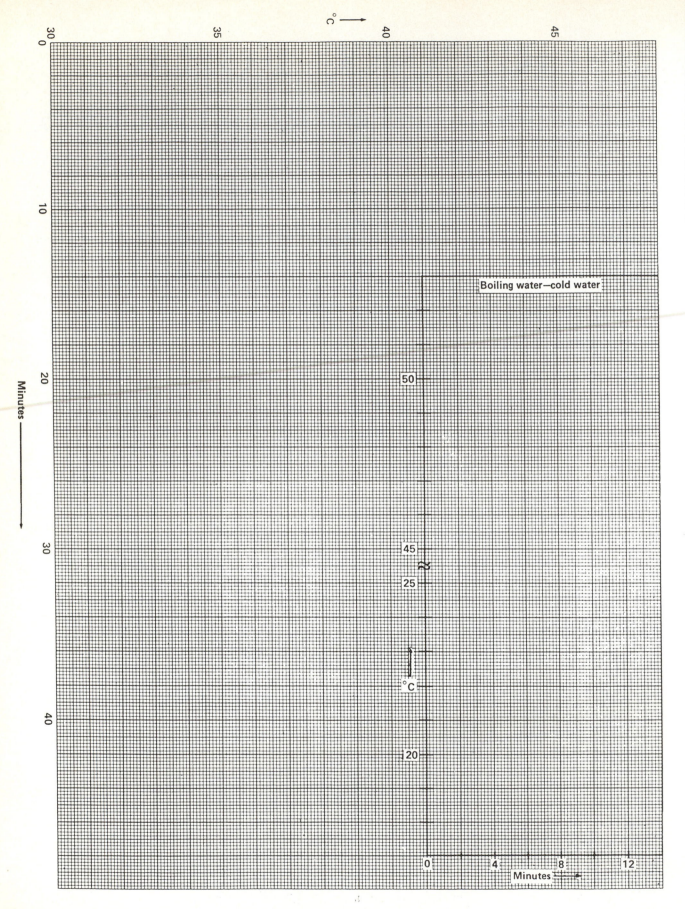

Boiling water—cold water

°C

Minutes

°C

Minutes

154

## Calculated Data with units

For the boiling water-room temperature water experiment

Mass of room temperature water ($m_{rT}$): _____ g

Mass of boiling water transferred $m_b$:    _____ g

Total mass of water in calorimeter ($m_w$): _____ g

Heat lost by the boiling water (give equation showing number used and resulting value):

_____

Heat gained by room temperature water (give equation showing number used and resulting value):_____

Heat gained by surroundings exclusive of water (give equation with numerical values):

_____

$K =$

For the thiosulfate experiment:

Mass of thiosulfate crystallized: _____

Mass of water ($m_w$):        _____

Temperature max:        _____

Equation, with numerical values shown, for calculation of $\Delta H_{cryst}^{spec}$:

_____

Equation for calculation of molar $\Delta H$:

_____

Value of experimental value of $\Delta H$ in units of J mol$^{-1}$:_____

Compare the enthalpy of crystallization of thiosulfate per gram and per mole with the corresponding data for water and sodium chloride.  Is sodium thiosulfate a better material to collect solar energy than water; than sodium chloride?  Why?  Would $\Delta H_{cryst}^{spec}$ be more appropriate for designing a solar energy collection system?

Date _____ Signature _____

# Investigation 15

## TETRAAMMINECOPPER(II) SULFATE

## INTRODUCTION

Elements can be classified in several ways: metals, metalloids, non-metals; solids, liquids, gases; insulators, conductors of electricity; soft, hard; reactive, unreactive. An often-used classification divides the elements into main-group elements and transition elements. The main-group elements have completely empty or completely filled d-subshells, whereas the transition elements have a partially filled d-subshell. In the transition elements (elements with atomic numbers 21–30, 39–48, 57, 72–80) the energies of the ns, np and (n–1)d subshells are close to each other. For this reason all of these orbitals can be used to form bonds. This ability leads to the formation of coordination compounds. In a coordination compound the central metal atom or ion is bonded to several molecules or anions, which are called ligands. A ligand has at least one lone electron pair that is used to form a coordinate covalent bond to the central atom or ion. The following formulas give the compositions and names of several coordination compounds. The

| [Mo(CO)_6] | Na[AgCl_2] | [Cr(NH_3)_6]Cl_3 |
|---|---|---|
| hexacarbonyl-molybdenum(0) | sodium dichloroargentate(I) | hexaamminechromium(III) chloride |

$$\begin{bmatrix} H_3N & & Cl \\ & Pt & \\ H_3N & & Cl \end{bmatrix}$$

cis–diamminedichloro-platinum(II)

$Na_3[TaF_7]$

sodium heptafluorotantalate(IV)

$K_4[Fe(CN)_6]$

potassium hexacyanoferrate(II)

central metal atom and its ligands are generally enclosed in brackets. Outside the brackets are written the cations or anions required to neutralize the charge on the "complex ion." The assembly of the central metal atom and its ligands can be anionic, cationic, or neutral. The large number of transition metals (many of which assume several valences), the even larger number of ligands, and the possibility of having more than one kind of ligand bonded to a central metal atom allow many different coordination compounds to be prepared. Chemists have synthesized several hundred thousand of these coordination compounds and more are made every day in academic and industrial laboratories.

Work on coordination compounds is of great economic importance. The petrochemical and other chemical industries rely on catalytic processes for their production. The catalysts are generally transition metal compounds that form coordination compounds with the chemicals to be transformed. The thermal cracking of petroleum fractions to gasoline and unsaturated hydrocarbons is an example of such a catalytic process that is carried out on a very large scale. Even small improvements in the effectiveness of a catalyst brings large economic rewards. Therefore, many industrial and academic chemists explore the preparation of coordination compounds and their chemical behavior.

Coordination compounds play a vital role in biology. Without metalloenzymes, proteins containing at least one metal ion, the chemical processes of life could not proceed. These metalloenzymes, coordination compounds with very complex, high-molecular-mass ligands, catalyze biochemical reactions. For instance, zinc is known to be present in more than 100 enzymes. The trace metals that are needed for the formation of metal-containing compounds by living systems are called essential trace elements. Copper, the subject of this experiment, is one of these essential elements.

Without the required amount of copper in the diet (2–3 mg Cu per day for an adult), severe anemia (iron deficiency in the blood) may develop, even though the diet contains more than sufficient iron for the production of hemoglobin. The blood contains a protein, ceruloplasmin, that complexes $Cu^{2+}$. Ceruloplasmin appears to play an essential role in triggering the release of iron needed for the production of hemoglobin from the liver. Hemocyanine, another copper complex, has the same role in the oxygen transport system of many invertebrates as hemoglobin, an iron complex, has in vertebrates. Copper-containing metalloproteins may have a relatively small molecular mass (25,000 amu, one $Cu^{2+}$ per molecule in cytochrome oxidase of the heart muscle) or a very high molecular mass (780,000 amu, 20 $Cu^{2+}$ per molecule in the hemocyanine of the lobster).

In this experiment you will synthesize tetraamminecopper(II) sulfate monohydrate as an introduction to the chemistry of coordination compounds.


## CONCEPTS OF THE EXPERIMENT

To prepare a coordination compound, an appropriate transition metal salt must be combined with the desired ligands under conditions that produce a pure product in high yield. The solvent, the temperature at which the reaction is carried out, the concentration of the reagents, and the presence or absence of light during the reaction often determine whether or not a synthesis of a coordination compound is successful.

In the preparation of tetraamminecopper(II) sulfate monohydrate, a solution of copper sulfate in water is treated with an excess of ammonia (the ligand). This excess shifts the equilibrium toward the product. The concentration of the reagents are chosen to make the resulting solution super-

$$[Cu(H_2O)_n]^{2+} + SO_4^{2-} \xrightarrow{\text{excess } NH_3} [Cu(NH_3)_4]^{2+}SO_4^{2-}\cdot H_2O \Downarrow$$

saturated with respect to the product. Therefore, the product will precipitate in the form of dark blue crystals. These crystals are then purified by dissolution in water and reprecipitation with ethanol. A small amount of concentrated ammonia is added to prevent the $[Cu(NH_3)_4]^{2+}$ from losing its ligands.

The formation of tetraamminecopper(II) is a ligand-exchange reaction. The copper ion in aqueous solution is surrounded by $H_2O$ molecules that are replaced by ammonia molecules.

## ACTIVITIES

- Perform the stoichiometric calculations to find the grams of $CuSO_4 \cdot 5H_2O$ and the volume of concentrated (14.8 M) ammonia required for the synthesis of 3.1 g of the copper complex under the assumption of a 90% yield.

- Measure out the reagents.

- Perform the synthesis.

- Purify your product.

- Calculate your yield.


## SAFETY

*Wear approved eye protection. A drop of concentrated ammonia in the eye can cause permanent damage. In case of contact with any reagents, wash immediately and thoroughly with water. Be cautious to keep the ethanol used away from flames.*


## PROCEDURES

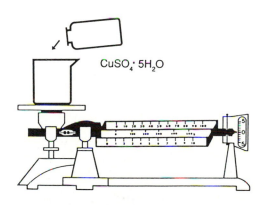

$CuSO_4 \cdot 5H_2O$

**1.** Without spillage, weigh the approximate mass of $CuSO_4 \cdot 5H_2O$ required for the preparation of 3.1 g of $[Cu(NH_3)_4]SO_4 \cdot H_2O$, assuming a 90-percent reaction yield, into a preweighed 100-mL beaker. Remove the beaker from the balance and add 15 mL of distilled water. Swirl to dissolve the salt. You may wish to warm the beaker gently to speed up dissolution. If you do, allow the solution to cool to room temperature before proceeding.

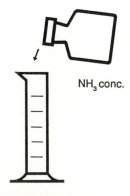

$NH_3$ conc.

**2.** If a graduated cylinder is not located in the fume hood for this purpose, take your 25-mL graduated cylinder and the beaker containing your $CuSO_4$ solution into the hood. Measure into the graduated cylinder twice the approximate volume of concentrated (14.8 M) ammonia calculated to convert all of your $CuSO_4 \cdot 5H_2O$ sample to the tetraammine complex. Close the ammonia bottle.

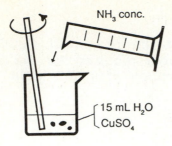

**3.** Slowly add the ammonia, with stirring, to the $CuSO_4$ solution. Continue stirring until all the initial precipitate of $Cu(OH)_2$, including any particles adhering to the beaker walls, has dissolved.

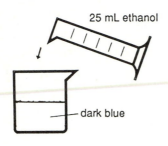

**4.** Very slowly, in small portions, pour 25 mL of ethanol into the deep blue solution. Stir the mixture thoroughly. Then set it aside for a few minutes to allow the crystals to settle. Prepare a "wash" solution by mixing 7 mL of ethanol and 5 mL of distilled water. *Laboratory ethanol (alcohol) usually contains toxic additives that make it non-potable (denatured). Denatured ethanol cannot be made potable by simple purification techniques.*

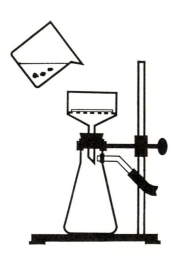

**5.** Assemble a vacuum filtration setup (Appendix F) and use it to isolate your crystals. Remember to moisten the filter paper and to decant most of the supernatant liquid through the funnel before transferring crystals from the beaker. Use four 3-mL portions of your "wash" solution to rinse the last of the crystals from the beaker into the funnel.

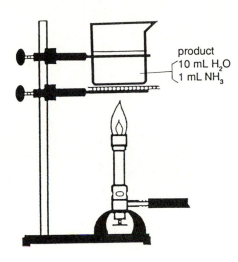

product
- 10 mL H$_2$O
- 1 mL NH$_3$

**6.** When all liquid has been pulled through the funnel, carefully break up the blue cake on the filter with a spatula. Don't tear the filter paper. Allow the aspirator to pull air through the crystals for 5 to 10 min to dry them. Then pull the hose off the vacuum flask and turn off the aspirator. Transfer your crystals completely onto a clean, *preweighed* watchglass. Reweigh to determine the product yield. Record the product mass and a description of the appearance of the crystals. This will be your yield after the first crystallization.

**7.** Place your product in a 100-mL beaker. Use 10 mL of distilled water to wash the watchglass. Add the wash-water to the beaker. Add 1 mL of concentrated ammonia. Dissolve the crystals by warming the beaker very gently. (Do not allow the solution to boil.)

**8.** Add ethanol slowly to the warm solution until you have a slightly cloudy solution or until you have added a total of 8 mL of ethanol. Then proceed as described in steps 4, 5, and 6 to obtain a recrystallized product.

Show your product to your instructor, who will direct you to the appropriate storage container. Ask your instructor to mark the product "grade" space on your Report Form. If your glassware has a dark film on it, clean with dilute acid as directed by your instructor.

recrystallized
product

**9.** On the basis of the mass of CuSO$_4\cdot$5H$_2$O used and the mass of [Cu(NH$_3$)$_4$]SO$_4\cdot$H$_2$O obtained after each crystallization, calculate the percentage yield for your preparation after each crystallization.

Use your calculations and notebook records to complete the Report Form.

# PRELAB EXERCISES

## INVESTIGATION 15:  TETRAAMMINECOPPER(II) SULFATE

Name: _____    ID No.: _____

Instructor: _____    Course/Section: _____

Date:_____

---

1.  Calculate the formula mass for each of the following compounds/ions:

$[Cu(NH_3)_4]^{2+}$

$[Cu(NH_3)_4]SO_4$

$[Cu(NH_3)_4]SO_4 \cdot H_2O$

2.  Write a balanced chemical equation for the preparation of $[Cu(NH_3)_4]SO_4 \cdot H_2O$ from $CuSO_4 \cdot 5H_2O$ and ammonia.

3.  Calculate the mass of $CuSO_4 \cdot 5H_2O$ required for the preparation of 3.1 g of $[Cu(NH_3)_4] SO_4 \cdot H_2O$ assuming a 90-percent reaction yield.

(over)

4.  Calculate the volume of concentrated (14.8 M) ammonia required for the preparation of 3.1 g of $[Cu(NH_3)_4]SO_4 \cdot H_2O$ assuming 90-percent yield.

5.  Calculate the percent yield for a reaction that gives 5.00 g of the complex and uses 5.50 g of $CuSO_4 \cdot 5H_2O$.

# REPORT FORM

## INVESTIGATION 15: TETRAAMMINECOPPER(II) SULFATE

Name: _____   ID No.: _____

Instructor: _____   Course/Section: _____

Partner's Name (if applicable): _____   Date: _____

---

Show all calculations on reverse side.

Mass of $CuSO_4 \cdot 5H_2O$ (procedure 1):

(beaker) _____ g   (beaker + compound) _____ g   (compound) _____ g

Volume of 14.8 M $NH_3$ required (procedure 2):

(theoretical volume _____ mL) x 2 = _____ mL

Product yield (first crystallization):               (second crystallization):

    Product + watchglass _____ g          Product + watchglass _____ g

           Watchglass _____ g                     Watchglass _____ g

    ————————————————                    ————————————————

             Product _____ g                        Product _____ g

               Yield _____ %                          Yield _____ %

Your description of product appearance after each crystallization

+------------------------------------------------------------------------+
| Instructor's evaluation:                                               |
|                                                                        |
| (1st crystallization)  ☐ Reagent grade  ☐ Tech grade  ☐ Unsatisfactory |
| (2nd crystallization)  ☐ Reagent grade  ☐ Tech grade  ☐ Unsatisfactory |
|                                                                        |
|                 Instructor's initials: _____                   |
+------------------------------------------------------------------------+

Date _____ Signature _____

# Investigation 16

---

## MOLECULAR MASSES OF PHARMACEUTICAL CHEMICALS

### INTRODUCTION

A drug is defined in pharmacology as a chemical substance of known structure that has therapeutic use. An enormous variety of synthetic drugs is now produced by pharmaceutical industries. Structural formulas for several familiar drugs are displayed below.

**Caffeine**

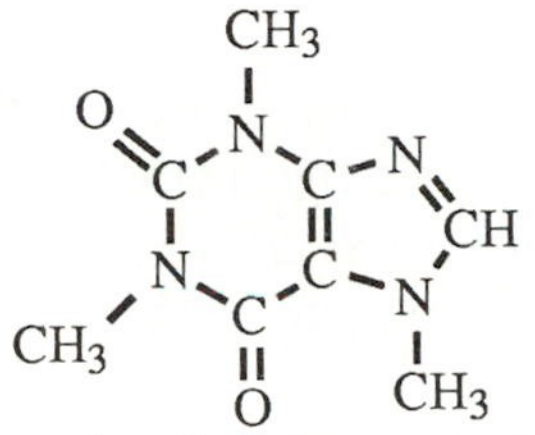

An ingredient of coffee, tea, and various "cola" beverages; used as a stimulant in several nonprescription medications, such as *No-Doz* and *Anacin*.

**Sulfanilamide**

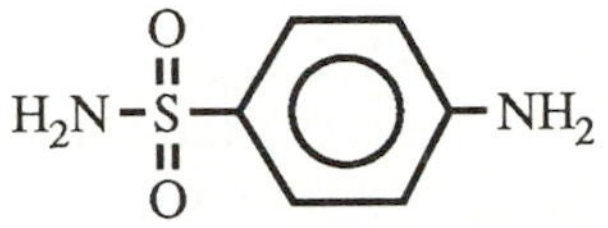

An antibiotic, one of the simpler *sulfa drugs*.

**Chlorpheniramine**

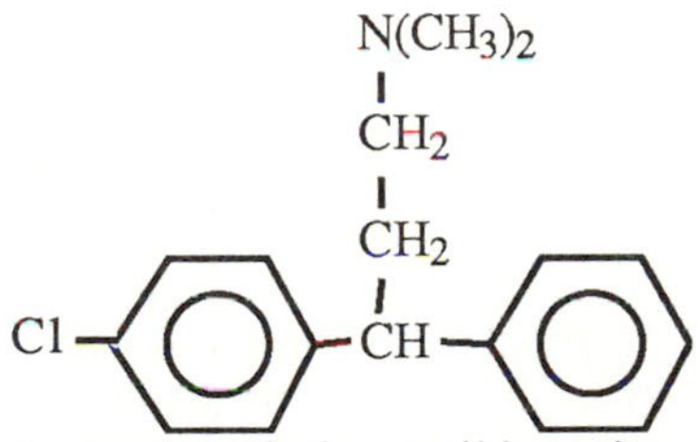

A nonprescription antihistamine.

**Benzocaine**

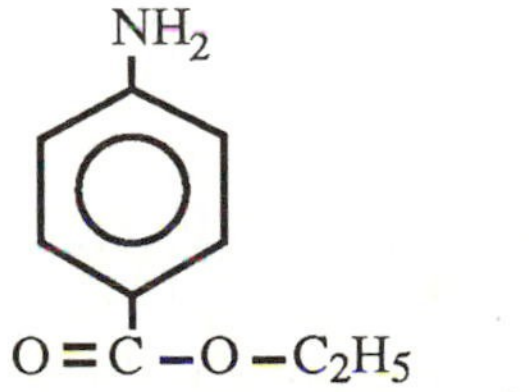

A local anesthetic, often used in topical ointments (e.g., in *Chiggerex*) and in solutions to reduce localized pain or irritation.

**Acetylsalicylic acid**

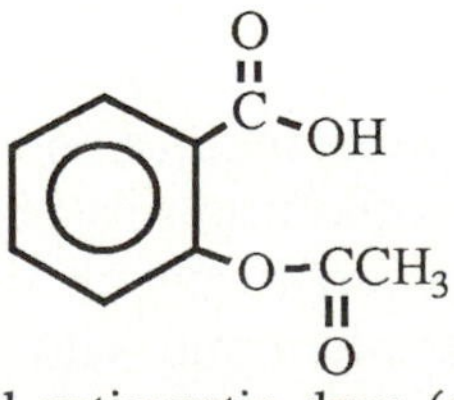

Analgesic and antipyretic drug (aspirin), most commonly used as the sodium or calcium salt.

**Camphor**

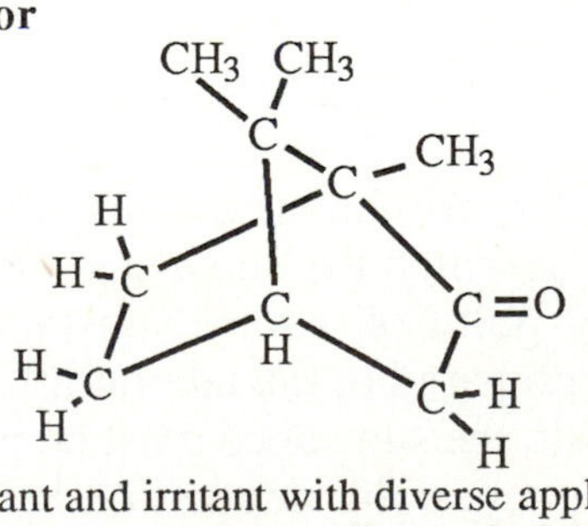

A stimulant and irritant with diverse applications ranging from mothproofing of fabrics to preparations of nasal decongestants.

**Scopolamine**

A nonprescription, sleep-inducing drug (e.g., in *Sleep-eze*).

**Phenolphthalein**

A laxative (e.g., in *Ex-Lax*), more familiar in the chemical laboratory as a pH indicator.

In spite of extensive research, we still know very little about the mechanisms of the actions of various drugs in the body. A number of theories exist, but none are entirely satisfactory. Therefore, we cannot design a drug for a specific purpose by selecting the structural features a molecule needs to have a particular biochemical function. Although we do know a few things about structure-function relationships – and more is being learned every day – this knowledge is not yet sufficient to permit the reliable design of drugs. Exhaustive testing of every new pharmaceutical chemical is necessary to determine its value as a drug and to learn about its adverse side effects that must be avoided or minimized.

Although some drugs, such as insulin, consist of very large molecules and others, such as cyclopropane (a general anesthetic), consist of very small molecules, most of the common drugs have molecular masses in the range of 100 to 400 amu. This experiment deals with the determination of the approximate molecular mass of a pharmaceutical chemical using the melting-point depression of a camphor sample, in which the "unknown" chemical has been dissolved.

## CONCEPTS OF THE EXPERIMENT

A pure substance – unless it decomposes on heating – melts and freezes at a characteristic temperature. This temperature is called the melting point when reached by heating a solid, and is known as the freezing point when approached by cooling a liquid. A melting point is determined by placing a small amount of a substance into a capillary tube and keeping the lower end of the tube next to the mercury reservoir of a thermometer that is immersed in a heating bath. The temperature of the heating bath is increased rapidly first and then very slowly (1°C per minute) as the melting point is approached. At the melting point the substance will begin to turn into a liquid. The temperature difference between the beginning of the melting process and its end (when the last solid has disappeared) should not exceed one to two degrees.

It is an experimental fact that the melting point of an impure substance (a mixture of two substances in which the substance present in much smaller amount is the impurity) is lower than the melting point of a pure substance. Melting points are often used to check the purity of substances prepared in the laboratory. If the melting temperature does not agree with the accepted melting point, the substance must be purified.

The freezing point depression increases with increasing concentration of the "impurity" (the solute) in the pure substance (the solvent). The concentrations are most conveniently expressed as moles of solute per one kilogram of solvent (molality). The same number of moles of a solute in one kilogram of different solvents will depress the melting point of each solvent to a different degree. This relationship is expressed by equation 1.

$$\Delta T_f = K_f \times m \qquad\qquad (1)$$

$\Delta T_f$ : freezing point depression of solvent (degree) [$T_f$(pure solvent) – $T_f$(mixture)]
$K_f$ : molal freezing point depression constant (degree kg mol$^{-1}$)
m : molality of solution (mol kg$^{-1}$)

Because the number of moles of a solute is related to its mass in grams by mass/molecular mass, the freezing point depression can be used to determine the molecular mass of the solute (equation 2) provided $K_f$ is known.

$$\text{Molecular mass} = K_f \frac{\text{grams(solute)}}{\Delta T_f \times \text{kg(solvent)}} \qquad\qquad (2)$$

The freezing point depression will be large and easily measured with precision and accuracy using common laboratory thermometers when $K_f$, the molal freezing point depression constant, is large. The value of this constant is equal to the freezing point depression of a one-molal solution. The molal freezing point depression constant for camphor (37.7°C mol$^{-1}$ kg) is one of the largest of the common solvents. For this reason, camphor is the solvent of choice for molecular mass determinations provided that the substances, the molecular masses of which are to be determined, are soluble in liquid camphor and do not react with camphor.

## ACTIVITIES

- Complete the PreLab Exercises before coming to the laboratory.

- Prepare a glass tube with a bulb at one end from soft-glass tubing.

- Prepare a solution of an unknown in camphor.

- Determine the melting points of pure camphor and an unknown/camphor mixture.

- Calculate the molecular mass of an unknown from freezing point depression data.

- Identify an unknown by comparing the experimental molecular mass with the molecular masses calculated for the pharmaceutical chemicals listed in the Introduction.

## SAFETY

*Under no circumstances are any of the chemicals used in this experiment to be tasted, swallowed, or taken from the laboratory. Many of these chemicals are toxic and all are hazardous when used without supervision. Be careful also in working with the hot-oil bath. In case of contact with hot oil, soak the affected area immediately with cold water and notify your instructor. Wear approved eye protection at all times.*

## PROCEDURES

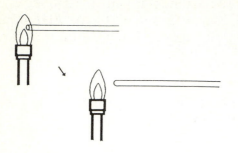

**1.** Use a hot burner flame to seal one end of a 10 to 30-cm length of 6 mm (inner diameter) soft-glass tubing. Rotate the tube slowly while heating. The hottest part of the flame is just above the inner blue cone.

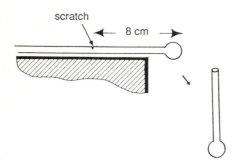

**2.** When the end of the tube is sealed and while the glass is still soft, remove the tube from the flame and in a controlled manner blow into the cool open end to form a bubble about 1.5 cm in diameter at the closed end.

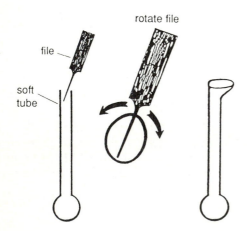

**3.** To cool the hot tube lean it against a ringstand base so that the hot end is above the base. *Do not place hot glass on the bench top.* When the tube has cooled, place it flat on the bench top with the bulb extending over the edge. With a file scratch the tube 8 cm from the sealed end. Carefully snap the tube at the scratch mark by placing your thumbs on the side behind the scratch and pushing away with your thumbs.

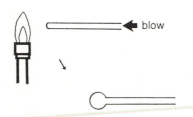

**4.** Carefully heat the open end of the tube until the top 1-cm has become soft. Take the tube from the flame and quickly flare the end into a small funnel by resting the file handle against the inner part of the tube just below the top soft part and rotating the file through a half circle. Allow the tube to cool. Weigh the cool tube on the analytical balance to ±0.1 mg.

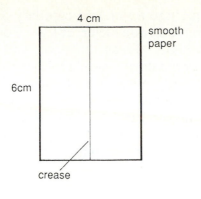

4 cm

smooth paper

6cm

crease

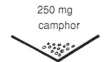

250 mg camphor

**5.** Weigh a creased piece of smooth paper (approximately 4 x 6 cm) to the nearest milligram. Place onto this paper approximately 250 mg camphor (acceptable range: 230 to 270 mg). Add or remove camphor as necessary to bring the mass of camphor into the acceptable range.

**6.** Your camphor sample should consist of small particles. If larger particles are present, divide them into smaller pieces with your spatula. Large pieces will not fit into the tube.

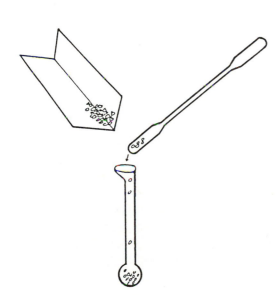

**7.** With your spatula transfer the camphor as quantitatively as possible into the bulb-tube. To assist the camphor in sliding into the bulb, you can gently tap the tube with your finger. When all the camphor is in the bulb, reweigh the tube to ±0.1 mg. The mass of camphor is obtained as the difference between the mass of tube plus camphor and the mass of the empty tube.

**8.** Obtain an unknown from your instructor. Weigh another piece of creased smooth paper to ±0.1 mg. Add an amount of unknown corresponding to approximately 10% (acceptable range: 5 to 15%) of the mass of camphor. Reweigh the paper plus the unknown to check that the mass is in the acceptable range. Transfer the unknown as quantitatively as possible into the tube containing the camphor. Reweigh the tube to ±0.1 mg and calculate the mass of the unknown.

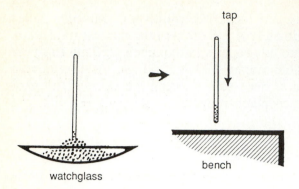

watchglass

bench

tap

**9.** Obtain or prepare a standard capillary melting-point tube (1 mm inner diameter, at least 6 cm long, closed at one end). Press the open end of the capillary tube into a small sample of camphor on a watchglass. Then tap the closed end of the tube against the bench top to help the crystals slide to the bottom of the tube. You should have a compact mass of crystals in the tube to a height of not more than 0.5 cm above the sealed end.

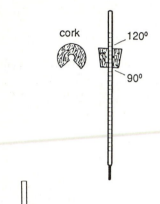

cork

120°

90°

**10.** Slide onto a 260-degree thermometer a large cork with a center hole and a pie-sized section cut away. Position the cork in the 90- to 120-degree region of the thermometer to keep the degree scale visible.

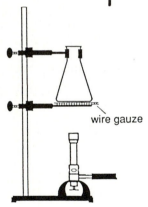

wire gauze

**11.** Attach a ring-clamp to a ringstand at a height that provides space for placement of a burner under the ring clamp. Place a wire gauze on the ring. Pour 75 mL of mineral oil into a 125-mL Erlenmeyer flask. Add one or two boiling chips to the oil. Attach a clamp securely – but not too tightly – to the neck of the flask. Place the flask on the wire gauze on the ring clamp and fix the clamp holding the flask to the ringstand. Check that the flask is securely fastened to the stand and cannot slip off the ring.

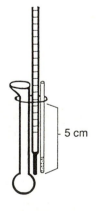

5 cm

**12.** With a rubber band tie the melting-point capillary and the melting-point bulb to the thermometer. The rubber band must be 5 cm above the lower end of the thermometer. Position the capillary and the bulb as shown in the illustration. The camphor and the camphor-unknown mixture must be next to the mercury reservoir of the thermometer.

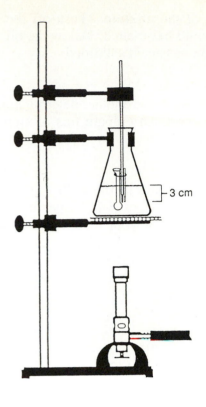

3 cm

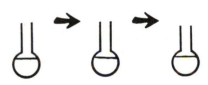

**13.** Attach a clamp to the cork positioned on the thermometer in the 90°-120° region. Carefully attach the other end of the clamp to the ringstand. Slide the thermometer clamp down the upright of the ringstand until the lower 3 cm of the thermometer are immersed in the oil. Tighten the clamp to the ringstand. Begin to heat the oil. Practice varying the air and gas controls of the burner to change the rate of heating. Also vary the location of the burner to find the best position for causing convection currents in the oil. You may heat rather rapidly to a temperature around 160°C, but above 170°C the temperature must not rise faster than 1°C per 30 seconds. Watch only the camphor in the capillary melting-point tube at this stage. Record the temperature at which liquid first appeared in this tube and the temperature at which the last crystals of the camphor melted. Record the temperature at which the last of the solid disappeared as the melting point of pure camphor. *If your heating rate was sufficiently low, the difference between the temperature at which the first liquid appeared and the temperature at which the last crystal melted will be 1 to 2 degrees.*

**14.** After you have determined the melting point of pure camphor, heat the oil to 180° but not higher than 185°. Hold this temperature until the camphor-unknown mixture has melted to a clear, homogeneous liquid. Turn off the burner as soon as the last solid has disappeared. Let the oil cool to approximately 120°. *If you heated your sample to a temperature higher than 185°C, some camphor may have sublimed to the cooler part of the tube. Check the top part of the bulb-tube for camphor crystals. If a considerable amount of camphor has sublimed, you may have to prepare another camphor-unknown mixture. Consult your instructor.*

**15.** The camphor-unknown mixture should have solidified by the time the oil temperature had dropped to 120°C. (If not, consult your instructor.) Reheat the oil at a rate of 5-10°C per minute and watch the camphor-unknown mixture in the bulb continuously. Record the temperature at which liquid first appears and the temperature at which the last cloudiness due to camphor crystals disappears. These two temperatures provide you with the *approximate* melting range of the mixture. Turn off the burner when the last solid has melted. Allow the oil to cool until the mixture has again solidified.

**16.** If you have to make another determination with a new mixture, you can prepare this mixture while the oil is cooling.

**17.** When the mixture in the bulb has solidified, reheat the oil at a rate of 5-10°C per minute to a temperature 10°C below your "approximate" melting range. Then reduce the heating rate to 1°C per minute and determine an *accurate* melting point of the mixture, defined as the temperature at which the last trace of solid disappears.

**18.** If time permits, complete a second determination using fresh samples of camphor and of the same unknown.

**19.** When you have finished your work, allow the oil to cool to room temperature while you clean up your work area. Dispose of the unused mixture and the melting tubes by placing them in the designated containers.

## CALCULATIONS

Be sure all data are recorded accurately in your permanent notebook. Using the melting points of the pure camphor and the camphor-unknown mixture, calculate the approximate molecular mass of the unknown compound. The molal freezing point constant ($K_f$) for camphor is 37.7°C kg (camphor) mol$^{-1}$ (solute).

Using your notebook records and calculations, complete the Report Form. By comparing your experimental molecular mass with the molecular masses calculated from the formulas given in the Introduction, suggest possibilities for the identity of your unknown. Your unknown may or may not be one of these compounds. Report your answer as that compound with a molecular mass closest to the value you determined.

# PRELAB EXERCISES

## INVESTIGATION 16: MOLECULAR MASSES OF PHARMACEUTICAL CHEMICALS

Name:_____

Instructor:_____   ID No.:_____

Course/Section:_____   Date:_____

---

1. The molecular formula gives the elements and their numbers present in one molecule of a substance. The atoms in a molecule are listed in the following sequence: C-H-all other elements in alphabetical order. For instance, the molecular formula for sucrose (sugar) is $C_{12}H_{22}O_{11}$. From the structural formulas in the Introduction find the molecular formulas for:

   Sulfanilamide: _____   Chlorpheniramine: _____

   Benzocaine: _____   Phenolphthalein: _____

   Scopolamine: _____   Acetylsalicylic acid: _____

   Camphor: _____   Caffeine: _____

2. The melting point of pure camphor is 177°C, its molal freezing point depression constant, $K_f$, is 37.7. Calculate the melting point of a solution prepared from 100 mg of camphor and 0.0001 moles of a solute.

3. In which area of the flame of a Bunsen burner is the temperature highest?

4. Define sublimation.

# REPORT FORM

## INVESTIGATION 16: MOLECULAR MASSES OF PHARMACEUTICAL CHEMICALS

Name: _____   ID No.: _____

Instructor: _____   Course/Section: _____

Partner's Name (if applicable): _____   Date: _____

Enter the molecular formulas and the molecular masses for the compounds named in the Table below and for additional compounds your instructor may identify.

| Compound | Molecular Formula | Molecular Mass | Compound | Molecular Formula | Molecular Mass |
|---|---|---|---|---|---|
| caffeine | | | scopolamine | | |
| sulfanilamide | | | acetylsalicylic acid | | |
| chlorpheniramine | | | camphor | | |
| benzocaine | | | phenolphthalein | | |
| | | | | | |
| | | | | | |

|  | First determination | Second determination |
|---|---|---|
| Melting point of pure camphor: | _____ °C | _____ °C |
| Melting point of camphor/unknown mixture: | _____ °C | _____ °C |
| Composition of the mixture: | | |
| Mass of camphor: | _____ mg | _____ mg |
| Mass of unknown: | _____ mg | _____ mg |

Approximate molecular mass of the unknown: _____        Unknown Number:_____

Possible identity of the unknown (from list above): _____

Answer all assigned problems.  Attach a separate sheet if necessary.

1. How would a trace of moisture in the camphor affect the melting point of pure camphor?  The molecular mass of your unknown?  (There is not a simple answer to this question.  You should consider the effect on $T_f$(solute) and $T_f$(solution) and whether or not $K_f$ will remain a constant.)

2. Camphor (200 mg)  and a substance having a molecular mass of 385 amu (15 mg) are mixed and melted to form a solution.  Calculate the melting point of the mixture.

Show calculations here.

Date _____ Signature _____

# Investigation 17

## INTRODUCTION

Most automobile engines employ a circulating-water cooling system to dissipate engine heat. Water, because of its relatively high specific heat [4.18 joule $g^{-1}$ $°C^{-1}$] and ready availability at low cost, makes a good coolant for the automobile radiator system.

The principal difficulty with coolant systems using water lies with the freezing point of water, 0°C (32°F). Much lower temperatures are encountered for extended periods during late autumn, winter, and early spring that would cause the water in the cooling system to freeze, burst hoses, crack the engine block, and make the car inoperable. With pure water as the coolant, automobiles could be operated only in regions in which the temperature does not fall below 0°C. Of course, water can be replaced by liquids with freezing points much lower than water. Methanol (freezing point −98°C) or ethanol (freezing point −114°C) could be used as coolants. Although these liquids would freeze nowhere except perhaps in the Antarctic, they are not very useful because of their low boiling points (methanol 65°C, ethanol 78°C) that favor loss by evaporation; because of the toxicity of methanol, consumption of which leads to blindness and death; and because of the attractiveness of ethanol as a legal intoxicant. Other liquids that might serve as coolants are too expensive.

The observation that water in which a substance is dissolved freezes at a lower temperature than pure water points the way toward an acceptable coolant. A substance, an "antifreeze" compound, must be added to the water. The resulting solution should have a freezing point lower than the lowest temperature encountered on the road. The antifreeze compound must not corrode the metal of the radiator, weaken the materials of which the hoses are made, or react with the alloy of which the engine block consists. The antifreeze now in use is 1,2-dihydroxyethane ($HOCH_2CH_2OH$), an alcohol that is also known as ethylene glycol or, simply, glycol. This compound has an almost ideal set of properties: it is non-corrosive to metal and rubber; it mixes with water in all ratios; it boils at 198°C, freezes at −13°C and is, thus, a liquid at room temperature; it has no odor, does not catch fire easily, and is quite inexpensive. However, ethylene glycol is toxic: ingestion of 100 mL of glycol causes death; repeated ingestions of small doses leads to brain and kidney damage. Glycol has a low vapor pressure (0.06 torr at 20°C). Therefore, very little glycol is in the air under normal conditions and available for inhalation. Glycol does not pose a hazard by the inhalation route at room temperature or below. How much glycol should be added to the cooling system to provide protection from freezing? This experiment will answer this question.

## CONCEPTS OF THE EXPERIMENT

When a substance (called a solute) is dissolved in a solvent, a solution is obtained. Experiments show that the solution freezes at a lower temperature than the pure solvent. From measurements of the freezing points of many solutions prepared from a large variety of solutes and solvents, a correlation was found between the concentration of solutes and the freezing points of the solutions (equation 1). What is $K_f$, the molal freezing point depression constant?

$$T_{f(solution)} - T_{f(solvent)} = \Delta T_f = K_f \times m \qquad\qquad (1)$$

$T_{f(solution)}$: freezing temperature of solution
$T_{f(solvent)}$: freezing temperature of solvent
$\Delta T_f$: freezing point depression
$K_f$: molal freezing point depression constant
m: molality, concentration of the solution in moles of solute per 1,000 g solvent.

Rearrangement of equation 1 provides the answer (equation 2). $K_f$ is the freezing point depression

$$K_f = \frac{\Delta T_f}{m} \qquad\qquad (2)$$

measured for a 1 molal solution. The units for $K_f$ are degree $mol^{-1}$ kg. Experiments with the same solvent but different solutes showed that $K_f$ does not depend on the nature of the solute. $K_f$ is a constant for each particular solvent. Different solvents have different values for $K_f$. $K_f$ for camphor is $-37.7(°C\ mol^{-1}\ kg)$, for water $-1.86(°C\ mol^{-1}\ kg)$. Equation 1 is valid for ideal solutions. When a solute, for instance sodium chloride, dissociates on dissolution ($Na^+$, $Cl^-$), the concentration to be used in equation 1 is the total concentration obtained by summing the concentrations of all the solute ions in solution.

Equation 1 can be used to calculate $K_f$ from the known concentrations and the measured freezing points of the solvent and the solution, to calculate the freezing point depression for a solution of a particular concentration when $K_f$ is known, or to calculate the molecular mass of a solute (Investigation 16). Values of $K_f$ for common solvents are listed in handbooks of chemistry. In principle, the freezing points of solutions of glycol in water can be calculated on the basis of experimental results obtained by earlier investigations. Equation 1 is valid only for ideal solutions in which interactions between solute molecules do not occur. This ideal situation is closely approximated in dilute solutions but not in concentrated solutions. The experimentally determined freezing points of concentrated solutions might be more reliable in evaluating the appropriateness of the composition of the cooling liquid for cold weather.

The cooling liquids are routinely checked for their composition by measuring the density of the liquid. Tables relating the measured density with the freezing point of the solution allow the protection against freezing to be estimated. For this reason, the densities (see Investigation 3) of the glycol-water mixtures are determined in this experiment and the density-freezing point relation displayed graphically. Antifreeze mixtures are prepared by mixing a certain volume of water with the required volume of the antifreeze compound. To make the results obtained in this experiment applicable to the practical task of preparing antifreeze mixtures, the dependence of the freezing points of the mixtures on their composition (percent by volume, percent by mass) is also plotted. The composition of a mixture in percent by volume is defined by equation 3 under the reasonable assumption that the volumes of glycol and water are additive. For instance, a mixture prepared from 10 mL water and 5 mL glycol is assumed to have a volume of 15 mL. This additivity of volumes does not apply to all mixtures but holds for water-glycol mixtures sufficiently well for the purpose of this experiment.

$$\text{percent glycol by volume} = \frac{\text{volume of glycol}}{\text{volume of glycol + volume of water}} \times 100 \qquad (3)$$

Similarly defined is the concentration in terms of percent by mass (equation 4). The mixtures of

$$\text{percent glycol by mass} = \frac{\text{mass of glycol}}{\text{mass of glycol + mass of water}} \times 100 \qquad (4)$$

glycol with water are prepared by allowing the desired volumes of the components to flow from burets into a flask. The masses of these mixtures are determined. The volumes and the masses allow the densities to be calculated (equation 5). The freezing point of these mixtures can be found

$$\text{density of glycol solution} = \frac{\text{mass of glycol} + \text{mass of water}}{\text{volume of glycol} + \text{volume of water}} \tag{5}$$

by cooling samples of the mixtures into which a thermometer has been placed until the mixtures begin to freeze. The crystals that form in solutions of glycol in water (more water is present than glycol: therefore, water is the solvent) are ice crystals. As ice forms, the remaining solution becomes more concentrated and its freezing point decreases with increasing concentration as predicted by equation 1. The freezing point of the mixture is taken as the temperature at which the first crystal is observed. Mixtures cooled to low temperatures generally become rather viscous, are reluctant to freeze, and often cool below their freezing point. To overcome this difficulty, mixtures are cooled to temperatures well below their freezing point and allowed to freeze. The solid is then heated carefully and the temperature determined at which the last crystal melts. This temperature is the same as the freezing temperature.

Temperatures are measured with thermometers. The common thermometer has mercury as the thermometric liquid. Mercury freezes at −39°C, a temperature that might be reached in this experiment. Frozen mercury is not a useful thermometric liquid. Therefore, thermometers with ethanol as thermometric liquid that function in the temperature range −50°C to +50°C will be used. Although thermometers are made to show the correct temperature, for careful work the thermometers are calibrated by immersing them into liquids of well defined freezing points. Water, glycol, and carbon tetrachloride will be used as the standards in this experiment. A calibration plot for the thermometer is then prepared by graphing the thermometer reading at the freezing temperatures of the standards versus the reported freezing points of the standards. The thermometer readings obtained in this experiment are then converted into corrected temperatures using this graph.

## ACTIVITIES

- Complete the PreLab Exercises before coming to the laboratory.
- Prepare several glycol-water mixtures by measuring appropriate volumes of the components from burets into a flask.
- Determine the densities of the glycol-water mixtures.
- Calculate the percent-by-volume and the percent-by-mass composition of these mixtures.
- Calibrate the thermometer (range -50°C to +50°C) with three standards.
- Prepare six glycol-water mixtures for freezing point determination.
- Determine the freezing points of these mixtures.
- Perform the necessary calculations and prepare the required graphs.
- Complete the Report Form.

## SAFETY

*Wear approved eye protection. Avoid contact with or inhalation of the vapors of the organic liquids used. Be particularly cautious with "dry ice". Contact with the skin for 2 seconds or more can cause severe freeze burns. Under no circumstances should you "play" with "dry ice." Never seal "dry ice" in a closed system. Gas pressures may cause an explosion.*

## PROCEDURES

### A.  DETERMINATION OF DENSITIES OF GLYCOL-WATER MIXTURES

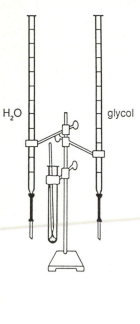

**A-1**. Obtain two burets and make sure that they are clean (Appendix D). Attach them with clamps to a ringstand. Rinse and fill one buret with distilled water. Rinse the other twice with 3-mL portions of glycol and then fill it with glycol. Label both burets. Then clamp a large test tube containing a wad of paper to the ringstand. Use this test tube as a storage place for the thermometer when it is not in use. Be careful with the thermometer. It is both fragile and expensive.

**A-2**. As a check on your technique, first determine the density of water. Weigh (Appendix A) a clean, dry 125-mL Erlenmeyer flask to the nearest 0.05 g. Drain just enough water from the buret into a waste flask to expel air from the buret tip and to drop the liquid level just below the zero mark. Read and record the volume to the nearest 0.02 mL. Then drain about 50 mL of water into the weighed flask. Record the buret reading and weigh the flask plus water to the nearest 0.05 g. Calculate the density of the water. If your value differs from 0.99 g mL$^{-1}$, consult your instructor. Discard the water and dry the flask, inside and out, with a clean towel.

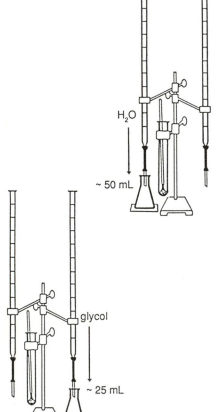

**A-3**. Reweigh the dried flask and measure into it from the buret about 25 mL of ethylene glycol. Record the volume to the nearest 0.02 mL. Weigh the flask plus contents.

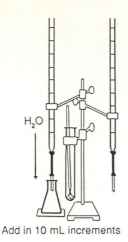

H₂O

Add in 10 mL increments

**A-4**. Add to the glycol in the flask approximately 10 mL of distilled water. Record the volume of the added water to 0.02 mL and calculate the total volume of liquid in the flask. Weigh the flask containing the water-glycol mixture to ±0.05 g. Add another 10-mL volume of water and proceed as described. Repeat this procedure until a total of 50 mL of water have been added. Calculate the density of each mixture.

*You may want to make a copy of the "Density Data" table from your Report Form in your notebook to facilitate the recording of your data.*

## B. CALIBRATION OF THE THERMOMETER

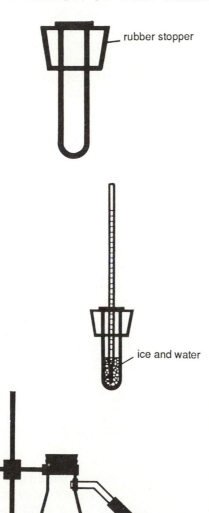

rubber stopper

ice and water

dry ice

**B-1**. Secure a No. 6 rubber stopper with a hole large enough to tightly hold a small test tube. Lubricate the stopper and the test tube with a few drops of glycol, protect your hands with a towel, and insert the test tube through the stopper nearly to the rim of the tube (Appendix G).

Form a team of three students. Each team member will prepare one of the calibration samples. All team members will use all three samples for calibrating their thermometers (Procedures B-3 to B-9).

**B-2**. Fill the test tubes that have been inserted into rubber stoppers as directed below, then stopper and label the tubes.
STUDENT 1: Crush some ice to a fine powder. Fill the test tube about halfway with the crushed ice. Then add about 1 mL of distilled water.
STUDENT 2: Pour 2 mL of pure glycol into the test tube.
STUDENT 3: Pour 2 mL of carbon tetrachloride into the test tube.

**B-3**. Take the ice-water sample, insert the thermometer *carefully* into the slush to a depth of about 3 cm. When the ethanol column has stabilized, read and record the temperature. Remove the thermometer and wipe it dry. Store it in the test tube on the ringstand until you need it again.

**B-4**. Clamp your suction flask to a ringstand. Then connect the flask to an aspirator. Place a few chunks of dry ice into the flask and stopper it.
**Caution: Do not handle dry ice with your bare hands. Use a dry towel.**

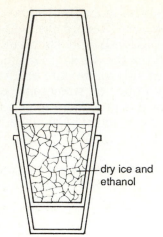

dry ice and
ethanol

dry ice

**B-5**. Obtain three styrofoam cups. Insert one cup into another and fill the upper cup nearly to the brim with powdered dry ice. Obtain about 60 mL of ethanol and pour it slowly onto the dry ice. Stir the mixture with a glass rod. When the very cold slurry evolves $CO_2$ only slowly, cover the mixture with the inverted third cup.

**B-6**. Obtain one of the liquid calibration samples and *carefully* slide the clean, dry thermometer into the test tube. Remove the cover of the dry ice-ethanol slurry and push the test tube into the slurry. Keep the test tube there until the sample has frozen completely. Then remove the test tube, cover the slurry, and ....

**B-7**. ....place the test tube in the neck of the suction flask. Turn on the aspirator. As the sample begins to thaw a little, stir the sample *carefully* with the thermometer using a <u>slow *circular* (*not* up-and-down) motion</u>. Observe the thermometer column. The temperature will rise slowly. Record as the melting point the temperature at which the last crystals melt to form a clear liquid.

Relieve the vacuum by removing the hose carefully from its connection to the aspirator. *Then* turn off the aspirator.

**B-8**. Remove the thermometer, wipe it dry, and return it to its storage tube until you need it again. Stopper the sample test tube and exchange samples with one of your team members. Using the new sample, repeat procedures B-6 and B-7. Continue until you have measured the melting points of all three samples.

**B-9**. In your notebook, plot a calibration graph using your experimentally determined melting points versus literature values (0°C for water, −13°C for ethylene glycol, and −23°C for carbon tetrachloride). You will use this graph to correct temperature readings obtained in part C of this experiment. Complete your calibration graph before starting Part C.

**B-10**. *After all team members have finished with all samples*, the ice-water mixture is to be poured down the drain. Each of the other samples is to be poured into its *own labeled* waste container. These liquids can be "recycled." Clean and dry the test tubes.

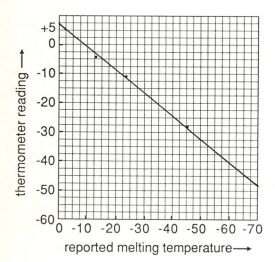

## C. MELTING (FREEZING) POINTS OF GLYCOL-WATER MIXTURES

You will work as a group on procedures C-1 and C-2. Your instructor will assign glycol-water mixtures to be prepared by various members of the group. Then each of you will determine the melting point of at least one of these mixtures.

### Table 17-1.   Glycol-water mixtures

| Mixture No. | Glycol (mL) | Water (mL) | Mixture No | Glycol (mL) | Water (mL) |
|---|---|---|---|---|---|
| 1 | 3.64 | 16.00 | 4 | 10.42 | 8.54 |
| 2 | 7.28 | 12.00 | 5 | 16.72 | 1.60 |
| 3 | 9.10 | 10.00 | 6 | 17.44 | 0.80 |

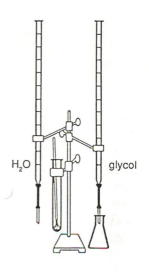

H₂O              glycol

←2 mL

**C-1.** Fill the "distilled water" buret to above the top graduation. Then drain water carefully until the liquid level is below the zero mark. To conserve glycol, add just enough to the "glycol" buret to provide 10 mL more than the amount calculated for the glycol-water mixture assigned to you.

Adjust the volume to below the closest whole milliliter mark by draining a little glycol into a waste flask. Record the initial buret readings to ±0.02 mL. Drain the volumes of water and of glycol for your assigned mixture into a clean, dry 125-mL Erlenmeyer flask. After you have drained the necessary volumes as accurately as possible, wait 1 minute and then check the buret readings again. If they have changed by more than 0.02 mL (due to liquid slowly draining down the buret walls), add the necessary remaining increments to the flask. Record both final buret readings.

**C-2.** Carefully swirl the flask to mix the liquids until all evidence of density gradients has disappeared and a clear homogeneous solution is obtained. Label the flask with the composition of the mixture and your name and the names of your partners. Stopper the flask.

The remaining steps are to be performed independently by each of you using portions of the prepared glycol-water mixtures. You will use the equipment and procedures described in part B of this experiment.

**C-3.** Check that all equipment is ready, that the test tube is clean and dry, and that you still have enough dry ice in the suction flask and styrofoam cups. Then transfer approximately 2 mL of a glycol-water mixture into the test tube.

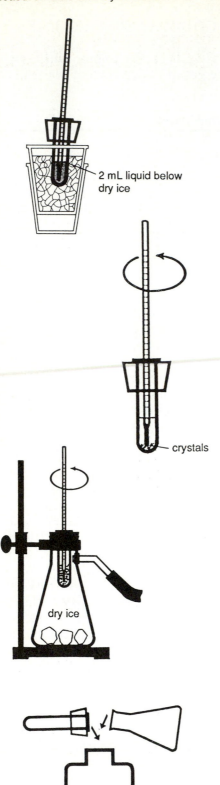

2 mL liquid below dry ice

crystals

dry ice

glycol waste

**C-4.** Slide the thermometer carefully into the test tube and place the tube in the dry ice-ethanol slurry. Stir the mixture slowly and *carefully* using a circular motion. Avoid mixing in air.

**C-5.** Mixtures of high glycol content will become quite viscous. Crystallization is typically slow, usually beginning near the bottom of the test tube where the temperature is lowest. Check periodically for signs of crystal formation. Rotating the thermometer around its own axis in *gentle* contact with the bottom of the tube may facilitate crystallization if the mixture has supercooled.

**C-6.** When the mixture has completely solidified, remove the test tube from the cooling bath, insert it into the top of the suction flask, and turn on the aspirator. When the solid begins to melt, stir slowly and carefully with a *circular* motion (not up-and-down). Avoid stirring in air. Air bubbles trapped in the viscous liquid cause cloudiness that makes it difficult to determine the melting point accurately. Record the temperature at which the last crystals disappear. Refreeze the mixture and repeat the melting point determination.

Use your thermometer calibration graph to correct the measured melting temperatures. Enter the temperatures in your notebook and enter your data in the *section* Report Form posted by your instructor for inclusion in the total data collection. Secure a copy of the final records of this total collection from your instructor.

**C-7.** Pour all glycol-water mixtures into the labeled "waste glycol" container.

**C-8.** Perform all your calculations with special attention to expressing the results with significant figures only. Plot the data using the graph paper provided. Complete the Report Form.

# PRELAB EXERCISES

## INVESTIGATION 17: ANTIFREEZE MIXTURES

Instructor:_____

Course/Section:_____

Name:_____

ID No.:_____

Date:_____

1. Define:

    Miscible liquids

    Molal freezing point depression constant

    Molality

    Freezing point lowering

2. Write a chemical formula for:

    Methanol

    Ethanol

    Ethylene glycol (1,2-dihydroxyethane)

    Glycol

(over)

3. If 20.05 mL of water (density = 0.992g mL$^{-1}$) and 25.12 mL of glycol (density = 1.11g mL$^{-1}$) are mixed, what will be the density of the resulting solution? The volumes are considered to be additive. How many significant figures are justified in the answer to this problem? If 20.05 mL of water and 25.12 mL of glycol are mixed and the resulting solution occupies 45.00 mL, would the non-additive nature of the volumes change the answer?

4. If the molal freezing point depression constant of water is -1.86°C kg (solvent) mole$^{-1}$ (solute), what should be the freezing point of a 1.0 molal solution of glycol in water? Assume ideal solution behavior.

# REPORT FORM

## INVESTIGATION 17: ANTIFREEZE MIXTURES

Name: _____ ID No.: _____

Instructor: _____ Course/Section: _____

Partner's Name (if applicable): _____ Date:_____

Density Data
Mass of empty flask: _____ g

Table 17.2 Density Data

| Glycol (mL) | | Water (mL) | | Mass | | Total liquid volume* | Density (g mL$^{-1}$) | Percent glycol by | |
|---|---|---|---|---|---|---|---|---|---|
| approx. | actual | approx. | actual | flask + liquid | liquid | | | volume | mass |
| 0 | 0.00 | 50 | | | | | | | |
| 25 | | 0 | 0.00 | | | | | | |
| 25 | | 10 | | | | | | | |
| 25 | | 20 | | | | | | | |
| 25 | | 30 | | | | | | | |
| 25 | | 40 | | | | | | | |
| 25 | | 50 | | | | | | | |

*Assume that volumes are additive.

CALCULATION OF DENSITY
(Show setup for one of the glycol-water mixtures.)

Assigned glycol-water mixture (procedure C): _____
Instructor's initials: _____

## CALIBRATION OF THERMOMETER

melting point of water:      accepted value _____ °C
                                      thermometer reading _____ °C

melting point of glycol:      accepted value _____ °C
                                      thermometer reading _____ °C

melting point of $CCl_4$:      accepted value _____ °C
                                      thermometer reading _____ °C

Table 17.3   Melting temperatures of glycol-water mixtures

| Mixture No. | Glycol (mL) | Water (mL) | Percent by volume glycol | water | Percent by mass glycol | water | Melting point °C experiment | corrected |
|---|---|---|---|---|---|---|---|---|
| 1 | 3.64 | 16.00 | | | | | | |
| 2 | 7.28 | 12.00 | | | | | | |
| 3 | 9.10 | 10.00 | | | | | | |
| 4 | 10.42 | 8.54 | | | | | | |
| 5 | 16.72 | 1.60 | | | | | | |
| 6 | 17.44 | 0.80 | | | | | | |
| 7 | pure glycol Part A | | | | | | | |
| 8 | pure water Part A | | | | | | | |

## SAMPLE CALCULATION

Calculate the percent glycol by mass and volume for the mixture assigned to you.



Prepare all the graphs and attach them to the Report Form.

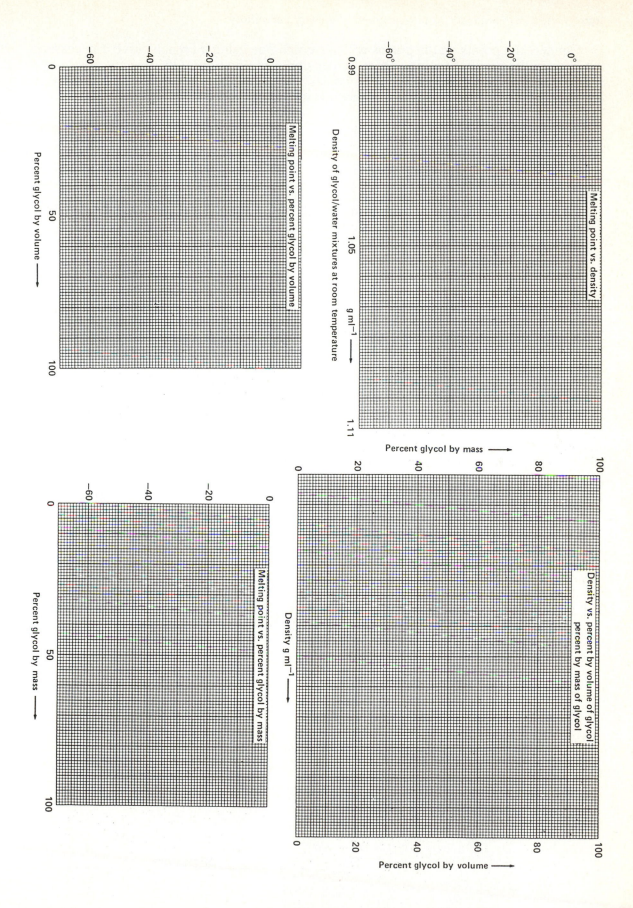

Melting point vs. density

Density of glycol/water mixtures at room temperature

0.99

1.05

g ml⁻¹ ——

1.11

Percent glycol by mass ——

Melting point vs. percent glycol by volume

Percent glycol by volume ——

Density vs. percent by volume of glycol
percent by mass of glycol

Melting point vs. percent glycol by mass

Percent glycol by mass ——

Density g ml⁻¹ ——

Percent glycol by volume ——

Connect the points on the graphs with a smooth line using a French curve if necessary. Extend the lines in the graphs "melting point vs. density", "melting point vs. percent glycol by volume", and "melting point vs. percent glycol by mass" until they meet at a minimum. From these graphs find the glycol-water mixture that has the lowest melting point:

Lowest melting point: _____ °C
Density of this mixture: _____ g mL$^{-1}$
Percent glycol by volume: _____ %
Percent glycol by mass: _____ %

The cooling system of a car requires 4 gallons of liquid. Calculate the gallons of water and gallons of glycol needed to fill this cooling system and protect it to a temperature of 0°F. Attach your work and report your answers below.

0°F = _____ °C

Composition of mixture freezing at 0°F:        _____ % glycol by volume
                                                _____ % water by volume
Needed for 4 gallon cooling system:            _____ gallon(s) water
                                                _____ gallon(s) glycol

Questions:
1. Criticize the following statement: Because a mixture containing about 60 percent glycol by mass will protect your radiator to -50°C, fill your radiator with pure glycol to protect your cooling system to temperatures well below -50°C.

2. Explain how a measurement of the density of the liquid in your automobile radiator could be used to find the lowest temperature at which your automobile's cooling system is protected against freezing.

Date _____ Signature _____

# Investigation 18

## MOLECULAR MASS OF A FATTY ACID

## INTRODUCTION

Fatty acids are monoprotic aliphatic saturated or unsaturated carboxylic acids. Some of the properties, the structure, and the composition of these acids are discussed in Investigations 7 and 8. Fatty acids have great economic and biochemical importance. Most of the fatty acids are produced from fats and oils that are naturally occurring esters containing glycerol (1,2,3-trihydroxypropane) as the alcohol component and long-chain carboxylic acids as the acid components. Saponification of fats and oils (Investigation 7) yields the fatty acids and glycerol. Fatty acids find many uses; for instance, in the manufacture of soap, as additives to increase the viscosity of cosmetics, as chemicals for the treatment of textiles, and as additives to lubricants, rubbers, and dyes. The reduction of fatty acids produces long-chain alcohols that are needed for the manufacture of biodegradable detergents. The human and animal body can make fatty acids from ingested organic compounds. These reactions converting excess food into fatty acids serve to store energy in the form of long-chain carboxylic acids that are present in the body as esters with glycerol (fat). When the body needs energy, it "burns" these fatty acids to produce carbon dioxide and water. The energy liberated in this room-temperature combustion is used to sustain life processes.

Table 18-1 Some fatty acids from natural fats and oils

| "Common" (trivial) name | Formula | m.p., °C | Obtained by hydrolysis of glyceryl esters found in |
|---|---|---|---|
| butyric | $CH_3(CH_2)_2CO_2H$ | −6 | butter |
| caproic | $CH_3(CH_2)_4CO_2H$ | −3 | butter and goat fat |
| caprylic | $CH_3(CH_2)_6CO_2H$ | 16 | butter, goat fat, and coconut oil |
| capric | $CH_3(CH_2)_8CO_2H$ | 31 | butter, goat fat, and coconut oil |
| lauric | $CH_3(CH_2)_{10}CO_2H$ | 44 | coconut oil and palm oil |
| myristic | $CH_3(CH_2)_{12}CO_2H$ | 54 | butter, coconut oil, palm oil, beef tallow, and cod liver oil |
| palmitic | $CH_3(CH_2)_{14}CO_2H$ | 63 | all of the above plus corn oil, cottonseed oil, lard, olive oil, peanut oil, and linseed oil |
| stearic | $CH_3(CH_2)_{16}CO_2H$ | 70 | all of the above |
| oleic | $cis\text{-}CH_3(CH_2)_7CH{=}CH(CH_2)_7CO_2H$ | 16 | all of the above |
| linoleic | $cis,cis\text{-}CH_3(CH_2)_4CH{=}CHCH_2CH{=}CH(CH_2)_7CO_2H$ | −5 | all of the above, but to the largest extent in vegetable oils |
| linolenic | $cis,cis,cis\text{-}$ $CH_3CH_2CH{=}CHCH_2CH{=}CHCH_2CH{=}CH(CH_2)_7CO_2H$ | −11 | soybean oil and linseed oil |

Whether fatty acids are used in industrial processes or studies by biochemists, these compounds must be identified and characterized. Examples of fatty acids from natural fats and oils, their names, their formulas, and their melting points are listed in Table 18-1.

Because fatty acids are acidic, they will react with bases. This neutralization reaction can be used to find the molecular mass of a fatty acid, or the average molecular mass of a mixture of different fatty acids. In this experiment an unknown fatty acid is identified by it molecular mass determined by titrating the unknown with an aqueous sodium hydroxide solution of known molarity.

## CONCEPTS OF THE EXPERIMENT

The fatty acids used in this experiment will be monoprotic carboxylic acids that have one acidic hydrogen atom per molecule. The general formula for such an acid is $C_nH_{2n+1}CO_2H$. The neutralization reaction occurring between a fatty acid and sodium hydroxide is described by equation 1 or in net-ionic form by equation 2. When a sodium hydroxide solution is added to a

$$C_nH_{2n+1}CO_2H + NaOH \longrightarrow C_nH_{2n+1}CO_2Na + H_2O \qquad (1)$$

$$\text{fatty acid} \qquad \text{base} \qquad\qquad \text{soap} \qquad\qquad \text{water}$$

$$C_nH_{2n+1}CO_2H + OH^- \longrightarrow C_nH_{2n+1}CO_2^- + H_2O \qquad (2)$$

solution of the fatty acid, the $OH^-$ from the added base and the $H^+$ ions from the fatty acid combine to form undissociated water. When all the acid has been neutralized at the endpoint of the titration, additional base will increase the hydroxide ion concentration because $H^+$ ions are no longer present to combine with $OH^-$. Phenolphthalein, an organic molecule, reacts with these excess hydroxide ions and forms a molecule that absorbs green light and makes the solution purple-red. Phenolphthalein, before its reaction with $OH^-$, does not absorb any visible light and is, therefore, colorless. The appearance of the pink color indicates that the neutralization of the acid is complete and that the endpoint of the reaction has been reached.

When the molarity of the sodium hydroxide solution is known and the volume of the sodium hydroxide solution that was needed to reach the endpoint was carefully measured by dispensing the solution from a buret, the moles of sodium hydroxide consumed in the reaction can be calculated (equation 3). Because each mole of sodium hydroxide consumed in the neutralization reaction

$$\text{moles (NaOH)} = M(\text{molarity}) \times V(\text{volume in liters}) \qquad (3)$$

(equation 1) requires one mole of fatty acid, the number of moles of sodium hydroxide calculated according to equation 3 is equal to the number of moles of carboxylic acid neutralized. The mass of the fatty acid used in the titration is known because the sample was weighed on an analytical balance before it was titrated. The mass of the fatty acid and the number of moles of fatty acid allow the molecular mass to be calculated (equation 4). When the molecular mass is known, the

$$\text{molecular mass (fatty acid)} = \frac{\text{mass of fatty acid}}{\text{number of moles of fatty acid}} \qquad (4)$$

"n" in $C_nH_{2n+1}CO_2H$ (or $C_nH_{2n+2}CO_2$) can be calculated using equation 5. Try your hand (and mind) in deriving this equation. When you have identified "n", you can write a formula for your

$$n = \frac{\text{molecular mass - 46}}{14} \qquad (5)$$

unknown and use Table 18-1 to find the name for the fatty acid. Because errors are always associated with experimental work, the calculated "n" very likely will not be an integer. You will have to round your answer. When attempting to identify your unknown, keep in mind that natural fatty acids almost always have an even number of carbon atoms in the molecule.

One additional question needs to be considered. Why must the sodium hydroxide solution be standardized? The balance is precise and accurate enough to produce a solution of well defined molarity. Sodium hydroxide is a rather reactive substance. If you keep a pellet of sodium hydroxide in moist air, you will notice a liquid film forming on the surface of the pellet. Sodium hydroxide is deliquescent (it absorbs water from the air) and reacts with carbon dioxide present in the air. For these reasons, sodium hydroxide is never pure when handled in air. Therefore, sodium hydroxide solutions are prepared by weighing out an approximate amount of solid sodium hydroxide and dissolving it in an approximate volume of distilled water. Graduated cylinders serve well for measuring the volume. The use of a volumetric flask is not necessary. The exact molarity of this solution is determined by standardization with an acid that can easily be obtained in high purity, will not absorb water, and will not react with carbon dioxide or other constituents of the atmosphere. Potassium hydrogen phthalate (KHP), the mono-potassium salt of the diprotic phthalic acid, is such a substance. The amount required for the standardization of a sodium hydroxide solution is weighed to 0.1 mg on an analytical balance. The sample is dissolved in hot water, phenolphthalein is added, and sodium hydroxide solution added until the endpoint is reached. At this point equimolar amounts of acid and base have reacted (equation 6).

potassium hydrogen
phthalate (KHP)

When "m" grams of the potassium hydrogen phthalate (KHP) of molecular mass $MM_P$ containing $m/MM_P$ moles of KHP required V mL of the sodium hydroxide solution of molarity $M_{NaOH}$, equation 7 will allow the molarity to be calculated. Rearrangement of this equation to isolate $M_{NaOH}$ will lead to an expression relating the molarity to the volume of base consumed, to the

$$\frac{m}{MM_P} = \frac{V(mL) \times M_{NaOH}}{1000} \tag{7}$$

mass of KHP titrated and to the molecular mass of KHP. The standardized sodium hydroxide solution can now be used to titrate the unknown fatty acid. The results of all calculations must be expressed with significant figures only.


## ACTIVITIES

- Complete the PreLab Exercises before coming to the laboratory.

- Prepare 300 mL of a sodium hydroxide solution of approximately 0.1 molar concentration.

- Standardize the sodium hydroxide solution using potassium hydrogen phthalate.

- Determine the molecular mass of an unknown fatty acid.

- Identify the fatty acid.

## SAFETY

*Wear approved eye protection at all times in the laboratory. Sodium hydroxide is a strong base; handle it carefully and avoid contact with your skin. If contact has occurred, wash with plenty of tap water. Use caution when heating ethanol, a flammable liquid.*

## PROCEDURES

### A. PREPARATION OF 300 mL OF A 0.1 M NaOH SOLUTION

**A-1**. Retrieve from problem 2 of the PreLab Exercises the value for the volume of 2.5 M sodium hydroxide solution needed to make 300 mL of a 0.1 M solution.

**A-2**. Take a clean, but not necessarily dry, 50- or 25-mL graduated cylinder to the reagent area and obtain the volume of 2.5 M sodium hydroxide calculated for the preparation of the needed 0.1 M NaOH solution. Add the 2.5 M NaOH solution to a 500-mL Erlenmeyer flask containing 100 mL distilled water. Swirl the flask carefully. Use 25 mL of water to rinse the graduated cylinder. Add this wash to the flask. Then add an additional 175 mL of distilled water. Mix the contents thoroughly by carefully swirling. Stopper the flask and label it "0.1 M NaOH."

**A-3**. Check that your 50-mL buret is clean, free of grease, and draining properly (Appendix D). Rinse the buret three times with 5 mL of your 0.1 M NaOH solution. Then fill the buret with the NaOH solution above the 0-mL mark. Drain some of the solution from the buret to expel air from the buret tip. Allow the liquid level to drop just below the zero mark

### B. STANDARDIZATION OF THE 0.1 M NaOH SOLUTION

**B-1**. Retrieve from your PreLab Exercise the mass of potassium hydrogen phthalate needed to neutralize 30 mL of a 0.1 M NaOH solution. Should you have doubts about your calculation, check with your instructor.

**B-2**. Fold a clean piece of smooth paper in the middle. Weigh this piece of paper to 0.1 mg. Take this paper to your instructor to obtain a sample of potassium hydrogen phthalate approximately of the amount you need. Take the paper with the sample to an analytical balance and determine its mass to 0.1 mg. Calculate the mass of the potassium hydrogen phthalate on your paper. This mass will not be exactly the amount you need. The mass can be ±4.0 mg of the calculated amount. If the mass is outside this limit, add or remove some of the phthalate. *Remember that the balance must be arrested before you remove anything from the pan.*

**B-3**. After you have determined the mass of phthalate to 0.1 mg, arrest the balance and carefully remove the paper from the balance pan. Transfer the weighed phthalate into a clean 250-mL Erlenmeyer flask.

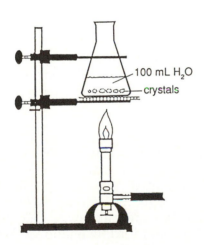

100 mL H₂O

crystals

**B-4**. Add approximately 100 mL distilled water to the flask. Heat the flask carefully until all the phthalate has dissolved. Then remove the flask from the flame.

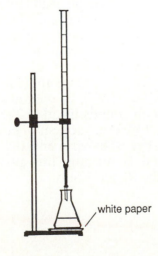

white paper

**B-5**. Add three drops of phenolphthalein indicator solution to the flask and set it on a white paper below the buret. Read and record the volume level in the buret to the nearest 0.01 mL. Titrate the phthalate solution to a permanent pale pink endpoint. Record the final buret reading. Clean the flask and rinse it with distilled water. Repeat the standardization if time allows.

## C. TITRATION OF AN UNKNOWN FATTY ACID

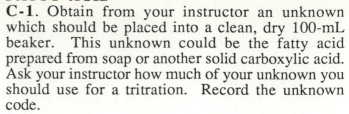

**C-1.** Obtain from your instructor an unknown which should be placed into a clean, dry 100-mL beaker. This unknown could be the fatty acid prepared from soap or another solid carboxylic acid. Ask your instructor how much of your unknown you should use for a tritration. Record the unknown code.

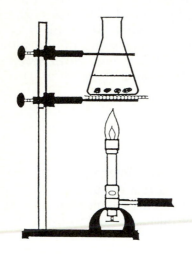

**C-2.** Weigh out the suggested amount of your unknown on the analytical balance to 0.1 mg using a creased piece of paper as described in Procedure B-2. The amount of unknown you use may deviate by ±20 mg from the suggested amount. Transfer the sample quantitatively into a clean 250-mL Erlenmeyer flask. Add 50 mL of 95-percent ethanol. Heat the mixture carefully over a Bunsen flame until the unknown has dissolved. Add 20 mL more ethanol if the substance has not dissolved when the ethanol boils. *Caution: Ethanol is flammable!*

**C-3.** When all the unknown has dissolved, add three drops of phenolphthalein to the solution and titrate with your NaOH solution to a pale pink endpoint. Record the initial and final buret readings. Calculate the formula mass of your unknown. Repeat the titration with another sample of the same unknown.

**C-4.** If you titrated your unknown more than once, calculate the average molecular mass of your unknown. Obtain from your instructor the molecular masses or the average molecular masses other students found for the same unknown. Calculate the average molecular mass for this unknown from the pooled data, calculate the standard deviation, and the relative standard deviation. On the basis of the average molecular mass from the pooled data, suggest a formula for your unknown assuming that the unknown has the general formula $C_nH_{2n+1}COOH$. Consult Table 18-1 for the formulas of fatty acids. Is your unknown identical to any of the compounds listed in Table 18-1? If more than one formula appears possible for your unknown, suggest experiments that would allow you to distinguish among the possible formulas. Complete the Report Form.

# PRELAB EXERCISES

## INVESTIGATION 18: MOLECULAR MASS OF A FATTY ACID

Name:_____

Instructor:_____     ID No.:_____

Course/Section:_____     Date:_____

1. Define:

    Molarity

    Acid

    Base

    Titration

2. Calculate the volume of 2.5 M NaOH solution needed to prepare 300 mL of a 0.10 M NaOH solution. (Record your answer here and in your notebook.)

3. Calculate the mass of solid sodium hydroxide needed to prepare 300 mL of a 0.1 M solution. If one pellet of sodium hydroxide weighs 0.170 g, how many pellets should be dissolved?

(over)

4. Write a balanced equation for the neutralization of potassium hydrogen phthalate ($C_8H_5O_4K$) with sodium hydroxide. Calculate the mass of potassium hydrogen phthalate that neutralizes 30 mL of a 0.1 M NaOH solution. (Record your answer here and in your notebook.)

5. A sample of a liquid carboxylic (fatty) acid weighing 0.2131 g required 36.5 mL of a 0.0973 M aqueous sodium hydroxide solution to make the pink color of phenolphthalein appear. Calculate the molecular mass of the acid, find the formula assuming the acid belongs to the series of carboxylic acids of general formula $C_nH_{2n+1}COOH$, and identify the acid by name.

6. Draw a structural representation of potassium hydrogen phthalate that shows all its hydrogen atoms.

# REPORT FORM

## INVESTIGATION 18: MOLECULAR MASS OF A FATTY ACID

Name: _____   ID No.: _____

Instructor: _____   Course/Section: _____

Partner's Name (if applicable): _____   Date: _____

---

Show all calculations on attached pages.

### Solution Preparation and Standardization.

Volume of 2.5 M NaOH needed for 300 mL of 0.1 M solution: _____ mL

Mass of $C_8H_5O_4K$ needed to neutralize 30 mL of 0.1 M NaOH: _____ g

| Phthalate sample | | | Phthalate titration | | |
|---|---|---|---|---|---|
| | Trial 1 | Trial 2 | | Trial 1 | Trial 2 |
| Sample + paper ____ g | | ____ g | Final volume NaOH ____ mL | | ____ mL |
| Paper ____ g | | ____ g | Initial volume NaOH ____ mL | | ____ mL |
| Phthalate ____ g | | ____ g | Vol. NaOH consumed ____ mL | | ____ mL |

Average standardized concentration of NaOH solution: _____ M

### Determination of Molecular Mass of a Fatty Acid

Unknown code: _____

| Fatty acid sample | | | Fatty acid titration | | |
|---|---|---|---|---|---|
| | Trial 1 | Trial 2 | | Trial 1 | Trial 2 |
| Sample + paper ____ g | | ____ g | Final volume NaOH ____ mL | | ____ mL |
| Paper ____ g | | ____ g | Initial volume NaOH ____ mL | | ____ mL |
| Fatty acid ____ g | | ____ g | Vol. NaOH consumed ____ mL | | ____ mL |
| | | | Moles NaOH consumed _____ | | _____ |
| | | | Moles $H^+$ neutralized _____ | | _____ |
| | | | Molecular mass of fatty acid _____ | | _____ |

Average Molecular Mass: _____ g mole$^{-1}$

## Calculation of Formula Mass

Group results: _____ g mole$^{-1}$

_____ g mole$^{-1}$

_____ g mole$^{-1}$

_____ g mole$^{-1}$

_____ g mole$^{-1}$

Your average formula mass: _____ g mole$^{-1}$

Average: _____ g mole$^{-1}$

Standard deviation: _____

Relative standard deviation: _____

The "unknown" acid was probably: _____ (name)

Formula of named acid: _____

Molecular mass of named acid: _____

Other possibilities, if any, are:

Suggestions for additional experiments, if needed:

Questions:

1.  The general formula for a diprotic saturated carboxylic acid is $C_nH_{2n}(CO_2H)_2$.  Derive an expression similar to equation 5 for the relationship between n and MM for this class of compounds.

2.  KHP is considered a primary standard.  Why shouldn't a solution of NaOH which was standardized yesterday be used as a standard solution today?

Date _____ Signature _____

# Investigation 19

## RAYON:  A VALUABLE POLYMER

### INTRODUCTION

Molecules come in all sizes.  The hydrogen molecule, $H_2$, with molecular mass of 2 amu is the smallest molecule.  Inorganic molecules belong – with few exceptions – to the class of small molecules.  They rarely consist of more than twenty or thirty atoms.  Among organic compounds, molecules of large size are frequently encountered.  A carbon atom can bind four other atoms and is capable of forming strong bonds with other carbon atoms.  In this way long chains of carbon atoms can be obtained in which each carbon atom is bonded to two other carbon atoms.  When the two valences (orbitals) on each carbon atom that are not needed for binding carbon atoms are used to hold two hydrogen atoms, a long-chain hydrocarbon ① is obtained.  Large molecules with

$$CH_3 \cdots + \underset{\underset{H}{|}}{\overset{\overset{H}{|}}{C}} - \underset{\underset{H}{|}}{\overset{\overset{H}{|}}{C}} + \underset{\underset{H}{|}}{\overset{\overset{H}{|}}{C}} - \underset{\underset{H}{|}}{\overset{\overset{H}{|}}{C}} + \underset{\underset{H}{|}}{\overset{\overset{H}{|}}{C}} - \underset{\underset{H}{|}}{\overset{\overset{H}{|}}{C}} + \underset{\underset{H}{|}}{\overset{\overset{H}{|}}{C}} - \underset{\underset{H}{|}}{\overset{\overset{H}{|}}{C}} + \underset{\underset{H}{|}}{\overset{\overset{H}{|}}{C}} - \underset{\underset{H}{|}}{\overset{\overset{H}{|}}{C}} + \underset{\underset{H}{|}}{\overset{\overset{H}{|}}{C}} - \underset{\underset{H}{|}}{\overset{\overset{H}{|}}{C}} + \cdots CH_3 \qquad (1)$$

①
long chain hydrocarbon

molecular masses of millions of atomic mass units are very common in nature.  For instance, wood consists of large-size molecules built from glucose; skin, hair, and fingernails are based on amino acids; our genes have sugar phosphates as building blocks; and most rocks are condensed silicates (large-size inorganic molecules) at least formally derived from ortho-silicic acid, $H_4SiO_4$.

Most large molecules are built from repeating units.  The long-chain hydrocarbon shown above can be considered to have the $CH_2$-$CH_2$ unit as the building block.  If the methyl ($CH_3$) groups at the two ends of the chain are neglected, the formula of the long-chain hydrocarbon can be written as $(CH_2CH_2)_x$ with x equal to about 100,000 for a material known as polyethylene.  Such materials are now made synthetically from ethylene, $CH_2=CH_2$, by polymerization.

Organic chemists had frequently encountered such polymers when they explored the chemistry of unsaturated organic compounds such as ethylene.  These polymers had made their presence known by viscous "gunks" at the bottom of the reaction flasks.  These "gunks" were hard to remove from the flasks and were rather unpopular with students.  Because professors did not clean flasks and had no first-hand experience with "gunk", they refused to believe in large-size molecules.  When Staudinger began to investigate these polymeric "gunks" during the first part of this century, he was advised by the experts in organic chemistry to use his time for more productive pursuits.  Out of Staudinger's work grew the science and the industrial endeavors concerned with polymers.  Today, life as we know it would come to a sudden halt without the huge variety of polymers that serve as materials of construction, as fibers for the production of clothing, and special materials for a multitude of special applications.

Before the idea of large-size molecules was accepted and synthetic polymers became available, attempts were made to convert natural polymers such as cellulose into products useful as fibers for the production of textiles. One of the first successes was the conversion of cellulose from cotton residues into a modified cellulosic fiber known as "Rayon."  This fiber had a luster similar to silk, but was much less expensive than natural silk, a polyamide fiber.

This experiment demonstrates the most important steps in the production of Rayon by the "cuprammonium" process. This process is used today to make hollow fibers for blood dialysis in artificial kidneys.

## CONCEPTS OF THE EXPERIMENT

Cellulose ② is a major constituent of wood and cotton fibers and is one of the most abundant organic compounds on earth. The molecular mass of native cellulose is several hundred thousand amu. This fibrous material is useful for the production of paper, but not very suitable in its native state as a fiber for the production of textiles. Cellulose is not soluble in common solvents. In

Cellulose

Repeating unit
(monomer)

Tetraamminecopper(II) Dihydroxide

1857 an aqueous ammoniacal solution of tetraamminecopper(II) hydroxide ③ was discovered to dissolve cellulose. Further investigations of this system eventually revealed that unique fibers are produced when the cellulose solution is squirted into an acidic solution. The processes that lead to the dissolution of cellulose in the very basic solution of tetraamminecopper(II) dihydroxide are not known in detail. However, the regenerated cellulose molecules have a lower molecular mass than the molecules of native cellulose. Therefore, the dissolution process must be aided by cleavage of large cellulose molecules (300,000 amu) into smaller molecules (60,000 amu). The chains are broken (hydrolyzed) at the "glycosidic" C-O-C bond in reactions with water molecules (equation 1).

In addition, in the very basic solution some of the alcoholic OH groups in cellulose will lose a proton and will be converted into negatively charged alkoxide groups that might interact with the tetraamminecopper(II) cations. The cellulose chain becomes electrically charged and capable of interacting strongly with the water molecules in the solvent through formation of hydrogen bonds.

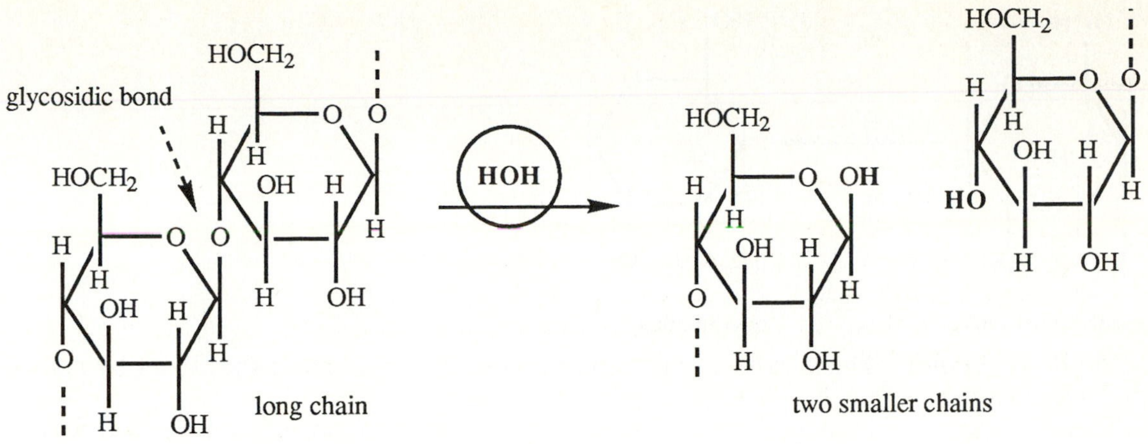

glycosidic bond

HOCH$_2$

HOCH$_2$

long chain

HOH

HOCH$_2$

HOCH$_2$

two smaller chains

(1)

Lower molecular mass through chain cleavage and increased interactions with the solvent increase the solubility of cellulose (equation 2).

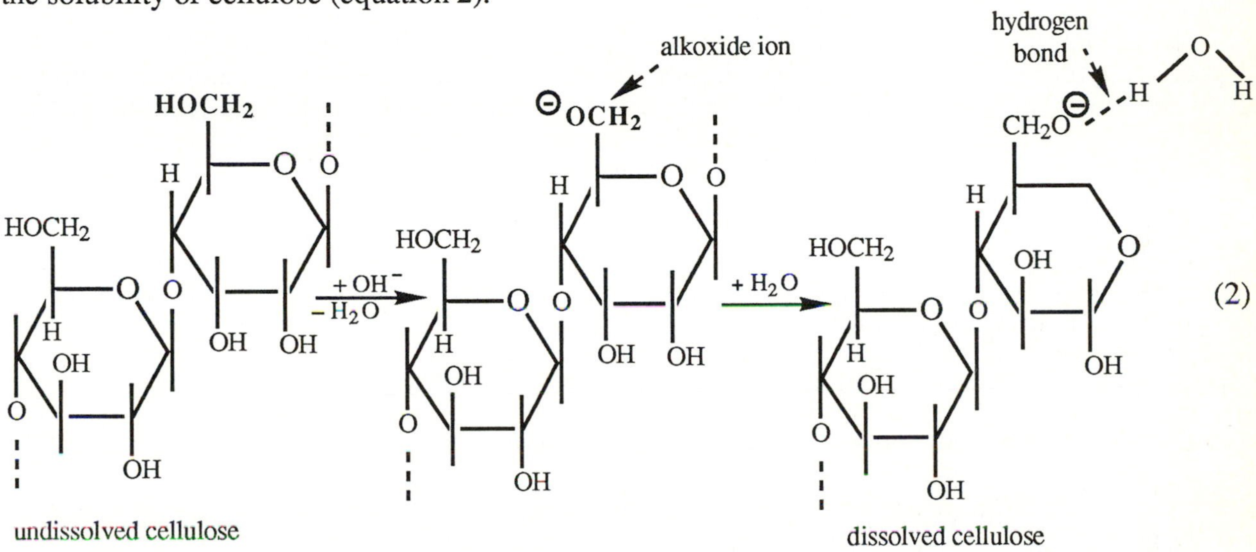

alkoxide ion

hydrogen bond

HOCH$_2$

HOCH$_2$

$^\ominus$OCH$_2$

HOCH$_2$

CH$_2$O$^\ominus$

HOCH$_2$

$\xrightarrow[-H_2O]{+OH^-}$

$\xrightarrow{+H_2O}$

(2)

undissolved cellulose

dissolved cellulose

To precipitate the cellulose, the viscous blue solution is drawn into an aqueous solution of sulfuric acid through a narrow nozzle. As soon as the basic solution reaches the acid, the ammonia solution is neutralized, ammonia is converted to ammonium sulfate, the tetraamminecopper(II) cation is decomposed to copper sulfate and ammonium sulfate, and all alkoxide groups on cellulose molecules are protonated and converted to neutral OH groups (equations 3a, 3b, 3c, and 3d). The cellulose becomes insoluble and precipitates in the form of fibers. Ammonium sulfate and copper sulfate adhere to the precipitated fibers. The copper sulfate on the fibers gives the precipitated

$$OH^- + H^+ \longrightarrow H_2O \qquad (3a)$$

$$2NH_3 + H_2SO_4 \longrightarrow (NH_4)_2SO_4 \qquad (3b)$$

$$[Cu(NH_3)_4]^{2+} + 3H_2SO_4 \longrightarrow 2(NH_4)_2SO_4 + CuSO_4 + 2H^+ \qquad (3c)$$

dissolved cellulose                precipitated cellulose

(3d)

cellulose a light blue color. These salts and the color can be removed from the fibers by rinsing them with distilled water.

The fibers prepared in this manner do not have all the characteristics of commercial Rayon. Several steps of the industrial process (formation of derivatives, aging, stretching) were not included in this experiment. These omissions affect the quality of the Rayon fibers also known as "artificial silk".

The tetraamminecopper(II) hydroxide solution is prepared in two steps. In the first step copper(II) hydroxide is precipitated by addition of the stoichiometric amount of aqueous ammonia to an aqueous solution of copper sulfate (equation 4). The copper(II) hydroxide precipitate must

$$CuSO_4(aq) + 2NH_3 + H_2O \longrightarrow Cu(OH)_2(s) + (NH_4)_2SO_4(aq) \qquad (4)$$

be washed entirely free of sulfate because the sulfate will interfere with subsequent reactions. The presence of sulfate in the washings is detected by the formation of a white barium sulfate precipitate upon addition of a solution of barium chloride to the acidified washings (equation 5). When the copper(II) hydroxide is sulfate-free, it is treated with an excess of concentrated

$$(NH_4)_2SO_4(aq) + BaCl_2(aq) \longrightarrow BaSO_4(s) + 2NH_4Cl(aq) \qquad (5)$$

aqueous ammonia to form a deep-blue solution of tetraamminecopper(II) dihydroxide (compound ③).

## ACTIVITIES

- Complete the PreLab Exercises before coming to the laboratory.

- Prepare a solution of tetraamminecopper(II) hydroxide.

- Prepare a solution of cellulose.

- Prepare Rayon fibers from the ammoniacal cellulose solution.

## SAFETY

*In this experiment you will handle corrosive ammonia and sulfuric acid solutions. Be cautious and wear approved eye protection at all times. Contact lenses should not be worn when working with ammonia. Ammonia vapors may penetrate the space between the contact lens and the eye and may cause damage to the eye. Perform all experiments with ammonia under a well ventilated hood. Avoid inhaling ammonia vapors. Be careful when acids and bases are combined in a neutralization reaction. Large amounts of heat are generated in such*

*reactions.   Burns   to   hands   may   result.    In   case   of   accidental   contact   with*
*chemicals,   wash   the   affected   area   immediately   and   thoroughly   with   water.*

## PROCEDURES

**1**.  On a triple-beam balance, weigh out about 7.5 g of $CuSO_4 \cdot 5H_2O$ crystals.  Use a clean mortar and pestle to grind the crystals into a fine powder.

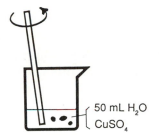

50 mL H₂O
CuSO₄

**2**.  Transfer the powdered $CuSO_4 \cdot 5H_2O$ into a clean 250-mL beaker, add 50 mL of distilled water, and stir the mixture vigorously until the copper sulfate has dissolved.

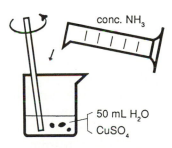

conc. NH₃

50 mL H₂O
CuSO₄

**3**.  Take the beaker with the copper sulfate solution into a fume hood.  Measure exactly 4 mL of concentrated (~15 M) aqueous ammonia from the storage bottle into a graduated cylinder. Pour the ammonia solution slowly into the gently stirred copper sulfate solution to precipitate bluish-white $Cu(OH)_2$.

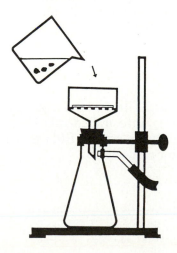

**4**.  Allow the $Cu(OH)_2$ precipitate to settle for a few minutes.  During this time set up the vacuum filtration apparatus and select a filter paper that will fit the Buchner funnel.  Weigh the filter paper on the triple-beam balance to the nearest 0.1 g.  Place the weighed filter paper into the Buchner funnel.  Pour a small volume of the clear liquid above the settled precipitate onto the filter paper to make the paper adhere to the perforated plate in the funnel.  Pour the supernatant liquid first and then the precipitate onto the filter paper.

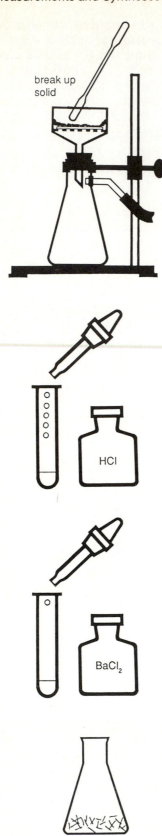

break up
solid

HCl

BaCl₂

**5**. Rinse the beaker with small portions of distilled water to transfer any residual solid to the filter. Then wash the solid on the filter with small portions of distilled water. The total volume of distilled water used for washing should be about 200 mL. After each wash is poured onto the filter, carefully break up the solid and stir it around. Then use strong suction to pull the wash water through the filter.

**6**. Disconnect the vacuum, carefully remove the Buchner funnel, and pour the wash liquid from the flask. Rinse the flask with a small portion of distilled water, then reassemble the filtration equipment. Wash the solid one more time with 2 to 3 mL of distilled water. Pour this rinsing from the flask into a test tube and add six drops of dilute HCl.

**7**. Add six drops of barium chloride solution to the acidified rinse solution. If a heavy white precipitate ($BaSO_4$) is observed, you have not washed the $Cu(OH)_2$ sufficiently free from sulfate anions and further washings must be performed until repeated tests indicate a negligible $SO_4^{2-}$ contamination. Then lift the filter paper from the funnel with the help of your spatula. Fold the filter paper into quarters to keep the precipitate on the inside of the folded filter.

**8**. Transfer the folded filter paper containing the sulfate-free $Cu(OH)_2$ into a 125-mL Erlenmeyer flask. Use your spatula to tear the filter paper inside the flask into small pieces.

conc. NH₃

**9**. Working under the hood, measure out 30 mL of concentrated aqueous ammonia. Pour the ammonia into the flask containing the paper and $Cu(OH)_2$. Add to this mixture 0.5 g of additional paper (or 0.5 g of "untreated" cotton) that has been torn into tiny pieces. Stopper the flask securely and swirl the mixture *gently* for 5 min. Set the flask aside, returning to swirl it periodically while you work on Procedures 10-12.

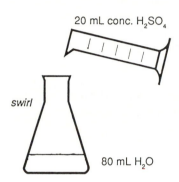

20 mL conc. H₂SO₄

swirl

80 mL H₂O

**10**. Carefully pour 20 mL of conc. $H_2SO_4$ into a 50-mL graduated cylinder. Prepare a dilute $H_2SO_4$ solution by *very carefully and slowly* pouring the 20 mL of concentrated (18 M) $H_2SO_4$ into 80 mL of distilled water contained in a 250-mL Erlenmeyer flask while swirling the flask gently to mix the liquids. Cool the final solution to room temperature by running cold tap water over the outside of the flask while swirling the flask gently. (Be careful when you clean the graduated cylinder you used for measuring the concentrated sulfuric acid.)

tip opening
~ 0.6 mm

**11**. Using *safe* and correct techniques, cut, constrict, bend and fire polish the glass tubing as needed for the apparatus shown. *If you have any questions about working with glass tubing, consult your instructor* before *starting work. Be careful to avoid burning yourself* or *the bench top with hot glass.* After the prepared tubing has cooled, lubricate the holes in the stopper and the part of the tubing to be inserted with glycerol. Protect your hands with a towel and insert the tubing through the stopper with a gentle twisting motion. *Do not exert undue force.* If you encounter problems, consult your instructor. Complete the assembly of the apparatus and have the setup checked by your instructor.

**12**. Check the mixture in your reaction flask. If it has not been recently disturbed, you should observe an intensely colored excess of tetraamminecopper(II) hydroxide above the more viscous mixture containing the complexed cellulose. The mixture should have "aged" for about 30 min and all paper (and cotton, if used) should have dissolved before the mixture is ready for spinning the Rayon fiber. Swirl the mixture gently. It should be fairly viscous ("syrupy") but not so thick as to pose any problems by plugging the constricted glass tubing. If the mixture appears to be too "gummy," add a few more milliliters of concentrated ammonia. Swirl the mixture carefully during addition of the ammonia.

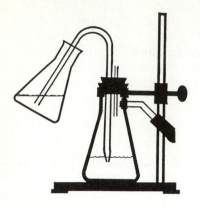

**13**.  You should now be ready to spin your Rayon fiber.  Pour the prepared dilute $H_2SO_4$ solution into the suction flask to a depth such that the constricted glass tubing will extend below the surface of the acid.

*Caution:  Be certain you are using* diluted $H_2SO_4$. *Concentrated $H_2SO_4$ may react explosively with the Rayon mixture.*

Remove the stopper from the flask containing the Rayon mixture and, holding the flask as illustrated, turn on the water aspirator.  Place a fingertip over the short piece of glass tubing and adjust finger pressure so that the Rayon mixture is drawn into the dilute $H_2SO_4$ at a rapid, even rate.  Disconnect the flask from the aspirator tubing immediately on finishing the spinning to avoid pulling too much air through the mixture or having water "back up" into the flask.

sink

**14**.  Allow the Rayon "threads" to remain in contact with the dilute $H_2SO_4$ until the blue color has leached from the fibers.  Then hold your Buchner funnel over a sink and pour the contents of the suction flask through the funnel.  Rinse the collected fibers on the funnel with tap water.

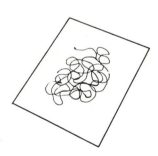

**15**.  After the wash water has drained from the fibers, transfer the moist fibers onto a preweighed paper and weigh the paper and fibers on a triple-beam balance.  Record the mass of Rayon obtained and have your instructor inspect your product.

# PRELAB EXERCISES

## INVESTIGATION 19:  RAYON:  A VALUABLE POLYMER

Name:_____

Instructor:_____   ID No.:_____

Course/Section:_____   Date:_____

---

1. Give an example of a small inorganic and organic molecule, a natural inorganic and organic large-size molecule, and a synthetic organic polymer.  For each example give the name of the compound, the formula, and the molecular mass.

2. Write the formula for cellulose, for ammonia, for tetraamminecopper(II) hydroxide, and for sulfuric acid.

(over)

3. Write a balanced equation for the reaction between tetraamminecopper(II) hydroxide and an aqueous solution of sulfuric acid.

4. Procedures 1 and 3 require that 7.5 g of $CuSO_4 \cdot 5H_2O$ be reacted with 4.0 mL of 15 M ammonia. Calculate the moles of $CuSO_4$ and $NH_3$ used in this reaction. Calculate the molar ratio $CuSO_4/NH_3$ for this reaction.

5. Write a reaction for the detection of sulfate ion with a solution of barium chloride.

# REPORT FORM

## INVESTIGATION 19: RAYON: A VALUABLE POLYMER

Name: _____     ID No.: _____

Instructor: _____     Course/Section:_____

Partner's Name (if applicable): _____     Date: _____

---

Show all calculations on the back of this form.

## Experimental Data

Mass of $CuSO_4 \cdot 5H_2O$ used: _____ g     Rayon + towel: _____ g

    Mass of single filter paper: _____ g     <u>       Towel: _____</u> g

    Mass of additional paper: _____ g     Rayon (moist): _____ g

Calculate the yield of Rayon under the assumption that your fibers are dry.

_____ % yield

If your product contained 65% of the cellulose you started with, which percentage of your weighed product would be water, assuming that all other reactants were rinsed away? (Show your calculations.) _____% water.

## Observations

Instructor's evaluation of Rayon:     ☐ good     ☐ average     ☐ poor _____ initials

Description of Rayon obtained:

Assume that the Rayon you have produced has a molecular mass of 60,000. Calculate the number of monomer units (the degree of polymerization) for this molecule.

Molecular mass: _____     Mass of Monomer: _____

Degree of Polymerization: _____

Date _____ Signature _____

# Investigation 20

## ANALYSIS OF AN UNKNOWN OXALATE SAMPLE

## INTRODUCTION

The analytical chemist is frequently called upon to determine the concentration of a solute in solution or to find the percentage composition of a solid. In chemical laboratories solutions of acids, bases, oxidizing agents, or reducing agents for which the concentration of the solute must be known precisely and accurately are used every day. For instance, laboratories associated with water treatment plants determine the hardness of water (the concentration of calcium and magnesium ions) by reacting a water sample with a solution of disodium ethylenediamine tetraacetate, the chemical oxygen demand by titration with a potassium dichromate solution, the alkalinity with a solution of an acid, and the acidity with a solution of a base. Although many analytical procedures are now automated and carried out with sophisticated instruments, the time-honored acid–base and redox titrations are still useful.

In titration procedures a solution of a reagent (the titrant) of known concentration is added to a sample until the endpoint is reached. At the endpoint, the titrant has reacted with all of the substance (the analyte) to be determined in the sample. The amount of analyte in the sample can then be calculated from the volume of the titrant consumed, the concentration of the titrant solution, and the stoichiometric relation between titrant and analyte. To obtain reliable results, the concentration of the reagent solution must be known. Such solutions with well-defined concentrations generally cannot be made by weighing the required amount of the reagent and dissolving it in water to the desired volume. Most of the substances cannot be obtained in sufficiently high purity. For instance, sodium hydroxide pellets are hygroscopic and react with carbon dioxide from the air making sodium hydroxide samples impure. Solutions of other reagents, such as potassium permanganate, do not hold the concentration because reactions with oxidizable substances decrease the concentration with time. Fortunately, a few substances can be prepared in high purity, are stable in the solid state and in solution, and do not react with atmospheric agents (moisture, carbon dioxide, oxygen, organic dust). These so-called primary standards are used to precisely and accurately determine the concentrations of the reagent solutions. The standardized solutions are then employed to analyze a variety of samples. Potassium hydrogen phthalate is an example of a primary standard for base solutions, and sodium oxalate is a primary standard for potassium permanganate solutions.

This experiment familiarizes you with the standardization of a potassium permanganate solution and the use of the standardized solution for the determination of the percentage of oxalate in an unknown.

## CONCEPTS OF THE EXPERIMENT

An aqueous solution of potassium permanganate reacts with an acidified aqueous solution of sodium oxalate according to equation 1. This redox reaction is catalyzed by $Mn^{2+}$ ion, a product

$$2KMnO_4 + 5Na_2C_2O_4 + 8H_2SO_4 \xrightarrow{Mn^{2+}} 2MnSO_4 + 10CO_2 + K_2SO_4 + 5Na_2SO_4 + 8H_2O \quad (1)$$

of the reaction. Because $Mn^{2+}$ ion is not present in the solution at the beginning of the reaction, $MnSO_4$ is added. The endpoint of this reaction is reached when one drop of $KMnO_4$ solution is not decolorized any more and imparts a light-pink color to the titrated solution. When a $KMnO_4$ solution is to be standardized, the required amount of pure sodium oxalate is weighed to $\pm 0.1$ mg, dissolved in water, acidified with sulfuric acid, and titrated with the $KMnO_4$ solution to the endpoint (V mL $KMnO_4$ solution used). The number of moles of oxalate are calculated from the mass of oxalate used and its formula mass (equation 2).

$$\text{moles } (Na_2C_2O_4) = \frac{\text{mass}}{\text{formula mass}} = a \tag{2}$$

The balanced reaction (equation 1) indicates that five moles of oxalate require 2 moles of $KMnO_4$. Therefore, the "a" moles of sodium oxalate (equation 2) will have reacted with 2a/5 moles of $KMnO_4$. The 2a/5 moles of $KMnO_4$ were present in V mL $KMnO_4$ solution. The molarity of a solution is defined as the number of moles of solute in one liter of solution. The molarity of the $KMnO_4$ solution is thus calculated according to equation 3.

$$M_{KMnO_4} = \frac{2a}{5} \times \frac{1}{V(mL)} \times 1000 = \frac{2 \text{ mass}_{Na_2C_2O_4}(g) \times 1000(mL/L)}{5 \text{ formula mass}_{Na_2C_2O_4}(g \text{ mol}^{-1}) \times V_{KMnO_4}(mL)} \tag{3}$$

Once the $KMnO_4$ solution is standardized and, thus, its concentration precisely and accurately known, the percentage of sodium oxalate and of the oxalate anion can be determined by dissolving an appropriate amount of an unknown weighed to $\pm 0.1$ mg in water, acidifying the solution, and titrating this solution with the $KMnO_4$ solution to the endpoint. Equation 3 can be solved for the mass of sodium oxalate. This mass can then be calculated from the experimental quantities.

To gain confidence in the results, the standardization and the titration of the unknown are carried out several times. The values are averaged and the standard deviation and the relative standard deviation calculated. Relative standard deviations of one or less than one percent are characteristic of precise results under the circumstances the experiment has been performed. Precision is not necessarily associated with accuracy.

## ACTIVITIES

- **Complete the PreLab Exercises before coming to the laboratory.**

- **Prepare a $KMnO_4$ solution with approximately 0.020 moles $KMnO_4$ per liter of solution.**

- **Titrate a weighed sample of pure sodium oxalate to determine the concentration of the prepared $KMnO_4$ solution.**

- **Use the $KMnO_4$ solution to determine the composition of an unknown oxalate sample.**

## *SAFETY*

*Wear approved eye protection at all times. Avoid contact of the $KMnO_4$ with organic material or concentrated $H_2SO_4$. $KMnO_4$ may produce an explosive mixture with organic material or concentrated $H_2SO_4$. If any of the reagents used in this Investigation come in contact with your skin, wash the affected areas*

**immediately** *and thoroughly with water.* **Notify your instructor.** *The reagents do not present a hazard unless they are misused.*

## PROCEDURES

### A. PREPARATION OF THE KMnO₄ SOLUTION

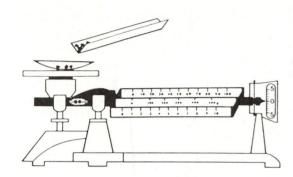

**A-1.** Weigh a clean, dry watchglass on the triple-beam balance to the nearest 0.01 g. Record this mass in your notebook. Advance the balance riders by the amount of $KMnO_4$ calculated for the preparation of 300 mL of 0.020 M $KMnO_4$ solution. Carefully add the crystals of $KMnO_4$ to the watchglass until the beam rises. Determine and record the mass of the watchglass and $KMnO_4$.

**A-2.** Fill a graduated cylinder with 300 mL distilled water. Quantitatively transfer the $KMnO_4$ crystals into a 500-mL Erlenmeyer flask. Use some of the water to rinse the last crystals off the watchglass and into the flask. Pour the rest of the water directly into the Erlenmeyer flask. Swirl the deep-purple solution for five minutes. Let the solution stand for a few minutes. Then decant the solution slowly into another large flask or beaker. Do not agitate the solution during this process. When the last milliliters of the solution are transferred, check for undissolved $KMnO_4$ crystals in the flask from which you are pouring. If undissolved crystals are seen, return about 200 mL of the solution into the original flask and swirl until dissolution is complete. Combine the solutions. Swirl and stir.

*The homogeneity of the KMnO₄ solution, i.e., the same concentration throughout the solution, is very important for the success of this experiment.*

### B. STANDARDIZATION OF THE KMnO₄ SOLUTION

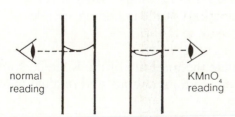

normal reading       KMnO₄ reading

**B-1.** Obtain a buret. Check that it works properly and is clean (see Appendix D). Rinse the buret twice with 5-mL portions of the $KMnO_4$ solution. Expel air from the buret tip and drain the buret until the solution level is *below* the zero mark. Because you cannot see the bottom of the meniscus, you will need to make all readings at the top of the solution level.

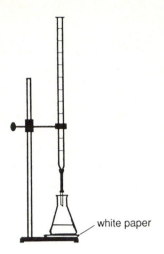

white paper

**B-2.** Using the analytical balance, weigh an amount of sodium oxalate within ±15 mg of the quantity calculated to react with 30 mL of a 0.02 M $KMnO_4$ solution. CAUTION: *Sodium oxalate is toxic if ingested!* Record the mass of sodium oxalate to the nearest 0.1 mg. Transfer the weighed sodium oxalate to a clean 500-mL Erlenmeyer flask. Use a total of 200 mL of distilled water to transfer and dissolve the sodium oxalate. It may not all dissolve until it is heated in the next step.

Add 10 mL of 9 M $H_2SO_4$ to the sodium oxalate water mixture. Protect your thumb and the fingers with which you will hold the hot flask with small pieces of rubber tubing cut open lengthwise. Heat the flask while swirling it until its temperature is approximately 70°C. When all the sodium oxalate has dissolved, set the hot flask on a towel near the $KMnO_4$ buret.

**B-3.** To your sodium oxalate solution add 10 drops of the $MnSO_4$ solution provided by your instructor. Record the level of the $KMnO_4$ solution in the buret. Begin to titrate by adding a drop of $KMnO_4$ solution to the hot oxalate solution. Swirl until the solution has become colorless. Then add the next drop. As the $Mn^{2+}$ concentration builds up, the reaction speeds up and you can add the titrant faster. If the solution becomes brownish, you added $KMnO_4$ too quickly or the solution cooled too much (or you didn't add the sulfuric acid). If necessary, reheat the solution. Titrate the sodium oxalate solution to the pale-pink endpoint.

**B-4.** Repeat the standardization twice more either at this time or after the completion of your work with the unknown.

## C. TITRATION OF AN UNKNOWN OXALATE SAMPLE

**C-1.** Repeat Procedures B-1 through B-4 with 0.6 g to 0.8 g of an unknown containing sodium oxalate (weighed to the nearest 0.1 mg). Record all weighings and buret readings. Repeat for a total of three determinations of the unknown sample.

**C-2.** Calculate the molarity of the $KMnO_4$ solution and the percent sodium oxalate and oxalate anion in the unknown.

# PRELAB EXERCISES

## INVESTIGATION 20:   ANALYSIS OF AN UNKNOWN OXALATE SAMPLE

Instructor:_____

Course/Section:_____

Name:_____

ID No.:_____

Date:_____

1. Define:

    Meniscus

    Homogeneous solution

    Standardization

2. Balance the following half-reaction:

    $MnO_4^- + H^+ \rightarrow Mn^{2+} + H_2O$

3. Calculate the mass of $KMnO_4$ required for the preparation of 300 mL of 0.020 M $KMnO_4$ solution.

(over)

4.  Balance the following equations:

   a. $MnO_4^- + H_2C_2O_4 + H^+ \rightarrow Mn^{2+} + CO_2 + H_2O$

   b. $MnO_4^- + C_2O_4^{2-} + H^+ \rightarrow Mn^{2+} + CO_2 + H_2O$

   c. $MnO_4^- + CaC_2O_4 + H^+ \rightarrow Mn^{2+} + Ca^{2+} + CO_2 + H_2O$

   d. $MnO_4^- + K_2C_2O_4 + H^+ \rightarrow Mn^{2+} + K^+ + CO_2 + H_2O$

5.  Calculate the amount of sodium oxalate required to consume 30 mL of 0.020 M $KMnO_4$.

# REPORT FORM

## INVESTIGATION 20: ANALYSIS OF AN UNKNOWN OXALATE SAMPLE

Instructor:_____

Course/Section:_____

Name: _____

ID No.: _____

Date:_____

## Preparation of KMnO₄ Solution

Mass of $KMnO_4$ + watchglass: _____ g

Mass of watchglass: _____ g

Mass of $KMnO_4$ used: _____ g

Volume of water used: _____ mL

## Standardization

|  | Determination | | |
|---|---|---|---|
|  | 1 | 2 | 3 |
| Mass of watchglass + $Na_2C_2O_4$: | _____ g | _____ g | _____ g |
| Mass of watchglass: | _____ g | _____ g | _____ g |
| Mass of $Na_2C_2O_4$ used: | _____ g | _____ g | _____ g |
| Final buret reading: | _____ mL | _____ mL | _____ mL |
| Initial buret reading: | _____ mL | _____ mL | _____ mL |
| Volume of $KMnO_4$ solution used: | _____ mL | _____ mL | _____ mL |
| Molarity of the $KMnO_4$ solution: | _____ M | _____ M | _____ M |

Average: _____ M   Std. deviation:_____ Relative std. deviation:_____ %

Attach a copy of all calculations including balanced equations.

## Unknown Determinations

Unknown # _____

|  | Determination | | |
|---|---|---|---|
|  | 1 | 2 | 3 |
| Mass of watchglass + unknown: | _____ g | _____ g | _____ g |
| Mass of watchglass: | _____ g | _____ g | _____ g |
| Mass of unknown used: | _____ g | _____ g | _____ g |
| Final buret reading: | _____ mL | _____ mL | _____ mL |
| Initial buret reading: | _____ mL | _____ mL | _____ mL |
| Volume of $KMnO_4$ solution used: | _____ mL | _____ mL | _____ mL |

Grams of $Na_2C_2O_4$ in the sample of
unknown:_____ g  _____ g  _____ g
Grams of $C_2O_4^{2-}$ in the sample of
unknown:_____ g  _____ g  _____ g

Attach all calculations, including balanced equations.

Percent $Na_2C_2O_4$ ($^w/_w$) in unknown
sample:_____ %  _____ %  _____ %
Percent $C_2O_4^{2-}$ ($^w/_w$) in unknown
sample:_____ %  _____ %  _____ %

For $C_2O_4^{2-}$ results:

Average: _____ %   Std. deviation:_____   Relative std. deviation:_____ %

Attach all calculations.

My results for percent $C_2O_4^{2-}$ in the unknown is limited to _____ significant figures because in the determination of _____ only _____ significant figures were available (or justifiable). The reported data and values represent work performed by me on _____.
                                                                                                              date

Questions:
1. Pure sodium oxalate (0.4554 g) consumed 26.73 mL of a $KMnO_4$ solution. Calculate the molarity of the $KMnO_4$ solution.

2. Using your textbook, find the definition of normality of an oxidizing or reducing agent. Write an expression that relates the normality and the molarity of the $KMnO_4$ used in a reaction in which the $MnO_4^-$ is reduced to $Mn^{2+}$.

Date _____    Signature _____

# Investigation 21

## DETERMINATION OF IRON IN PYRITES

### INTRODUCTION

Modern societies cannot survive without metals. One of the most important metals is iron from which a large variety of alloys is made. Iron and other metals are produced from ores. Ores are assemblages of minerals containing one or several metals in deposits of sufficient size to justify a mining operation. These deposits must be located and the metal concentrations in them determined. During the expansion of the United States toward the west, prospectors roamed the hills and the mountains for signs of ore deposits on the surface. Geologists and sophisticated geochemical prospecting methods have replaced the prospector. The concentrations of the metals in the ore deposits were determined by assayers often using gravimetric techniques on the samples brought by the prospectors from the field. Now instrumental methods have replaced gravimetric and titrimetric techniques. However, titrimetric methods are much less expensive than instrumental methods with regard to the required equipment. For this reason, the titrimetric determination of metals in ore samples is still used.

The gravimetric determination of iron in an iron ore will proceed in the following manner. The weighed amount of an ore sample is dissolved by heating it with concentrated nitric acid. When all the sample has dissolved, the solution is transferred to a volumetric flask and diluted with distilled water. An aliquot (e.g., 50.0 mL) of this solution is pipetted into a beaker. A solution of sodium hydroxide is slowly added until the mixture in the beaker becomes basic and brown iron trihydroxide, $Fe(OH)_3$, precipitates. When all the iron has precipitated, the mixture is filtered through an ash-free filter paper. The brown hydroxide on the filter is washed free of soluble salts with distilled water. The filter paper with the iron (III) hydroxide is placed into a porcelain crucible that has been heated until it attained constant mass. The crucible is then heated with a Bunsen burner until the filter paper is completely burned and the water is driven from the hydroxide converting it to Fe(III) oxide, $Fe_2O_3$. The crucible is placed into a furnace maintained at 300°C for two hours. After the crucible has cooled to room temperature, its mass is determined on an analytical balance. The heating and weighing procedure is repeated until the mass of the crucible does not change any more. From the mass of the $Fe_2O_3$ and the mass of the ore sample, the percent iron in the sample is calculated. This gravimetric procedure based on mass determinations is very precise and accurate but also very time-consuming.

Titrimetric procedures, although ultimately also based on mass determinations, require much less time. The ore sample must be dissolved as in the gravimetric method. However, the content of iron in an aliquot of the solution is determined by addition of a solution of potassium permanganate that oxidizes Fe(II) to Fe(III). For a titrimetric method to work properly, the reaction between the analyte [Fe(II)] and the titrimetric reagent ($KMnO_4$) must proceed stoichiometrically, that is in a fixed ratio by mass as expressed by a balanced equation. In addition, the endpoint, at which all the analyte has reacted, must be clearly noticeable. From the volume of the titrimetric reagent consumed and the concentration of the reagent, the amount of the analyte can be calculated and the percent of iron in the ore sample obtained. Although the solution of the titrimetric reagent is prepared by weighing the required amount and dissolving it to an appropriate volume, only one weighing is necessary. Many titrations can be carried out once the solution is prepared. Titrimetric determinations are less time-consuming than gravimetric determinations and can be automated much easier.

In this experiment the percent iron will be determined with a titrimetric procedure based on the oxidation of Fe(II) by $KMnO_4$ to Fe(III).

## CONCEPTS OF THE EXPERIMENT

Divalent iron present, for instance, in the green salt $FeSO_4 \cdot 7H_2O$ [iron(II) sulfate heptahydrate] or $(NH_4)_2Fe(SO_4)_2 \cdot 6H_2O$ [ammonium iron(II) sulfate hexahydrate] is easily oxidized in aqueous solution acidified with sulfuric acid by solutions of potassium permanganate to iron(III). Potassium permanganate is reduced to Mn(II) in this reaction. The stoichiometry of an example reaction is given by equation 1 which is written in molecular form. The equation in the

$$10FeSO_4 + 2KMnO_4 + 8H_2SO_4 \longrightarrow 5Fe_2(SO_4)_3 + 2MnSO_4 + 8H_2O + K_2SO_4 \quad (1)$$

net-ionic form containing only the species that take part in the reaction is more useful for calculating the amount of iron in the ore sample (equation 2). The endpoint in this reaction is

$$5Fe^{2+} + MnO_4^- + 8H^+ \longrightarrow 5Fe^{3+} + Mn^{2+} + 4H_2O \quad (2)$$

conveniently indicated by a faint pink color of the solution caused by the first drop of $KMnO_4$ solution added in excess of the required volume. Because $KMnO_4$ solutions have a deep purple color, one drop that is not reduced to the almost colorless $Mn^{2+}$ is sufficient to signal that the endpoint has been reached.

A solution of Fe(III) formed during the titration is yellow. At the endpoint the yellow color from Fe(III) would make it difficult to clearly observe the faint pink color. Fortunately, Fe(III) reacts with phosphoric acid to form colorless Fe(III)-phosphate complexes. The addition of the Reinhardt-Zimmermann reagent, which contains phosphoric acid, has the purpose of making the endpoint easy to recognize. This reagent also contains manganese(II) sulfate. The oxidation of Fe(II) to Fe(III) by $KMnO_4$ is catalyzed by $Mn^{2+}$. To have the reaction proceed quickly even at the beginning of the titration, $Mn^{2+}$ is added in the form of the Reinhardt-Zimmermann reagent. This catalyst, the substance that speeds up the reaction, is formed during the titration by reduction of $MnO_4^-$ to $Mn^{2+}$. The $MnSO_4$ provides the needed catalyst at the beginning of the titration.

The iron(II) solution that is titrated with $KMnO_4$ must be prepared by dissolving a weighed sample of pyrite. Pyrite is $FeS_2$, although quartz and silicates may also be present in the ore samples. When the pyrite is treated with a mixture prepared from concentrated nitric and concentrated hydrochloric acid in a ratio of 1 volume $HNO_3$ to 2 volumes HCl, the pyrite dissolves forming Fe(III) salts, sulfate, and some elemental sulfur. The acid mixture used is called "aqua regia" (royal water) because it is the only acid that will dissolve elemental gold. After the dissolution of the $FeS_2$ is complete, all the excess acid must be removed by careful evaporation of the mixture to dryness. Any acid remaining in the sample will interfere with subsequent operations. The dry residue will consist of Fe(III) salts, mostly $FeCl_3$, and any inert, undissolved quartz, silicates and sulfur. The Fe(III) salts are dissolved in 6 M hydrochloric acid.

Because the determination of iron by titration with $KMnO_4$ requires a solution of Fe(II), the Fe(III) solution must be reduced to Fe(II). This reduction is accomplished by the dropwise addition of a solution of $SnCl_2$ in hydrochloric acid to the Fe(III) solution. As the Fe(III) is reduced (equation 3), the yellow color of the solution characteristic of Fe(III) fades. Solutions of

$$2Fe^{3+} + SnCl_2 + 4Cl^- \longrightarrow 2Fe^{2+} + SnCl_6^{2-} \quad (3)$$

Fe(II) are only faintly green and dilute solutions are almost colorless. When the solution has become colorless (check against a white background), all the Fe(III) has been reduced. Although a few drops of $SnCl_2$ solution added in excess do no harm, a larger excess must be avoided. An excess of $SnCl_2$ in the Fe(II) solution would react with $KMnO_4$ and produce incorrect high results for iron upon titration. Therefore, the excess $SnCl_2$ must be converted to $SnCl_4$, a tetravalent tin compound that is unreactive toward $KMnO_4$. This conversion is accomplished by addition of a solution of $HgCl_2$ (equation 4). The mercurous chloride, $Hg_2Cl_2$, is rather insoluble and

$$SnCl_2 + 2HgCl_2 \longrightarrow SnCl_4 + Hg_2Cl_2\downarrow \tag{4}$$

precipitates in the form of very fine, white and shiny crystals. The solution is now ready to be titrated. When too much $SnCl_2$ is added, some of the $HgCl_2$ will be reduced to elemental mercury producing a grayish precipitate. The potassium permanganate will then react not only with Fe(II) but also with elemental mercury and give erroneous results.

A potassium permanganate solution is strongly oxidizing and not very stable because oxidizable dust, always present in the laboratory atmosphere, falls in the solution, reacts with $KMnO_4$, and reduces the $KMnO_4$ concentration in the solution. For this reason, $KMnO_4$ solutions are standardized just before or after use. Standardization is the process of determining the exact concentration of a solution. Sodium oxalate, $Na_2C_2O_4$, a substance that can be obtained in high purity, is used for the standardization of the $KMnO_4$ solution. An appropriate amount of sodium oxalate ($m_{so}$ grams) is weighed out and dissolved in warm water. Sulfuric acid and Reinhardt-Zimmermann reagent are added because the oxidation of oxalate requires an acidic medium and is catalyzed by $Mn^{2+}$. The warm solution is then titrated with $KMnO_4$ solution to the faint pink endpoint ($V_{P-so}$ mL consumed). Oxalate is oxidized to $CO_2$ and $MnO_4^-$ reduced to $Mn^{2+}$ (equation 5). When the solution turns brown because of precipitation of dark brown manganese dioxide,

$$5Na_2C_2O_4 + 2KMnO_4 + 8H_2SO_4 \longrightarrow 10CO_2 + 2MnSO_4 + K_2SO_4 + 5Na_2SO_4 + 8H_2O \tag{5}$$

$MnO_2$, the solution was not warm enough, the titration was performed too fast, or the acid had not been added. The experiment can be saved by heating the solution until the brown color has disappeared indicating that all $MnO_2$ has been converted to $Mn^{2+}$. The titration can then be continued.

The molarity of the $KMnO_4$ solution can now be calculated from the mass of sodium oxalate used, the molecular mass of sodium oxalate, and the volume of $KMnO_4$ solution consumed. The moles of sodium oxalate titrated are obtained by dividing the mass of sodium oxalate by its molecular mass (equation 6). The balanced equation (5) demands that each mole of sodium

$$\text{\# of moles of } Na_2C_2O_4 = \frac{m_{so}}{MM_{so}} \tag{6}$$

oxalate consumes 0.4 moles of $KMnO_4$. The moles of $KMnO_4$ used in the titration are given by equation 7. The molarity of a solution is defined as the number of moles of solute per one liter of

$$\text{\# of moles of } KMnO_4 \text{ consumed} = 0.4 \times (\text{\# of moles of } Na_2C_2O_4) = \frac{2}{5}\frac{m_{so}}{MM_{so}} \tag{7}$$

solution. Because $V_{P-so}$ mL of $KMnO_4$ solution were needed to reach the endpoint, this volume must contain the number of moles of $KMnO_4$ given by equation 7. To find the molarity, the number of moles of $KMnO_4$ in 1000 mL (1.0 liter) of solution must be calculated (equation 8).

$$M_{KMnO_4} = \text{\# of moles } KMnO_4 \text{ in 1000 mL} = \frac{2\,m_{so}\,1000}{5\,MM_{so}V_{P-so}} \tag{8}$$

This $KMnO_4$ solution is used to titrate the Fe(II) solution obtained from the ore. The following data are available.

$m_{Py}$: mass of pyrite dissolved (in grams)
$V$: volume of volumetric flask = total volume of Fe(II) solution (in mL)
$V_{Fe}$: volume of Fe(II) solution titrated (in mL)
$V_{P-Fe}$: volume of $KMnO_4$ solution consumed to reach the endpoint in the titration of $V_{Fe}$ mL of Fe(II) solution (in mL)
$M_{KMnO_4}$: molarity of $KMnO_4$ solution.

The number of moles of $KMnO_4$ consumed in the titration of $V_{Fe}$ mL of Fe(II) solution is given by the expression in equation 9. According to the balanced equation 2, each mole of $KMnO_4$ oxidizes

$$\text{\# of moles of } KMnO_4 = M_{KMnO_4} \times \frac{V_{P-Fe}}{1000} \tag{9}$$

5 moles of Fe(II). The moles of Fe in $V_{Fe}$ mL of Fe(II) solution can be calculated according to equation 10. The total number of moles of Fe in the solution of $V$ mL (the volume of the

$$\text{moles Fe} = 5 \times M_{KMnO_4} \times \frac{V_{P-Fe}}{1000} \tag{10}$$

volumetric flask) will be obtained by using equation 11. The mass of iron in the volumetric flask

$$\text{total moles of Fe} = 5 \times M_{KMnO_4} \times \frac{V_{P-Fe}}{1000} \times \frac{V}{V_{Fe}} \tag{11}$$

is the product of the atomic mass of iron and the moles of iron. The percent of iron in the pyrite sample can then be calculated according to equation 12.

$$\% \text{ Fe} = 5 \times M_{KMnO_4} \times \frac{V_{P-Fe}}{1000} \times \frac{V}{V_{Fe}} \times 55.84 \times \frac{100}{m_{Py}} \tag{12}$$

## ACTIVITIES

- Complete the PreLab Exercises before coming to the laboratory.

- Prepare a 0.02 M $KMnO_4$ solution.

- Dissolve a weighed amount of a pyrite sample.

- Prepare the reagents needed for the titration of Fe(II) by $KMnO_4$.

- Reduce the Fe(III) in the solution prepared from the ore to Fe(II).

- Titrate the Fe(II) solution with the $KMnO_4$ solution.

- Standardize the $KMnO_4$ solution with sodium oxalate.

- Calculate the percent iron in your ore sample.

- Complete the Report Form.

## SAFETY

*During this experiment you will work with very corrosive acids and toxic substances. Wear approved eye protection at all times. Perform all operations generating corrosive fumes in a well ventilated hood with the door to the hood closed as much as possible. Avoid contact of the permanganate solution with organic material. Never mix $KMnO_4$ with concentrated sulfuric acid. This mixture can explode violently. Do not spill any solutions or solids. In case of contact with any of the reagents, wash the affected areas immediately and thoroughly with water.*

## PROCEDURES

Part of this experiment (Procedures B-5 and C-1 through C-4) will be carried out by a team to conserve time and reagents. Your instructor will form 6-member teams from students occupying neighboring work areas. When your team has been established, meet and assign the tasks outlined in Procedures B-5, C-1, C-2, C-3, and C-4 to members of your team.

## A. PREPARATION OF THE KMnO₄ SOLUTION

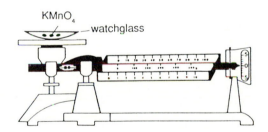

KMnO₄ — watchglass

**A-1.** If the $KMnO_4$ solution to be used in this experiment is supplied, take a clean and dry 250-mL Erlenmeyer flask to the storage bottle. Obtain approximately 150 mL of the $KMnO_4$ solution. Stopper the flask. Record the approximate molarity of this solution.

**A-2.** If you must prepare the $KMnO_4$ solution, take a clean, dry watchglass to the triple-beam balance and weigh the watchglass to the nearest 0.1 g.

**A-3.** Advance the balance masses by the amount calculated for the preparation of 150 mL of a 0.02 M $KMnO_4$ solution. Carefully transfer the crystals of $KMnO_4$ onto the watchglass until the balance returns to equilibrium.

**A-4.** Carefully transfer all the crystals on the watchglass, without spilling any, into a clean 250-mL Erlenmeyer flask. Add 150 mL distilled water to the crystals. Rinse any crystals clinging to the watchglass into the flask with the help of your squeeze bottle. Swirl the flask until all the crystals have dissolved. To determine that there are no undissolved crystals, check the bottom of the flask for any remaining solids. Stopper the flask.

# B. DISSOLUTION OF THE PYRITE SAMPLE

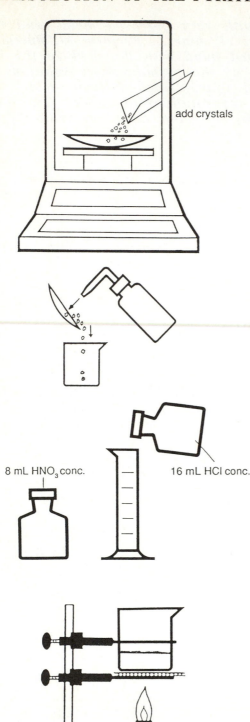

add crystals

8 mL HNO$_3$ conc.

16 mL HCl conc.

**B-1.** Weigh a clean, dry watchglass on an analytical balance to the nearest 0.1 mg. Take a 3x5 cm piece of smooth paper to your instructor. Obtain a pyrite sample. You will need approximately 1 gram. Transfer a sufficient amount of this sample onto the weighed watchglass to have the mass of pyrite between 0.9 and 1.0 g. Weigh the watchglass with the pyrite on an analytical balance to the nearest 0.1 mg.

**B-2.** Transfer the pyrite quantitatively into a clean 100-mL beaker. Rinse the last crystals of pyrite from the watchglass into the beaker with not more than 5 mL of water.

**B-3.** In the hood assigned to your team, set up a ringstand with two ring clamps. Place a wire gauze on the bottom ring. Connect a Bunsen burner to a gas outlet. Take the beaker with your pyrite sample, a 25-mL graduated cylinder, a bottle with concentrated hydrochloric acid, and a bottle of concentrated nitric acid into the hood. *The preparation of "aqua regia" and the dissolution of your pyrite sample must be carried out in a well ventilated hood.*

**B-4.** To prepare "aqua regia", carefully pour 8 mL of concentrated nitric acid into a clean and dry 25-mL graduated cylinder. Then carefully add concentrated hydrochloric acid until the total volume of the mixture is 24 mL. Slowly pour this mixture (aqua regia) into the beaker containing the pyrite. Take some tap water in an Erlenmeyer flask to the hood. Fill the graduated cylinder that held the aqua regia with tap water to dilute the small amount of strong acid still in the cylinder. You may pour this solution into the sink while the tap water is running.

**B-5.** Set the beaker with the pyrite/aqua regia mixture on the wire gauze supported by the ring clamp. Adjust the burner to give a low flame. Heat the mixture to keep it boiling gently. Adjust the flame if the mixture boils too vigorously. When all your team members have their mixtures boiling gently, the cooperative part of this experiment begins.

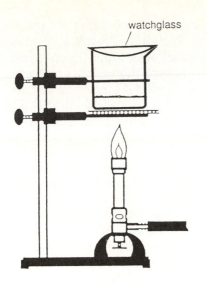

watchglass

**B-6.** Team member #1 remains to watch over the six beakers in the hood adjusting the burners - if necessary - to ensure that all the mixtures boil gently. During this process the front of the hood must be almost completely closed to have proper ventilation and keep the brown nitrogen dioxide fumes from escaping into the laboratory. When the liquid volumes in the beakers have dropped to 5-10 mL, all team members must be called back to their beakers.

**B-7.** Cover the beaker with a watchglass. Very carefully continue heating the beaker to minimize splattering until a dry, yellow residue consisting of iron(III) salts and some undissolved silicates remains. When the solid in the beaker is dry, turn off the burner and let the beaker cool to room temperature. Then take the beaker covered with the watchglass from the hood to your workplace.

## C. PREPARATION OF SPECIAL REAGENTS

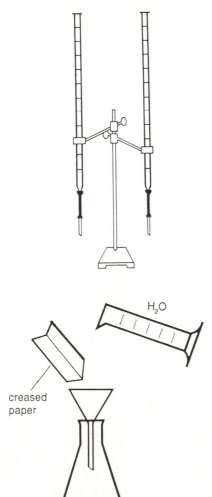

H₂O

creased paper

While the pyrite/aqua regia mixtures boil gently in the hood watched closely by team member #1, the other team members set up and, if necessary, clean two burets for each team member and prepare a 5% $HgCl_2$ solution, a 1 M $SnCl_2$ solution, and a $MnSO_4$ solution in dilute sulfuric acid (Reinhardt-Zimmermann solution).

**C-1.** Team members #2 and #3 check 2 burets per team member for proper drainage. Fill each buret with distilled water. Allow the water to flow out through the buret tip. Check the walls of the buret for water droplets. When you observe water droplets clinging to the inside wall of a buret, clean it as described in Appendix D. Mount two clean burets for each team member as shown. Other team members who have finished their tasks should help with the cleaning of the burets if necessary.

**C-2.** Team member #4 prepares 120 mL of a 5-percent (mass by volume) $HgCl_2$ solution. Weigh a creased piece of smooth paper to the nearest 0.1 g on a triple-beam balance. Advance the weights by 6 grams. Transfer $HgCl_2$ crystals onto the paper until the balance returns to equilibrium. Quantitatively transfer the 6 g of $HgCl_2$ into a clean 250-mL Erlenmeyer flask. Add 120 mL of distilled water and warm the flask with occasional swirling until all the $HgCl_2$ has dissolved. Let the solution cool to room temperature. Label the flask "5% $HgCl_2$".

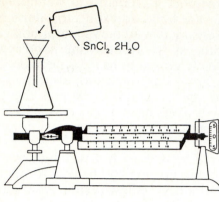

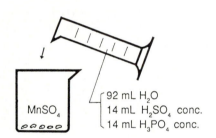

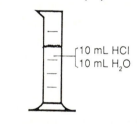

**C-3.** Team member #5 prepares 50 mL of a 1 M $SnCl_2$ solution. Weigh a 125-mL Erlenmeyer flask and a powder funnel on a triple-beam balance to the nearest 0.1 g. Advance the weights to correspond to the amount of $SnCl_2 \cdot 2H_2O$ required for 50 mL of a 1.0 M $SnCl_2$ solution. Carefully add $SnCl_2 \cdot 2H_2O$ until the balance returns to equilibrium. Remove the flask from the balance. Place the flask inside a hood. Pour 20 mL of concentrated hydrochloric acid from the storage bottle into a 25-mL graduated cylinder. Use these 20 mL of acid to rinse any $SnCl_2$ adhering to the powder funnel into the flask. Heat the mixture cautiously with swirling until all $SnCl_2$ has dissolved. Then add 30 mL of distilled water. Swirl the solution until it has become homogeneous. Take this solution back to your work area. Label the flask "1 M $SnCl_2$".

**C-4.** Team member #6 prepares 120 mL of Reinhardt-Zimmermann reagent. Weigh a 250-mL beaker on the triple-beam balance to the nearest 0.1 g. Advance the weights by 8.4 grams. Add $MnSO_4 \cdot 4H_2O$ crystals to the beaker until the balance returns to equilibrium. Remove the beaker from the balance. Pour 92 mL distilled water into the beaker. Stir until all the manganese sulfate has dissolved. Then slowly pour 14 mL of concentrated sulfuric acid into the stirred solution. Then add 14 mL of concentrated phosphoric acid. Stir the mixture until a homogeneous solution is formed. Label the beaker "R-Z Reagent".

## D.  REDUCTION OF Fe(III) OBTAINED BY DISSOLUTION OF THE PYRITE SAMPLE TO Fe(II)

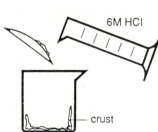

**D-1.** Prepare 20 mL of a 6 M hydrochloric acid by pouring 10 mL of distilled water and ice into a 25-mL graduated cylinder and adding concentrated (12 M) hydrochloric acid to the cylinder until the liquid level reaches the 20-mL mark.

**D-2.** Take the watchglass covering the beaker with the dry, yellow residue from the dissolution of pyrite, hold the watchglass over the beaker, and rinse the underside with small volumes of the 6 M hydrochloric acid to transfer any solids clinging to the watchglass into the beaker. Similarly, rinse any solids on the inside walls of the beaker down to the

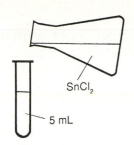

SnCl₂

5 mL

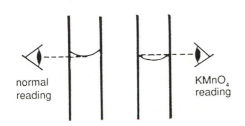

normal
reading

KMnO₄
reading

bottom. Pour any remaining acid into the beaker. Set the beaker on a wire gauze supported by a ring clamp above a burner. Warm the mixture and stir it with a short glass rod until the solid has dissolved. Some inert silicate particles may remain undissolved. Heat the yellowish brown solution until it starts to boil. Then turn off the burner. Leave the beaker on the ring clamp and the glass rod in the beaker.

**D-3.** In a clean test tube obtain 5 mL of the 1 M $SnCl_2$ solution. With a medicine dropper add this solution dropwise to the hot solution of Fe(III). Stir the solution in the beaker with the glass rod. Add the $SnCl_2$ solution dropwise until the yellow color of Fe(III) has disappeared. Avoid adding an excess of $SnCl_2$. Let the colorless solution now containing Fe(II) cool to room temperature.

**D-4.** While the solution is cooling, thoroughly stir the $KMnO_4$ solution with a glass rod and rinse one of the two burets twice with 5-mL portions of the $KMnO_4$ solution. Then fill the buret with the $KMnO_4$ solution to a level above the zero mark. Drain enough solution into a beaker to expel all air from the buret tip and drop the solution level to below the zero mark. Because the $KMnO_4$ solution has an intense color, the bottom of the meniscus is not visible; therefore, all volume readings with $KMnO_4$ solution are made at the top of the solution level.

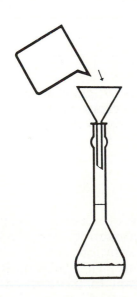

**D-5.** Place a clean funnel into the neck of a clean, but not necessarily dry, 100-mL volumetric flask. When the Fe(II) solution has cooled to room temperature, transfer the solution quantitatively into the volumetric flask. Rinse the beaker and the glass rod several times with small portions of distilled water. Pour all rinses through the funnel into the volumetric flask. Rinse the funnel with distilled water from a wash bottle, and then remove the funnel from the flask. Fill the flask with distilled water almost to the mark. Add water dropwise until the bottom of the meniscus is at the same height as the mark on the neck of the flask. Stopper the flask securely, invert it, and shake it to mix the contents thoroughly.

# E. TITRATION OF THE Fe(II) SOLUTION PREPARED FROM PYRITE WITH POTASSIUM PERMANGANATE

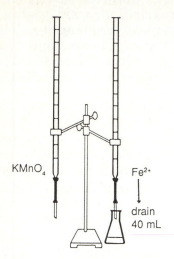

KMnO₄

Fe²⁺

↓ drain
40 mL

**E-1.** Rinse your second buret with two 5-mL portions of the Fe(II) solution. Then fill the buret with this solution. Drain enough of the solution into a beaker to expel all air from the buret tip and drop the liquid level to below the zero mark. Read and record the liquid levels in the Fe(II) buret and in the KMnO₄ buret. Then drain approximately 40 mL of the Fe(II) solution into a clean 500-mL Erlenmeyer flask. Wait thirty seconds. Then read and record the liquid level in the Fe(II) buret.

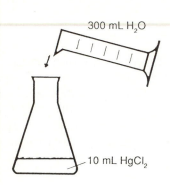

300 mL H₂O

10 mL HgCl₂

**E-2.** Obtain in a clean 25-mL graduated cylinder 10 mL of the 5% HgCl₂ solution. Pour this solution into the Fe(II) solution in the 500-mL Erlenmeyer flask. After a few seconds a fine, white precipitate of Hg₂Cl₂ should form. Let the mixture stand for two minutes.

Should the precipitate be grey or black, you added too much SnCl₂ in step D-3. Continue with the experiment. However, your result will not be very accurate. You may want to repeat the procedure with another aliquot of your Fe(II) solution.

**E-3.** Add 300 mL distilled water to the mixture. Swirl the flask to mix the contents.

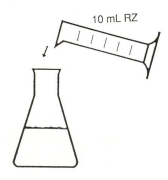

10 mL RZ

**E-4.** Obtain in a clean 25-mL graduated cylinder 10 mL of the Reinhardt-Zimmermann reagent. Add this reagent to your mixture. Swirl the flask to mix its contents.

**E-5.** Place the Erlenmeyer flask on a white piece of paper under the KMnO₄ buret. Check whether the level of the KMnO₄ solution agrees with the previously recorded value. Then add the KMnO₄ solution in a slow stream to the swirled Fe(II) solution. The additions can be fairly rapid as long as the KMnO₄ solution is quickly decolorized. When the color fades only slowly or persists, stop the addition of $KMnO_4$ and allow the $KMnO_4$ to react. When the mixture has become colorless, continue the titration dropwise. The closer your mixture is to the endpoint, the slower the dropwise addition of $KMnO_4$ must be. The endpoint is reached when one drop of the $KMnO_4$ solution gives your mixture a faint pink color that persists in the swirled mixture for 30 seconds. Record the liquid level in the $KMnO_4$ buret.

Should you have overshot the endpoint (your mixture is dark pink), you may add some Fe(II) solution from your Fe(II)-buret until your mixture has become colorless. Then titrate this colorless mixture carefully drop by drop to a faint pink endpoint. Record the liquid levels in both burets.

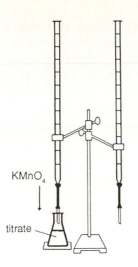

KMnO₄

titrate

**E-6.** Refill both burets. Clean your 500-mL Erlenmeyer flask. Perform another titration by repeating Procedures E-1 through E-5, if time permits.

## F. STANDARDIZATION OF THE $KMnO_4$ SOLUTION

**F-1.** Weigh the amount of sodium oxalate required for the reduction of 30 mL of a 0.02 M $KMnO_4$ solution to $Mn^{2+}$ on an analytical balance to the nearest 0.1 mg using a watchglass. To make the weighing operation easier and faster, you do not have to weigh out the calculated amount to ±0.1 mg. The amount you use for the standardization may vary ±1% from the calculated amount. However, the mass of the amount you use must be determined to ±0.1 mg. Transfer the sodium oxalate on the watchglass quantitatively into a clean 500-mL Erlenmeyer flask. Rinse any oxalate crystals remaining on the watchglass into the flask with a stream of distilled water from a squeeze bottle. Add distilled water to the flask until the volume of liquid in the flask is approximately 200 mL.

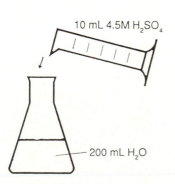

10 mL 4.5M H₂SO₄

200 mL H₂O

**F-2.** Prepare 10 mL of 4.5 M sulfuric acid by adding the required volume of distilled water and crushed ice to a 25-mL graduated cylinder. While you stir the water in the graduated cylinder with a glass rod, slowly pour the required volume of concentrated sulfuric acid into the water. Pour this dilute sulfuric acid into the Erlenmeyer flask containing the oxalate sample. Heat the flask with occasional swirling to approximately 70°C. The flask will feel almost too hot to hold. Two pieces of split-open rubber tubing stuck on your fingers will protect your fingers. Set the hot flask on a piece of white paper near the $KMnO_4$ buret.

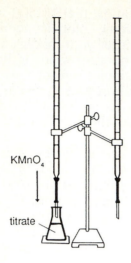

KMnO$_4$

titrate

**F-3.** Refill the KMnO$_4$ buret and drain enough solution to drop the level just below the zero mark. Record the buret reading. Add 10 drops of the Reinhardt-Zimmermann reagent to the oxalate solution. Titrate this solution slowly while it is still hot to a pale pink endpoint. Reheat the solution if it becomes too cool. Record the final buret reading.

**F-4.** Repeat the standardization of the KMnO$_4$ solution (Procedures F-1 through F-3) if time permits.

## G. CALCULATIONS

**G-1.** From the mass of sodium oxalate and the volume of KMnO$_4$ consumed, calculate the molarity of the KMnO$_4$ solution (Procedures F-1 through F-3). If you have more than one set of mass/volume data, average the calculated molarities and use the average molarity in all subsequent calculations.

**G-2.** From the volume of Fe(II) solution titrated and the volume of KMnO$_4$ solution required to oxidize Fe(II) to Fe(III), calculate the grams of Fe in the volume of Fe(II) you used (Procedures E-1 through E-5). Then calculate the grams of Fe(II) in 100 mL of the Fe(II) solution. If you have more than one set of data, calculate the average "grams of Fe in 100 mL Fe(II) solution". Use this average in all subsequent calculations.

**G-3.** From the amount of pyrite dissolved (Procedure B-1) and the "grams of iron in 100 mL Fe(II) solution" calculate the percent iron in your pyrite sample.

# PRELAB EXERCISES

## INVESTIGATION 21: DETERMINATION OF IRON IN PYRITES

Name:_____

Instructor:_____     ID No.:_____

Course/Section:_____     Date:_____

1. Balance the equation for the oxidation of $Fe^{2+}$ by $KMnO_4$ in aqueous acidic solution. *Show your work.* Also write the balanced equation in net-ionic form.

   $$KMnO_4 + FeSO_4 + H_2SO_4 \longrightarrow MnSO_4 + Fe_2(SO_4)_3 + K_2SO_4 + H_2O$$

2. Calculate the mass of $KMnO_4$ needed to prepare 150 mL of a 0.02 M $KMnO_4$ solution. Make sure that the equation in Exercise 1 is balanced correctly. $KMnO_4$ needed: _____ g.

3. Calculate the mass of $HgCl_2$ needed to prepare 120 mL of a 5% (mass/volume) solution of $HgCl_2$ in distilled water. A 5% mass-by-volume solution has 5.0 grams of a substance dissolved in 100 mL of water.

4.  Calculate the mass of $SnCl_2 \cdot 2H_2O$ needed to prepare 50 mL of a 1.0 M aqueous solution of $SnCl_2$.

5.  Balance the equation for the oxidation of sodium oxalate by potassium permanganate in an aqueous solution acidified with sulfuric acid. The products of the reaction are $Mn^{2+}$ and carbon dioxide. Write the balanced equation in molecular form and in net-ionic form.

6.  Calculate the mass of sodium oxalate, $Na_2C_2O_4$, needed to consume 30 mL of a 0.02 M $KMnO_4$ solution.

7.  Calculate the volume of concentrated (18.0 M) sulfuric acid and the volume of distilled water needed to prepare 10 mL of 4.5 M sulfuric acid under the assumption that the volume of the mixture is equal to the sum of the volumes of sulfuric acid and water that were mixed.

# REPORT FORM

## INVESTIGATION 21: DETERMINATION OF IRON IN PYRITES

Name: _____     ID No.: _____

Instructor: _____     Course/Section: _____

Partner's Name (if applicable): _____     Date: _____

---

## Analytical Data

### Pyrite

Pyrite + watchglass: _____ g

Watchglass: _____ g

Pyrite: _____ g

### Sodium oxalate

|  | 1st trial | 2nd trial |
|---|---|---|
| Oxalate + watchglass: | _____ g | _____ g |
| Watchglass: | _____ g | _____ g |
| $Na_2C_2O_4$: | _____ g | _____ g |

### Fe(II) Titration

1st trial

|  | $Fe^{2+}$ | $KMnO_4$ |
|---|---|---|
| Final volume: | _____ mL | _____ mL |
| Initial volume: | _____ mL | _____ mL |
| Volume used: | _____ mL | _____ mL |

2nd trial

|  | $Fe^{2+}$ | $KMnO_4$ |
|---|---|---|
| Final volume: | _____ mL | _____ mL |
| Initial volume: | _____ mL | _____ mL |
| Volume used: | _____ mL | _____ mL |

### Oxalate titration

|  | 1st trial | 2nd trial |
|---|---|---|
| Final $MnO_4^-$ volume: | _____ mL | _____ mL |
| Initial $MnO_4^-$ volume: | _____ mL | _____ mL |
| Volume $MnO_4^-$ consumed: | _____ mL | _____ mL |

Molarity of $KMnO_4$: _____ (1st trial) _____ (2nd trial) _____ (average)

Grams of Iron per 100 mL Fe(II) solution:

_____ (1st trial) _____ (2nd trial) _____ (average)

Percent Iron in Pyrite Sample: _____ %

Attach your calculations.

Signature _____

237

# Investigation 22

## THE KINETICS OF THE DECOMPOSITION OF HYDROGEN PEROXIDE

### INTRODUCTION

Chemical reactions are conducted to convert starting materials to products. Many of these reactions are carried out by people who are not chemists in every-day life. Examples of such reactions are the combustion of wood (an organic material containing carbon, hydrogen, and other elements) to carbon dioxide and water in a fireplace; the combustion of gasoline (a mixture of hydrocarbons, e.g., $C_7H_{16}$) also to carbon dioxide and water in an internal combustion engine; the reaction of sodium hydroxide with aluminum to generate hydrogen used to hydrolyze fats in an attempt to unclog a kitchen sink, or the removal of calcium and magnesium cations from hard water by a water-softening ion-exchange cartridge. About these reactions and the many other reactions that are being explored in laboratories and exploited for industrial purposes, three important questions can be asked:
- Can a particular reaction occur under defined conditions?
- How much of the starting material can be converted to products?
- If a reaction can occur, how fast will it proceed?

The first two questions are answered by the science of thermodynamics through consideration of the energy and entropy changes associated with a reaction. When the sign and the magnitude of the Gibb's free energy ($\Delta G$) of a reaction are known, the first two questions can be answered. When $\Delta G$ is negative for a reaction, the reaction will proceed spontaneously. However, a reaction with a negative $\Delta G$ can be practically useless if it proceeds very slowly. Thermodynamics does not deal with the timely progress of a reaction. The investigations of the speed (the rates) of reactions is the domain of kinetics. Some reactions are very fast and others are imperceptibly slow. Trinitrotoluene (TNT), an explosive, will react in a fraction of a second in the form of an explosion to produce carbon dioxide, water, nitrogen, and nitrogen oxides. When an aqueous solution of lead chloride is mixed with dilute sulfuric acid, a white lead sulfate precipitate is formed "instantaneously". The complete hydrolysis of fats to glycerine and fatty acids (Investigation 7) may take an hour. Rusting of iron, of which car bodies are made, fortunately takes years. The rate of a reaction is determined by experiment. The influence of the concentrations of the reagents and of the temperature on the rate of a reaction provides insight into the mechanism of a reaction and establishes the order of a reaction. In this experiment the dependence of the rate of decomposition of hydrogen peroxide under the catalytic influence of the iodide anion on the concentration of hydrogen peroxide and iodide will be investigated.

### CONCEPTS OF THE EXPERIMENT

Hydrogen peroxide decomposes to oxygen and water as shown in the balanced equation 1. Hydrogen peroxide is thermodynamically unstable with respect to water and oxygen and should

$$2H_2O_2 \longrightarrow 2H_2O + O_2 \uparrow \qquad (1)$$

readily and quickly decompose. However, experiments show that solutions of hydrogen peroxide in water decompose only very slowly because the unstable hydrogen peroxide molecule is kinetically protected by a high activation energy. This activation energy can be lowered by a catalyst. A catalyst is a substance that accelerates a reaction without being used up.

A reaction between two molecules is only possible when the two molecules collide. Each hydrogen peroxide molecule in the aqueous medium slithers randomly about the water molecules. When it meets another hydrogen peroxide molecule, the pair could react according to equation 1 if the collision is sufficiently energetic and the right parts of the two molecules meet. These conditions are rarely met and hydrogen peroxide decomposes very slowly. When a solution of potassium iodide is mixed with the hydrogen peroxide, oxygen is formed readily and quickly. Hydrogen peroxide molecules still collide in this iodide-containing solution; their collisions lead to reactions just as rarely as in the iodide-free system. The logical conclusion from this observation suggests that the iodide ion must initially react with the hydrogen peroxide molecule to produce an intermediate (equation 2 ) that in subsequent reactions must lead to the formation of oxygen and the regeneration of iodide. The decomposition of hydrogen peroxide is more complex than pictured by

$$H_2O_2 + I^- \longrightarrow H_2O + IO^- \tag{2}$$

equation 2. The decomposition proceeds by a series of reactions whose sum must agree with equation 1. Each one of these reactions occurs when two molecules (ions) collide. Collisions of three molecules (ions) do occur, but are so rare that they do not influence the overall reaction. Equation 2 demands that an iodide anion collide with a hydrogen peroxide molecule. Such collisions occur very frequently. When Avogadro's number of molecules of hydrogen peroxide and Avogadro's number of iodide anions are present in one liter of solution, one can expect a certain large number of $H_2O_2$–$I^-$ collisions to occur per second. If the rate of this reaction is defined as the number of $H_2O_2$–$I^-$ collisions per second that lead to the formation of $H_2O$ and $IO^-$, then the rate will be proportional to the total number of $H_2O_2$–$I^-$ collisions per second (equation 3). The total number of collisions will increase when the number of hydrogen peroxide molecules or

$$\text{rate (successful collisions per second )} \propto \text{total number of collisions per second} \tag{3}$$

the number of iodide anions in one liter of solution increases. When the number of molecules per liter of solution is divided by Avogadro's number, the moles of substance per liter are obtained. Moles per liter is a concentration. Equation 3 can now be expressed in terms of concentrations (equation 4).

$$\text{rate (moles of } H_2O_2 \text{ decomposed per second)} \propto [H_2O_2]\,[\,I^-] \tag{4}$$

To convert proportionality into equality, the specific rate constant k is introduced (equation 5).

$$\text{rate} = k\,[H_2O_2]\,[\,I^-] \tag{5}$$

Through transformation of equation 5 "k" is recognized as the rate of the reaction when the concentrations of both reactants are 1 molar.

When the reaction described by equation 1 is similarly considered, a different expression for the rate is obtained (equation 6). Nature is not bound to obey such "paper chemistry".

$$R = \text{rate} = k'\,[H_2O_2]^2 \tag{6}$$

To find out what the hydrogen peroxide really does when it decomposes, experiments must be carried out to establish how the rate depends on the concentration of each molecule or ion. Equation 5 should be written more generally (equation 7) until experimental results select one of the many possible equations.

$$R = \text{rate} = k\,[H_2O_2]^n\,[\,I^-]^m \tag{7}$$

Equation 6 is a special form of equation 7 with n=2 and m=0.

The goal of this Investigation is the establishment of values for n and m. These exponents are called the "orders" of the reaction with respect to hydrogen peroxide and iodide. The sum of the exponents is the overall order of the reaction. Once the order of the reaction is known, stepwise reactions can be postulated that must agree with the experimental rate law (equation 7). The rate of the overall reaction (equation 1) is determined by the slowest of the stepwise reactions. Even when such a scheme of "elementary" stepwise reactions agrees with the experimental rate law, the scheme may not describe what the hydrogen peroxide really does when it decomposes.

How is the rate of the decomposition measured? In the case of the decomposition of hydrogen peroxide, the volume of oxygen produced or the decrease of the concentration of hydrogen peroxide in the solution could be measured at various times after the reaction had been initiated. Experimentally, it is easier to measure the volume of oxygen evolved as a function of time. As the reaction proceeds, the hydrogen peroxide concentration in the solution will decrease and the volume of oxygen formed per unit time (the rate) will decrease. When all the hydrogen peroxide is decomposed, its concentration is zero, and no more oxygen is formed.

The primary data provided by the experiment are the volume readings every 20 seconds. The increase in volume at the barometric pressure (the water level in the buret and in the leveling bulb are kept at the same height) comes from the oxygen generated by the decomposition of the hydrogen peroxide and from the water vapor. Any closed space in contact with liquid water will contain water vapor that exerts its partial pressure. The partial pressure of water is only dependent on the temperature when the space is saturated with water vapor. The partial pressure of water at various temperatures can be taken from vapor pressure tables.

In an idealized experiment a volume $V_{O_2}$ of oxygen may be collected at the barometric pressure without allowing the water to saturate the volume $V_{O_2}$ with its vapor by evaporation from the liquid surface. As the oxygen is generated and collected in the buret, the water level in the buret will drop and the leveling bulb is concomitantly lowered to keep the water levels in the buret and in the bulb at the same height and, consequently, the oxygen under the barometric pressure $P_b$. After $V_{O_2}$ mL of oxygen have been collected, the water is allowed to evaporate and saturate the volume $V_{O_2}$ with water vapor until the water vapor pressure $P_w$ is reached. If this process is allowed to proceed without a change in volume (the buret reading remains constant), the pressure must increase to $P_b + P_w$. When the gas mixture (oxygen and water vapor) is allowed to return to the barometric pressure $P_b$, the gas mixture will expand to a total volume V, which is the sum of the volume $V_{O_2}$ occupied by the oxygen at barometric pressure and the volume $V_w$ occupied by the water vapor at atmospheric pressure.

In the real experiment $V_{O_2}$ and $V_w$ cannot be measured separately; only the total volume V is obtained. However, for the decomposition of hydrogen peroxide only the volume $V_{O_2}$ is relevant. The measured total volume must be corrected by subtracting from it $V_w$. The volume $V_w$ can be calculated using the ideal gas law $PV = nRT$. Applying this equation to the water vapor ($P_w$: partial pressure of water at the laboratory temperature; $n_w$: moles of water in total volume V), the following relations are obtained (equations 8, 9). Substitution for $n_w$ produces equation (10).

$$P_w V = n_w RT \qquad\qquad (8)$$

$$P_b V_w = n_w RT \qquad\qquad (9)$$

$$V_w = V\frac{P_w}{P_b} \qquad\qquad (10)$$

This equation is used to calculate the volume correction $V_w$ that must be subtracted from the total volume V to obtain the volume $V_{O_2}$ needed. A table of vapor pressure of water ($P_w$) as a function of temperature is located in Investigation 12.

When these corrected volumes of oxygen produced during the decomposition of hydrogen peroxide are plotted versus time elapsed since the beginning of the reaction, a curve similar to the one shown in Fig. 1 is obtained.

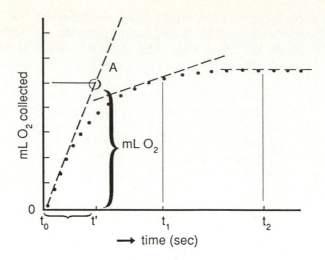

Figure 1:  Volume of oxygen collected during the decomposition of hydrogen peroxide

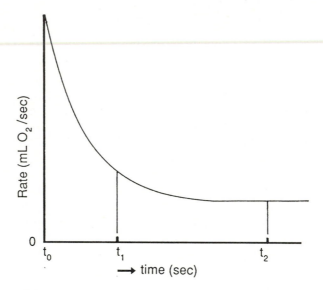

Figure 2:  The change of the rate of decomposition of hydrogen peroxide as a function of the time elapsed since the beginning of the reaction

From the experimental curve (Vol. $O_2$ collected vs. Time) the rate of the decomposition reaction (mL $O_2$ collected per second) can be obtained for any particular time as the slope of the tangent to the curve at that time.  Such tangents are shown in Fig. 1.  The tangent at $t_0$ (a straight line from Point 0 to Point A) is easy to draw and the slope (rate) easy to calculate because at the beginning of the reaction the curve is almost a straight line and no serious mistake is made when this part of the curve is approximated by a straight line.  The slope of the straight line that passes through the origin of the coordinate system (Point 0) gives the rate of the reaction at time $t_0$ and is obtained according to equation 11.  As shown in Fig. 1 the volume of oxygen at time t' is

$$rate_{(t=0)} = \frac{mL\ O_2\ from\ Volume\ axis\ at\ t'}{t'} \qquad (11)$$

found in the following manner:  Draw a perpendicular line to the time axis from time t'.  That line (parallel to the volume axis) intersects the tangent (line 0A) at Point A.  Draw a parallel to the time axis through Point A. The intersection of this parallel with the volume axis gives the volume of oxygen that is needed in equation 11. If the rates of the reaction are calculated for various times and then plotted versus time, Fig. 2 is obtained. Fig. 2 shows that the rate is highest at the beginning of the reaction, slows as the reaction proceeds, and is zero as soon as the reaction is complete.

Although the rates of a reaction under different conditions could be compared at any fixed time, it is experimentally convenient to select time zero for this purpose because the rate (the slope) can be easily determined and the reaction must be followed for a short time only. To find the order of the reaction, the rates at $t_0$ are determined for solutions of different concentrations for the reagents ($H_2O_2$, $I^-$). Usually, the concentration of one reactant is kept constant and the concentration of the other reactant is varied. The rates are determined, the initial concentrations are known, and transformed equation 7 is solved for the unknown exponents (equation 12).

$$R = k\,[H_2O_2]^n[I^-]^m \qquad\qquad [H_2O_2]\ \text{changed},\ [I^-]\ \text{constant} \qquad (12)$$

$$R' = k\,[H_2O_2']^n[I^-]^m$$

$$\frac{R}{R'} = \frac{[H_2O_2]^n}{[H_2O_2']^n} = \left\{\frac{[H_2O_2]}{[H_2O_2']}\right\}^n$$

$$\log(R/R') = n\,\log\,[H_2O_2]/[H_2O_2']$$

$$n = \frac{\log(R/R')}{\log\left\{[H_2O_2]\Big/[H_2O_2']\right\}}$$

A similar expression for m is obtained by keeping $[H_2O_2]$ constant and varying the iodide concentration.

Because of experimental errors the exponents n and m are rarely integers. Although fractional orders of reactions are known, in this experiment the exponents should be rounded to the nearest whole number. With the order of the reaction with respect to $H_2O_2$ and $I^-$ known, chemically-based intuition is needed to propose a "mechanism" for the decomposition of hydrogen peroxide that agrees with the experimentally established rate law.


## ACTIVITIES

- Complete the PreLab Exercises before coming to the laboratory.

- Prepare 200 mL of a 0.15 M KI solution.

- Determine the cumulative volumes of oxygen for the decomposition of 3% $H_2O_2$ in the presence of KI at three concentration ratios of $[H_2O_2]/[KI]$.

- Repeat these measurements several times.

- Correct the volume of $O_2$ collected with the volume occupied by water vapor and plot the corrected values versus time.

- Calculate the rates for these reactions at the beginning of the reactions.

- Determine the order of the reaction.

- Propose a mechanism for the reaction.

- Complete the Report Form.

## SAFETY

*Wear approved eye protection at all times. Avoid contact between skin and hydrogen peroxide. Be cautious with the apparatus for generating and collecting oxygen. Check that your apparatus is not a "closed system" in which build-up of pressure could cause breakage of glass and injury to bystanders.*

## PROCEDURES

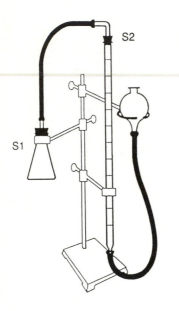

**1**. On a triple-beam balance weigh the amount of KI required to prepare 200 mL of a 0.15 M solution. Transfer the crystals quantitatively into a 250-mL Erlenmeyer flask and add 200 mL water. Stir to dissolve the KI and homogenize the solution.

**2**. Attach a buret clamp parallel to the direction of the base of a ringstand and about 30 cm above the base. Attach two other clamps near the top of the ringstand at angles of about 45 deg to the direction of the first clamp. Connect one end of a 60-cm length of rubber tubing to the lower end of a 50-mL buret and fasten the buret securely in the center clamp so that the lower end of the buret is about 15 cm above the ringstand base.

Fasten the 125-mL Erlenmeyer flask securely in the upper left clamp. Connect the two rubber stopper-glass tube assemblies S1 and S2 with a 60-cm length of rubber tubing. Insert the stoppers into the top of the buret and into the flask. Connect the tubing from the lower end of the buret to a leveling bulb. Fasten the bulb securely to the right-hand clamp. Make sure that the rubber tubing is not constricted by the clamp.

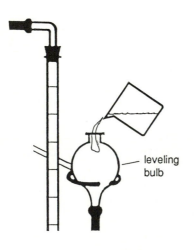

leveling bulb

**3**. Remove the stopper from the Erlenmeyer flask. Pour 125 mL of distilled water at room temperature into the securely clamped leveling bulb. Loosen the clamp holding the bulb. Lower and raise the bulb several times to expel all air from the tubing. When bubbles rising to the surface can no longer be observed, clamp the bulb in its original position.

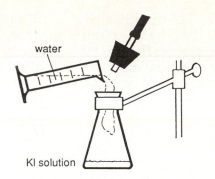

water

KI solution

**4.** Fill a clean and dry 25-mL graduated cylinder with the 0.15 M KI solution to the 20-mL mark. Pour the 20 mL of KI solution into the 125-mL Erlenmeyer flask. Fill the same graduated cylinder with distilled water at room temperature to the 20-mL mark. Add this water to the flask.

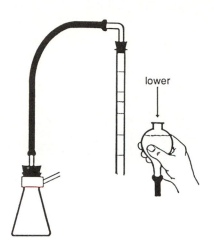

lower

**5.** Stopper the 125-mL Erlenmeyer flask tightly. Check the stopper in the buret for a tight fit. Lower the leveling bulb to a position approximately 10-cm below its original position. Observe the water level in the buret. The water level in the buret will drop and should stop at a level considerably above the level in the bulb. If the water level in the buret keeps dropping slowly, the system has a leak. Check the stoppers and the hose connections. Do not continue until your system is leak-free. Consult your instructor if you cannot eliminate your leaks.

*Plan Steps 6 and 7 carefully. Ask your neighbor at the laboratory bench to assist you during the measurement of the evolved oxygen.*

| Time | Volume Reading mL | Vol. Coll. | Vol. Corr. for $H_2O$ |
|---|---|---|---|
| 2:26:35 | 3.2 | 0 | 0 |
| 2:26:55 | | | |
| 2:27:15 | | | |

**6.** Pour a 3% solution of $H_2O_2$ at room temperature into a clean and dry 25-mL graduated cylinder until the 10-mL mark is reached. Your neighbor will remove the leveling bulb from its clamp and raise or lower the bulb to keep the water levels in the bulb and in the buret at the same height throughout Step 7. Your neighbor will announce the time at the beginning of the experiment and read the volume in the buret (with the water levels in the buret and the bulb at the same height) every 20 seconds. You will need to be prepared to enter the volume readings into a previously prepared table.

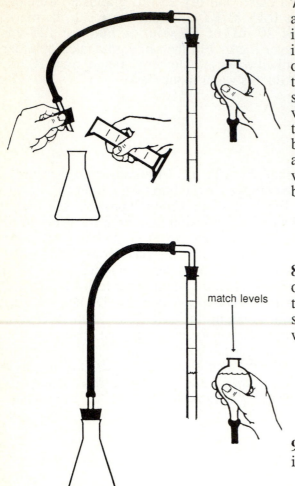

match levels

**7**. Unstopper the 125-mL Erlenmeyer flask, quickly add the 10 mL 3% $H_2O_2$ solution at once, and immediately stopper the flask tightly. This moment is the beginning of the experiment. Record the time of this moment as read by your assistant. Remove the Erlenmeyer flask quickly from its clamp and swirl the solution in the flask gently. Record the volume reading in the buret (with the water level in the leveling bulb exactly at the same height as in the buret) every twenty seconds as announced by your assistant. Continue swirling the flask and recording volumes until approximately 20 mL of oxygen have been released.

**8**. Place the leveling bulb in its clamp. Cautiously open the stopper in the Erlenmeyer flask. Remove the flask from the clamp, pour the solution down the sink and thoroughly rinse the flask with distilled water. Dry the inside of the flask with a towel.

**9**. Repeat Procedure 7 with the following solutions in the Erlenmeyer flask:

- 20 mL 0.15 M KI, 10 mL $H_2O$, 20 mL 3% $H_2O_2$
- 40 mL 0.15 M KI, no water, 10 mL 3% $H_2O_2$

If time permits, repeat the three experiments. Obtain and record the temperatures in the laboratory and the barometric pressure.

**10**. Calculate the volumes of gas (oxygen and water vapor) present in the buret at each time. Correct these volumes by subtracting the volumes occupied by the water vapor at the barometric pressure and the temperature of the laboratory. Plot the volumes of oxygen generated by the reaction versus time for each experiment. Determine the slope of each of the curves at time zero. Average the slopes from repeat experiments. Calculate the order of the reaction with respect to hydrogen peroxide and iodide using averaged slopes.

# PRELAB EXERCISES

## INVESTIGATION 22: THE KINETICS OF THE DECOMPOSITION OF HYDROGEN PEROXIDE

Name:_____

Instructor:_____     ID No.:_____

Course/Section:_____     Date:_____

1.  Calculate the mass of KI needed to prepare 200 mL of a 0.15 M solution of KI.

2.  A solution is prepared by mixing 40 mL 0.15 M KI, 20 mL of distilled water and 40 mL 0.9 M $H_2O_2$.  Calculate the concentrations of each of these substances in the final solution before any reaction takes place.

3.  The hydrogen peroxide in a solution containing 10 mL 3% $H_2O_2$, 10 mL $H_2O$, and 10 mL KI was completely decomposed to water and oxygen.  How many milliliters of dry oxygen are generated when the oxygen is collected under 1 atm pressure at 22°C?

(over)

4. During the decomposition of hydrogen peroxide, 10.0 mL oxygen saturated with water vapor were collected. The laboratory temperature was 22°C and the barometric pressure 755 torr. Calculate the volume occupied by the dry oxygen under these conditions.

5. The following data were collected in an experiment in which hydrogen peroxide was decomposed to water and oxygen. The data are given as mL $O_2$ evolved/time in seconds elapsed since the beginning of the experiment: 2.0/15, 4.0/30. 6.0/40. 8.0/65, 10.0/90, 12.0/110, 14.0/130, 16.0/150. 18.0/170. Plot these data, calculate the slope of the best straight line (You may want to use the least squares program in a calculator.) and give the numerical value and the units for the rate of formation of oxygen.

# REPORT FORM

## INVESTIGATION 22:  THE KINETICS OF THE DECOMPOSITION OF HYDROGEN PEROXIDE

Name: _____     ID No.: _____

Instructor: _____     Course/Section:_____

Partner's Name (if applicable): _____     Date: _____

| Sample | Initial Concentrations (mol/L) $H_2O_2$* | KI | Slope (mL $O_2$/min) for Trial 1 | 2 | 3 |
|---|---|---|---|---|---|
| 20 mL KI 20 mL $H_2O$ 10 mL $H_2O_2$ | | | | | |
| 20 mL KI 10 mL $H_2O$ 20 mL $H_2O_2$ | | | | | |
| 40 mL KI 0 mL $H_2O$ 10 mL $H_2O_2$ | | | | | |

*3% $H_2O_2 \equiv 0.9$ mol/L

Barometric Pressure: _____     Laboratory Temperature: _____     Partial Press. of Water: _____

1. Suggested Rate Law:  Rate of $O_2$ formation (mL/min) = k x [      ] [      ]
   Attach your calculations and rate plots.

2. Suggested mechanism: (Attach your answer and calculations.)

3. Calculate the grams (moles) of hydrogen peroxide decomposed, the grams (moles) of oxygen produced and the percent of hydrogen peroxide decomposed at the time the experiment is terminated. (Attach your answer and calculations.)

4. Calculate the number of collisions between $H_2O_2$ and $I^-$ per second that lead to a reaction (equation 2) and ultimately to the production of oxygen.  Use the experimental rate at $t_0$; assume the experiment is carried out at STP; pay attention to equations 1 and 2; use Avogadro's number in your calculations. (Attach your answer and calculations.)

Date _____ Signature _____

# Investigation 23

## INTRODUCTION

When we eat, the chewed food (carbohydrates, fats, proteins) is swallowed and enters the stomach. On arrival of the food in the stomach, an intricate sequence of events is triggered that initiates the digestion of the food. Among other substances, dilute hydrochloric acid is excreted. Acting on chunks of food, the acid helps to swell and loosen the fibrous portions for better contact with the aqueous liquid in the stomach. The hydrochloric acid deactivates the enzyme α-amylase that was mixed into the food during the chewing process and activates proteolytic enzymes. These enzymes that hydrolyze proteins (meat) to amino acids are secreted initially in inactive forms, probably to protect the walls of the stomach from being digested. Without activation of these enzymes by hydrochloric acid, the digestion of protein foods would not proceed.

For various reasons, not all of which are yet understood, the stomach may secrete too much hydrochloric acid causing a condition known as gastric hyperacidity resulting in discomfort familiar to most of us as indigestion. This condition is often induced by highly spiced or fat-rich food or by simply eating too much. Mild cases of gastric hyperacidity may be treated by taking "antacids". These antacids neutralize excess stomach acid. An excess of the neutralizing ingredients would upset the delicate pH regimen of the digestive enzyme systems. Therefore, the antacids must be taken only in controlled amounts. The antacids cannot contain strong bases such as sodium hydroxide that would produce a very alkaline pH in the stomach and cause serious problems when taken in excess. The antacids are designed to prevent such problems by consisting of substances that reduce excess stomach acid but keep the liquids in the stomach in the correct acidic pH region.

In this experiment the capacities of antacids to neutralize stomach acid are determined. These capacities are needed to establish the correct doses of antacids to be taken.

## CONCEPTS OF THE EXPERIMENT

The active ingredients in antacids are substances such as calcium carbonate, sodium hydrogen carbonate, and magnesium hydroxide. These substances react with hydrochloric acid as described by equations 1-4.

$$CaCO_3 + H^+ + Cl^- \longrightarrow Ca^{2+} + HCO_3^- + Cl^- \tag{1}$$

$$CaCO_3 + 2H^+ + 2Cl^- \longrightarrow Ca^{2+} + 2Cl^- + H_2O + CO_2 \uparrow \tag{2}$$

$$NaHCO_3 + H^+ + Cl^- \longrightarrow Na^+ + Cl^- + H_2O + CO_2 \uparrow \tag{3}$$

$$Mg(OH)_2 + 2H^+ + 2Cl^- \longrightarrow Mg^{2+} + 2Cl^- + 2H_2O \tag{4}$$

In all these reactions hydronium ions that are responsible for the acidity are used to form water or hydrogen carbonate ion. This decrease in the hydronium ion concentration in the stomach liquid caused by the swallowed antacid tablet relieves the hyperacidity and the pain. The "acid-neutralizing capacity" of an antacid, defined as the moles of HCl neutralized by one gram of the antacid, can be determined by reacting an exactly weighed amount of the antacid, m, with an excess of hydrochloric acid ($V_A$ mL of molarity $M_A$). The excess hydrochloric acid is then titrated with sodium hydroxide ($V_B$ mL of molarity $M_B$). The moles of hydrochloric acid consumed by m g of antacid are calculated using equation 5.

$$\text{moles HCl} = \frac{V_A M_A - V_B M_B}{1000} \tag{5}$$

The "acid-neutralizing capacity" (ANC) is given by equation 6.

$$\text{ANC} = \frac{\text{moles HCl}}{m} = \frac{V_A M_A - V_B M_B}{1000\,m} \tag{6}$$

The endpoint of the titration of the excess hydrochloric acid with sodium hydroxide (a titration of a strong acid with a strong base) is performed with bromophenol-blue, an indicator that changes color (yellow to blue) at pH 4.6. Although a strong acid-strong base titration could be carried out with any indicator changing color in the pH 4 to pH 10 range, the active ingredients of the antacid would begin to react with the added base at pH values approaching 7. For instance, magnesium ions would begin to form rather insoluble magnesium hydroxide and dissolved carbon dioxide would be transformed to $HCO_3^-$. These reactions would consume additional base and would cause the calculated acid-neutralizing capacity to be low.

The molarities of the hydrochloric acid and of the sodium hydroxide solution used in this experiment must be known accurately and precisely. Dilution of the concentrated hydrochloric acid and weighing out an exact amount of solid sodium hydroxide will not produce dilute solutions of exactly known molarities. Therefore, the dilute solutions must be standardized. A convenient standard acid is potassium hydrogen phthalate, $C_8H_5KO_4$ (KHP). This crystalline substance can be obtained in high purity, is stable, does not absorb water from the atmosphere, is soluble in water, and can be weighed accurately and precisely. KHP is a monoprotic acid and reacts with

sodium hydroxide as shown in equation 7. When m grams of KHP (molecular mass $M_{KHP}$) require $V_B$ mL of sodium hydroxide to reach the phenolphthalein endpoint (colorless to pink), the molarity of the base ($M_B$) can be obtained using equation 8. In this titration phenolphthalein must

$$M_B = \frac{1000\,m}{M_{KHP} V_B} \tag{8}$$

be used as the indicator (color change at pH 8.5) because the solution of completely neutralized KHP will be weakly alkaline. The sodium hydroxide solution of exactly known molarity can now be used to standardize the dilute hydrochloric acid. Phenolphthalein or bromophenol-blue could be used for this titration. With phenolphthalein as indicator, the endpoint is easier to see when the sodium hydroxide is added to the hydrochloric acid. When $V_A$ mL hydrochloric acid required $V_B$ mL of sodium hydroxide solution of molarity $M_B$ to reach the endpoint, the molarity of the acid $M_A$ can be calculated according to equation 9. To achieve precise results in these standardizations,

$$M_A = \frac{V_B M_B}{V_A} \tag{9}$$

the volumes of acid and base consumed should not be too small and not larger than one buret-volume. It is also advisable to repeat the standardizations and work with average molarities.

## ACTIVITIES

- Complete the PreLab Exercises before coming to the laboratory.

- Prepare 400 mL 0.1 M HCl from concentrated HCl.

- Prepare 400 mL 0.1 M NaOH from 2.0 M NaOH.

- Standardize the sodium hydroxide solution with potassium hydrogen phthalate.

- Standardize the hydrochloric acid with the standardized sodium hydroxide solution.

- Determine the moles of HCl neutralized by an exactly weighed amount of an antacid.

- Calculate the acid-neutralizing capacity of the antacid.

- Complete the Report Form.

## SAFETY

*Do not, under any circumstances, taste or swallow any of the antacids used for this experiment. Storage in the chemical stockroom may have resulted in contamination. Wear approved eye protection and use proper care in working with acids and alkalies.*

## PROCEDURES

### A. PREPARATION OF SOLUTIONS

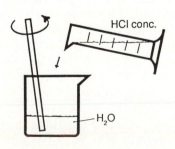

A-1. To prepare 400 mL of 0.1 M HCl, pour 200 mL of distilled water into a 600-mL beaker. In the hood fill a 10-mL graduated cylinder from the storage bottle for concentrated (12 M) HCl to the level needed to produce 400 mL of 0.1 M HCl. Pour the concentrated acid slowly into the stirred distilled water in the 600-mL beaker. Then add 200 mL distilled water. Stir the solution until it is homogeneous. Cover the beaker with a piece of paper and label the beaker "0.1 M HCl".

A-2. To prepare 400 mL of 0.1 M NaOH solution, pour 200 mL of distilled water into a 500-mL Erlenmeyer flask. Measure the volume of 2.0 M

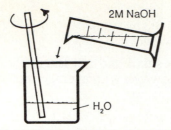

NaOH solution needed to prepare 400 mL of 0.1 M NaOH solution into a 25-mL graduated cylinder. Add 200 mL of distilled water. Mix the solution thoroughly by stirring with a clean glass rod. Stopper the flask and label it "0.1 M NaOH".

# B. STANDARDIZATION OF THE HCl AND NaOH SOLUTIONS

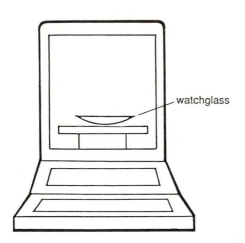

**B-1.** Obtain two burets. Fill them with distilled water. Drain the water through the stopcock. Observe the interior of the buret walls. The water should drain without leaving water droplets clinging to the wall. If you observe water droplets, clean the buret and repeat this procedure. Mount the burets as shown in the drawing.

**B-2.** Rinse and fill one buret with 0.1 M HCl and the other with 0.1 M NaOH. Label the burets. Drain enough solution from each buret into a beaker to expel air from the buret tips and to drop the liquid levels just below the zero marks.

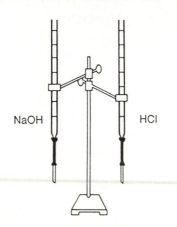

**B-3.** Obtain from your instructor a small vial containing potassium hydrogen phthalate. Weigh a clean and dry watchglass on an analytical balance to the nearest 0.1 mg. Transfer potassium hydrogen phthalate corresponding within ±10 percent to the amount needed for the neutralization of 30.0 mL 0.10 M NaOH solution. Weigh the watchglass and the acid phthalate to the nearest 0.1 mg. Record all the masses.

**B-4.** Carefully remove the watchglass from the balance. Transfer the acid phthalate quantitatively into a clean 250-mL Erlenmeyer flask. Rinse the watchglass with distilled water to wash the last traces of the phthalate into the flask.

**B-5.** Add distilled water to the flask until the flask contains approximately 100 mL of water. Heat the flask carefully over a Bunsen burner until all the

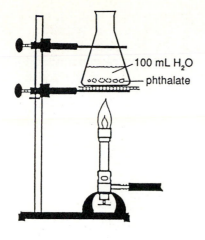

phthalate has dissolved. Set the flask on an insulating pad and allow it to cool until it is warm but not hot to the touch.

**B-6**. Add three drops of phenolphthalein indicator solution to the flask. Set the flask on a sheet of white paper below the tip of the "NaOH" buret. Read and record the volume in this buret to the nearest 0.01 mL. Titrate the phthalate solution with the NaOH solution to a permanent pale-pink endpoint. Record the final buret reading. Discard the solution in the flask, clean the flask, and rinse it with distilled water. Repeat this standardization procedure twice. Calculate the average molarity of the NaOH solution and the standard deviation.

**B-7**. Read and record the volume in the "HCl" buret. Drain approximately 30 mL of the hydrochloric acid into a clean 250-mL Erlenmeyer flask. Record the final buret reading to 0.01 mL. Add approximately 70 mL distilled water and 5 drops bromophenol-blue solution. Swirl the flask to mix the contents.

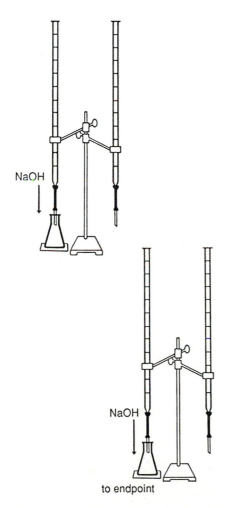

**B-8**. Refill the "NaOH" buret and drain sufficient liquid into a waste beaker to drop the liquid level just below the zero mark. Record the buret reading to 0.01 mL. Place the Erlenmeyer flask on a sheet of white paper below the tip of the "NaOH" buret. Titrate the HCl solution with the sodium hydroxide.

The endpoint is reached when the bromophenol-blue changes its "acidic" color (yellow at pH 3 and below) to its "basic" color (pure blue at pH 4.6 and above). Before the pure blue color is reached, a greenish-blue color prevails. To recognize the pure-blue endpoint, titrate a solution back and forth between the yellow and the pure-blue. Read the buret and record the reading to 0.01 mL. When the pure-blue endpoint has been reached, repeat the standardization of the HCl. Calculate the molarity of the HCl.

## C. NEUTRALIZING CAPACITY OF AN ANTACID

**C-1**. Obtain from your instructor a sample of an antacid. Ask your instructor for suggestions about the amount of sample to be used in the experiment and the volume of HCl needed to dissolve the sample.

tablet

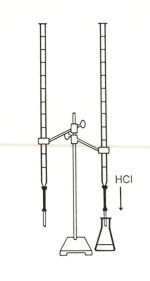

HCl

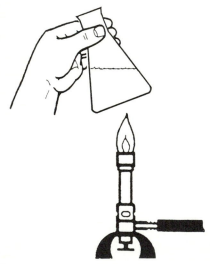

**C-2.** Weigh a clean and dry watchglass to the nearest 0.1 mg on an analytical balance. Then determine the mass of one antacid tablet. If necessary, break the tablet into pieces and add pieces to the watchglass until the mass suggested by your instructor is reached within ±10 percent. Weigh this amount to the nearest 0.1 mg. Record all masses. Transfer the weighed sample quantitatively into a clean 250-mL Erlenmeyer flask.

**C-3.** Refill the HCl buret and drop the liquid level just below the zero mark. Record the buret reading. Add to your antacid sample in the Erlenmeyer flask the volume of HCl solution suggested by your instructor. Record the buret reading to 0.01 mL. Add approximately 30 mL of distilled water and 5 drops of bromophenol-blue solution.

**C-4.** Carefully warm the flask over a low Bunsen flame. Swirl the contents until the antacid sample has disintegrated. Do not boil the solution. The sample may not completely dissolve. Inert binders and fillers may remain undissolved. The color of the mixture in the flask should remain yellow. Should the color change to blue, add HCl solution until the yellow color is restored. Then add an additional 5 mL of the HCl solution. Read and record the total volume of HCl added.

**C-5.** Allow the solution to cool. Refill the "NaOH" buret. Titrate the warm, yellow solution with the sodium hydroxide solution to the pure-blue endpoint. You have reached the endpoint when the blue color without greenish tint persists for at least 30 seconds. Record all buret readings to 0.01 mL. If time permits, repeat the titration with another sample of your antacid.

**C-6.** Calculate the neutralization capacity (the moles of HCl neutralized by 1.00 g of the antacid) of your antacid. The moles of HCl consumed by the amount of your antacid used in the experiment is the difference between the moles of HCl added to the sample and the moles of NaOH needed to reach the endpoint. The moles of HCl and NaOH are obtained as the products of the molarities of the solutions and the volumes (expressed in liters) used in the experiment.

# PRELAB EXERCISES

## INVESTIGATION 23:  SOME COMMERCIAL ANTACIDS

Name:_____

Instructor:_____    ID No.:_____

Course/Section:_____    Date:_____

1. Describe the correct way to prepare dilute acids and bases from concentrated solutions and distilled water.

2. Calculate the volume of 12 M HCl required for the preparation of 400 mL of 0.10 M HCl.

3. Calculate the volume of 2.0 M NaOH needed to prepare 400 mL 0.10 M NaOH.

4. Describe how a buret is checked for cleanliness (grease-free).  What must be done when the buret does not pass the check?

5. Calculate the mass of potassium hydrogen phthalate that will neutralize 30.0 mL 0.10 M NaOH. Express your answer with the correct number of significant figures.

6. Calculate the molarity of a dilute solution of HCl from the following data: 33.45 mL of 0.0985 M NaOH had to be added in a titration to 41.23 mL HCl to make the indicator change from its "acid" to its "basic" color. Express your answer with the correct number of significant figures.

# REPORT FORM

## INVESTIGATION 23: SOME COMMERCIAL ANTACIDS

Name: _____     ID No.: _____

Instructor: _____     Course/Section:_____

Partner's Name (if applicable): _____     Date: _____

Show calculations on attached pages.

## Standardization of NaOH solution

### Potassium Hydrogen Phthalate

Phthalate & Watchglass: _____ g

Watchglass: _____ g

Phthalate: _____ g

### Titration of phthalate

Final volume NaOH: _____ mL

Initial volume NaOH: _____ mL

NaOH consumed: _____ mL

### NaOH solution

Concentration: _____ M

Average concentration: _____ ± ___ M

## Standardization of HCl solution

### HCl solution

Final "HCl" buret reading: _____ mL

Initial "HCl" buret reading: _____ mL

Volume HCl solution: _____ mL

### Titration of HCl with NaOH

Final "NaOH" buret reading: _____ mL

Initial "NaOH" buret reading: _____ mL

Volume NaOH consumed: _____ mL

### HCl solution

Concentration: _____ M

Average concentration: _____ ± ___ M

If the standardizations were repeated, record the data on copies of the Report Form. Attach these copies to your report. Calculate averages and standard deviations.

## Antacid

Name of antacid used: _____

**Mass of one antacid tablet**

Tablet & Watchglass: _____ g

Watchglass: _____ g

Tablet: _____ g

**Mass of antacid sample used**

Sample and Watchglass: _____ g

Watchglass: _____ g

Sample: _____ g

**HCl added to antacid**

Final "HCl" buret reading: _____ mL

Initial "HCl" buret reading: _____ mL

Volume HCl added: _____ mL

Moles of HCl added: _____

**Titration of excess HCl**

Final "NaOH" buret reading: _____ mL

Initial "NaOH" buret reading: _____ mL

Volume NaOH consumed: _____ mL

Moles of NaOH consumed: _____

Moles of HCl consumed by antacid sample: _____

Moles of HCl consumed by one tablet of the antacid: _____

Moles of HCl consumed by one gram of the antacid (neutralizing capacity): _____ moles g$^{-1}$

Average neutralizing capacity: _____ ± _____ moles g$^{-1}$

If the antacid experiment was repeated, record the data on copies of the Report Form. Attach the copies to your report. Calculate the average and the standard deviation for the neutralizing capacity.

The hydrochloric acid concentration in a full, hyperacidic stomach is 0.03 M. The volume of the liquid in the stomach is 300 mL. How many tablets of your antacid would have to be taken to bring the stomach liquid to the more normal pH of 3.5? (Show your work!!)

Date _____ Signature _____

# Investigation 24

## DETERMINATION OF AN IONIZATION CONSTANT

## INTRODUCTION

Chemists classify acids as strong or weak. The strength of an acid in aqueous solution is expressed by its dissociation constant $K_a$ or its $pK_a$, the negative logarithm of the $K_a$. When an acid, HA, is dissolved in water, it will dissociate into hydronium ion, $H_3O^+$, and the anion, $A^-$ (equation 1). The hydronium ion is frequently written for the sake of brevity as $H^+$ in spite of the

$$HA \xrightleftharpoons{H_2O} H_3O^+ + A^-  \qquad (1)$$

fact that a "naked" proton does not exist in aqueous solution; $H^+$ is always associated with one or more water molecules. The acid dissociation constant and $pK_a$ are defined by equations 2 and 3, respectively. An acid is called strong when it is completely or almost completely dissociated. A

$$K_a = \frac{[H^+][A^-]}{[HA]} \qquad (2)$$

$$pK_a = pH + \log \frac{[HA]}{[A^-]} \qquad (3)$$

strong acid has a large positive value for $K_a$ and a negative value for $pK_a$. In a solution of a weak acid in water, only a small fraction of the acid present is dissociated. At equilibrium most of the weak acid molecules will be dissolved as undissociated HA. A weak acid has a low value for $K_a$ (e.g., $1.8 \times 10^{-5}$ for acetic acid) and a positive value for $pK_a$ (+4.76).

Acid strength is often confused with corrosive ability and unique chemical properties. For instance, many people consider hydrofluoric acid to be strong because it etches glass. Actually, hydrofluoric acid is a weak acid ($K_a = 6.9 \times 10^{-4}$). Its glass-etching ability is based on the formation of strong silicon-fluorine bonds in the reaction transforming $SiO_2$ to $SiF_4$ or $H_2SiF_6$ and not on its acid strength. Aqua regia, a mixture of concentrated nitric and hydrochloric acids, is considered "super strong" because it is one of the few reagents that will react with and dissolve gold and platinum. The driving force for these dissolution reactions is not acid strength but the strong oxidizing power of aqua regia and the formation of $AuCl_4^-$ and $PtCl_6^{2-}$ complex ions.

Acids are very important substances. They are used to dissolve metals and other solids; they are employed in huge quantities in the chemical industry; plants and animals need them for their life processes; organic (carboxylic) acids are building blocks of fats that make up cell membranes and are responsible for obesity, of esters that give fruits their characteristic aroma, and of buffer systems that keep our blood pH close to neutral. They are a destructive nuisance in the form of acid rain that attacks buildings and acidifies lakes with severe consequences for organisms.

One of the most important properties of an acid that is responsible for its chemical and biological behavior is its strength as expressed by the dissociation constant. This experiment is designed to show how the dissociation constant of a weak monoprotic acid can be determined.

## CONCEPTS OF THE EXPERIMENT

The dissociation constant $K_a$ can be calculated according to equation 2 when the concentrations of $[H^+]$, $[A^-]$, and $[HA]$ are known. The experiment is performed by dissolving an exactly weighed amount of an acid in distilled water and determining the volume, $V_i$, of the resulting solution. This solution is then titrated with a sodium hydroxide solution of exactly known molarity, $M_{NaOH}$, to the phenolphthalein endpoint. After each addition the pH of the solution is measured with a pH meter. The total number of moles of the acid HA can be calculated from the molarity of the sodium hydroxide solution and the volume of the sodium hydroxide solution expressed in liters needed to reach the endpoint (equation 4).

$$\text{moles HA} = M_{NaOH} \times V_{NaOH}^{end} \tag{4}$$

The initial volume of the acid solution, $V_i$, is determined as described in Procedure 4. The initial concentration of the acid, $[HA]_i$, before any sodium hydroxide has been added can now be obtained (equation 5). When a certain volume, $V_{NaOH}$, of sodium hydroxide solution has been

$$[HA]_i = \frac{\text{moles HA}}{V_i \text{ (in L)}} = \frac{M_{NaOH} V_{NaOH}^{end}}{V_i} \tag{5}$$

added, the total volume of the solution increases to $V_i + V_{NaOH}$, and a quantity of HA equivalent to the added NaOH is neutralized. The pH of the solution is measured with the pH meter and the hydronium ion concentration can be calculated from the pH value. Thus, the expression for $K_a$ (equation 6) has three unknowns ($K_a$, $[A^-]$, $[HA]$) and two more equations are needed to calculate a value for $K_a$. These two additional equations are generated by considering that the sum of the concentrations of all positively charged ions must be equal to the sum of the concentrations of all negatively charged ions (charge balance, equation 7), and the sum of the concentrations of the acid anion and the concentration of the undissociated acid must be equal to the total HA concentration (mass balance, equation 8).

$$K_a = \frac{[H^+][A^-]}{[HA]} \tag{6}$$

$$[H^+] + [Na^+] = [A^-] \tag{7}$$

$$[A^-] + [HA] = [HA]_{total} \tag{8}$$

By substituting the expression for $[A^-]$ (equation 7) into equations 6 and 8, the relations 9 and 10 are obtained.

$$K_a = \frac{[H^+]\left([H^+] + [Na^+]\right)}{[HA]} \tag{9}$$

$$[HA] = [HA]_{total} - [H^+] - [Na^+] \tag{10}$$

When the expression for $[HA]$ (equation 10) is introduced into equation 9, an expression for $K_a$ is obtained (equation 11) that consists entirely of experimentally known quantities: $[H^+]$ from the pH

$$K_a = \frac{\left[H^+\right]\left\{H^+ + Na^+\right\}}{[HA]_{total} - H^+ - Na^+} \tag{11}$$

measurement; $[HA]_{total}$ from the volume of sodium hydroxide solution, $V_{NaOH}^{end}$, needed to reach the endpoint of the titration, the molarity of the sodium hydroxide solution, $M_{NaOH}$, and the total volume of the solution after addition of a certain volume of sodium hydroxide, $V_{NaOH}$ (equation 12); and $[Na^+]$ from the volume of sodium hydroxide solution added, $M_{NaOH}$, the molarity of the sodium hydroxide solution and the total volume of solution (equation 13).

$$[HA]_{total} = \frac{V_{NaOH}^{end} \, M_{NaOH}}{V_i + V_{NaOH}} \tag{12}$$

$$[Na^+] = \frac{V_{NaOH} \, M_{NaOH}}{V_i + V_{NaOH}} \tag{13}$$

Introduction of equations 12 and 13 into the equation for $K_a$ (equation 11) produces equation 14 that can now be used to calculate $K_a$ for any pH measurement before the endpoint.

$$K_a = \frac{[H^+] \left\{ [H^+] + \dfrac{V_{NaOH} \, M_{NaOH}}{V_i + V_{NaOH}} \right\}}{\dfrac{V_{NaOH}^{end} \, M_{NaOH}}{V_i + V_{NaOH}} - [H^+] - \dfrac{V_{NaOH} \, M_{NaOH}}{V_i + V_{NaOH}}} \tag{14}$$

With no sodium hydroxide added ($V_{NaOH} = 0$), expression 14, simplifies to equation 15. The same expression is obtained from equation 6 by noting that under these circumstances $[H^+] = [A^-]$ whenever the pH is not close to 7. When the pH is close to 7, the self-dissociation of water would

$$K_a = \frac{[H^+]^2}{\dfrac{V_{NaOH}^{end} \, M_{NaOH}}{V_i} - [H^+]} \tag{15}$$

have to be taken into account. At the half-neutralization point, when $V_{NaOH}$ is $V_{NaOH}^{end}/2$, equation 14 transforms into expression 16. If the acid HA is sufficiently weak, $[H^+]$ at half neutralization is

$$K_a = \frac{[H^+] \left\{ [H^+] + \dfrac{V_{NaOH}^{end} \, M_{NaOH}}{2 \left( V_i + \dfrac{V_{NaOH}^{end}}{2} \right)} \right\}}{\left\{ \dfrac{V_{NaOH}^{end} \, M_{NaOH}}{2 \left( V_i + \dfrac{V_{NaOH}^{end}}{2} \right)} - [H^+] \right\}} \tag{16}$$

much smaller than $\left( V_{NaOH}^{end} \, M_{NaOH} \right) / 2 \left( V_i + V_{NaOH}^{end}/2 \right)$ and can be dropped from the sum and the difference in equation 16. The well known half-neutralization approximation equation $K_a = [H^+]$ or $pK_a = pH$ is the result of this simplification.

To calculate $K_a$ using equation 14, the volume of sodium hydroxide solution consumed to reach the endpoint must be known. This volume can be obtained from the initial buret reading and the buret reading corresponding to the color change of the indicator. When the titration curve is plotted, this volume can be determined graphically as shown in Figure 1. Extend the linear flat

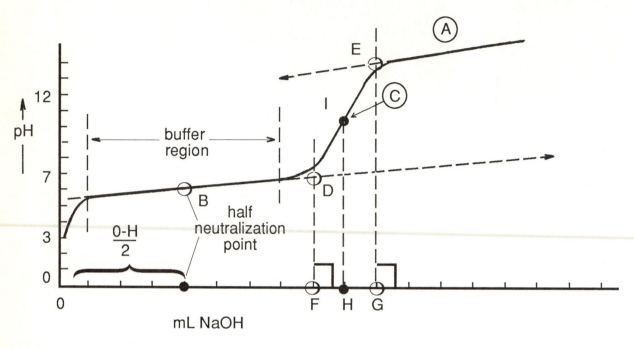

Figure 1.

regions A and B of the titration curve. Draw the best fitting straight line through the steep part C of the titration curve to obtain the intersection points D and E. Drop verticals from D and E onto the volume axis. These verticals will intersect the volume axis at points F and G. Halve the distance FG (point H) and draw a vertical at point H. The intersection of the vertical with the titration curve is the endpoint and the volume corresponding to point H is $V_{NaOH}^{end}$. The endpoint-volumes found by these two methods should not differ significantly. All the data are now available to calculate $K_a$. In precise and accurate work only values falling into the buffer region are used for the calculation of $K_a$. At the beginning of the titration only a small volume of sodium hydroxide has been added and the small volume consumed is associated with a large error. At the end of the titration, the two terms in the denominator of equation 14 having $V_{NaOH}^{end}$ and $V_{NaOH}$ become about equal and their difference becomes imprecise. It is instructive to calculate $K_a$ values from all the experimental data up to the endpoint. This task can be made easier by writing a small program for expression 14 to be executed by a programmable calculator or small computer. The average for $K_a$ and the standard deviation should be calculated using first all the values and then only the values in the buffer region. These averages and the values for $K_a$ obtained from the first pH measurement (no NaOH added) and from the pH at the half-neutralization point should be compared and the most precise value selected as the final result of this experiment.

## SAFETY

*Wear approved eye protection in the laboratory at all times. Exercise standard precautions in working with acids and bases.*

## ACTIVITIES

- Complete the PreLab Exercises before coming to the laboratory.

- Prepare carbon dioxide-free distilled water.

- Standardize the 0.1 M NaOH solution with potassium hydrogen phthalate.

- Titrate a sample of a weak acid with the standardized sodium hydroxide solution, measure the pH of the solution after each addition of sodium hydroxide solution, and construct the titration curve (pH versus mL NaOH) for your sample.

- Calculate $K_a$ values in several ways.

- Select the most reliable value for $K_a$.

- Complete the Report Form.

## PROCEDURES

Before starting your experiment you must know how to properly work with the pH meter in use in your laboratory. Your instructor might have passed out instruction sheets for you to study or will give a short lecture on this topic in the laboratory. You must know:

- how to protect the pH meter from electrical damage.

- how to protect the glass electrode from physical, chemical, and electrical damage.

- how to rinse and dry the glass electrode.

- when to rinse and dry the electrode.

- how to use the temperature dial on the pH meter.

- how to use the magnetic stirrer to stir a solution whose pH is to be determined.

- how to calibrate the pH meter.

- how to measure the pH of a solution.

- how to place the meter on "standby" when it will not be in use for a while.

- how to record the use of the pH meter and any malfunctions if such records are maintained.

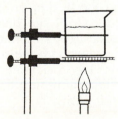

1. Pour approximately 450 mL distilled water into a 600-mL beaker. Heat the beaker with a Bunsen burner until the water boils. Keep the water boiling for several minutes. Then turn off the burner, cover the beaker with a watchglass, and let the water cool to room temperature. When you have exhausted this $CO_2$-free water supply, boil more water.

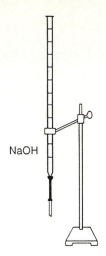

**2**. Take a clean and dry 250-mL Erlenmeyer flask to obtain approximately 150 mL 0.1 M NaOH solution. Keep your flask stoppered to prevent the carbon dioxide in the air from reacting with the sodium hydroxide. Set up a clean buret and rinse it twice with 5 mL of the NaOH solution. Fill the buret with the NaOH solution.

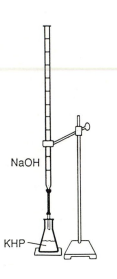

**3**. Weigh on the analytical balance to 0.1 mg the amount of potassium hydrogen phthalate (KHP) needed to neutralize approximately 35 mL 0.1 M NaOH solution. The actual amount taken may vary by ±10 percent from the calculated amount. Transfer the phthalate quantitatively into a 250-mL Erlenmeyer flask, add 100 mL distilled water (boiled to remove carbon dioxide), and heat the flask to dissolve the phthalate. Cool the flask until you can hold it, add 3 drops phenolphthalein solution, and titrate with sodium hydroxide to the pink endpoint. Record all masses and buret readings. Calculate the molarity of the sodium hydroxide solution. Repeat the standardization of the sodium hydroxide solution. Calculate the average for the NaOH molarity.

**4**. Obtain from your instructor approximately 1 g of a solid acid. Place this sample on a smooth piece of paper. Fold the paper to protect the acid from contamination. On the analytical balance weigh a creased piece of smooth paper to 0.1 mg. Add to the paper an amount of the solid acid between 0.35 and 0.45 g. Determine the mass of the sample and the paper to 0.1 mg. Weigh a clean and dry 500-mL beaker on a triple-beam balance to 0.1 g. Record the mass. Transfer the sample quantitatively into the weighed beaker, add 300 mL of boiled, distilled water, and gently heat the beaker until the acid has completely dissolved. Cool the beaker to room temperature. Add three drops of phenolphthalein solution. Weigh the beaker with the solution on the triple-beam balance to 0.1 g. Calculate the volume of your solution under the assumption that the density of the solution is 0.9980 g/mL.

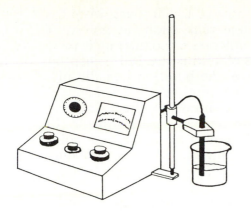

**5**. Standardize the pH meter assigned to you using the buffer solution(s) provided and following the instructions provided for the type of pH meter in use in your laboratory.

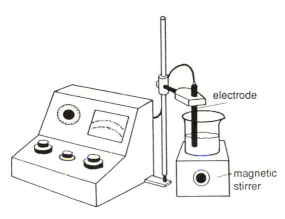

electrode

magnetic
stirrer

**6**. Make sure the switch is in the "off" position on the magnetic stirrer. Place the beaker with the acid solution on the stirrer. Carefully add a magnetic stirring bar to the solution. Turn the speed control on the stirrer to slow and switch on the stirrer. Adjust the speed control until the stirring bar turns properly at a reasonable rate in the center of the beaker. Turn the stirrer off. Carefully insert the pH electrode into the solution. The end of the electrode must be one to two cm below the liquid level and a safe distance above the stirring bar. Turn on the stirrer. Measure the pH of the solution. Record the value as the "initial pH".

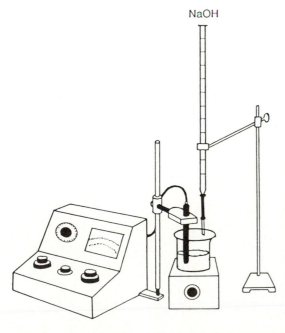

NaOH

**7**. Fill your buret with the standardized NaOH solution. Drop the level just below the zero mark. Record the buret reading. Arrange the buret to permit addition of the NaOH solution to the beaker. Have graph paper ready with a pH axis labeled from pH 0 to 13 and a volume axis labeled from 0 to 50 mL. Add the NaOH solution dropwise very slowly and watch the pH meter. When the pH has changed by approximately 0.2 pH units, stop the addition, allow the pH meter to stabilize, read the pH meter, and record the pH value and the buret reading. Continue adding NaOH solution and take measurements at such increments of NaOH volumes that will define the initial part of the titration curve. When the buffer region is reached, the pH should be measured after each addition of 2 mL of NaOH solution. Smaller volume increments must be used before and after the equivalence point. When you plot the titration curve as you titrate, comparison of your curve with the curve in Figure 1 will allow you to judge in which region of the titration curve you are. Make sure you do not overshoot the endpoint.

Measure the pH and read the buret when the indicator changes to the pink endpoint color. Then add appropriate volumes of sodium hydroxide solution until the pH of 11.6 is reached. Then turn off the stirrer, remove the buret, switch the pH meter to "standby", and remove the beaker. Rinse the electrodes and immerse them into a "storage solution" as directed by your instructor. Repeat this titration with another aliquot of your solid acid and adjust your volume increments to obtain a better-defined titration curve.

**8**. With the help of the equations derived in the section "Concepts of the Experiment" calculate:

- $K_a$ from the initial pH measurement (no NaOH added).

- $K_a$ from the half neutralization point by the graphical procedure (Figure 1).

- $K_a$ from equation 16 (half-neutralization).

- $K_a$ for four measurements in the buffer region, the average of these $K_a$ values and the standard deviation.

- $K_a$ for four measurements near the equivalence point, the average of these $K_a$ values and the standard deviation.

Select the most reliable and precise value for $K_a$.

# PRELAB EXERCISES

## INVESTIGATION 24: DETERMINATION OF AN IONIZATION CONSTANT

Name:_____

Instructor:_____     ID No.:_____

Course/Section:_____     Date:_____

1. Calculate the amount of potassium hydrogen phthalate needed for the neutralization of 35 mL 0.1 M NaOH solution.

2. An empty, clean, and dry 500-mL beaker weighed 190.5 g. After a solid acid and distilled water had been added to the beaker, the mass was 612.1 g. Calculate the volume of the solution under the assumption that the solution has a density of 0.9980 g/mL.

3. A sample of potassium hydrogen phthalate (0.3672 g) was titrated with a sodium hydroxide solution to the phenolphthalein endpoint. The initial buret reading was 0.26 mL; the final reading was 32.60 mL. Calculate the molarity of the sodium hydroxide solution. The potassium hydrogen phthalate has the formula $C_8H_5KO_4$.

(over)

4. Draw the titration curve for a weak acid/strong base and label the following points or regions:
   a) the initial pH of the weak acid solution     b) the equivalence point
   c) the two inflection points     d) the half-neutralization point
   e) the pH at half-neutralization     f) the buffer region
   g) the region of excess strong base

5. Draw the titration curve for a strong acid/strong base titration.

# REPORT FORM

## INVESTIGATION 24: DETERMINATION OF AN IONIZATION CONSTANT

Name: _____     ID No.: _____

Instructor: _____     Course/Section:_____

Partner's Name (if applicable): _____     Date: _____

---

| Standardization of NaOH: | 1st titration | 2nd titration |
|---|---|---|
| Mass of paper + KHP: | _____ g | _____ g |
| Mass of paper: | _____ g | _____ g |
| Mass of KHP: | _____ g | _____ g |
| | | |
| Moles of KHP: | _____ moles | _____ moles |
| mL NaOH consumed: | _____ mL | _____ mL |
| Molarity of NaOH: | _____ mol L$^{-1}$ | _____ mol L$^{-1}$ |
| Average Molarity of NaOH: | _____ mol L$^{-1}$ | |

| Titration of Solid Acid: | 1st titration | 2nd titration |
|---|---|---|
| Mass of paper + acid: | _____ g | _____ g |
| Mass of paper: | _____ g | _____ g |
| Mass of acid: | _____ g | _____ g |
| | | |
| Mass of beaker + acid + water: | _____ g | _____ g |
| Mass of beaker: | _____ g | _____ g |
| Mass of acid + water: | _____ g | _____ g |
| Volume of solution, $V_i$: | _____ mL | _____ mL |
| | | |
| mL NaOH to reach endpoint (graphically): | _____ mL | _____ mL |
| (color change): | _____ mL | _____ mL |
| Moles of acid titrated: | _____ moles | _____ moles |
| | | |
| Molecular mass of acid from mass and moles titrated: | _____ g/moles | _____ g/moles |

**Titration Data**

| Buret Reading mL | mL NaOH added ($V_{NaOH}$) | Total Vol of solution, mL ($V_i + V_{NaOH}$) | pH | [$H^+$] | $K_a$ | $pK_a$ |
|---|---|---|---|---|---|---|
|  |  |  |  |  |  |  |
|  |  |  |  |  |  |  |
|  |  |  |  |  |  |  |
|  |  |  |  |  |  |  |

$K_a$ from initial pH measurement: _____      $pK_a$: _____

$K_a$ from half-neutralization: _____      $pK_a$: _____

$K_a$ from buffer region (average/standard deviation): _____ ± ____      $pK_a$: _____ ± ____

$K_a$ from measurements near the equiv. point (average/standard deviation): _____ ± ____

$pK_a$: _____ ± ____

Most reliable value for $K_a$ with standard deviation: _____ ± _____

On a separate sheet of paper show:
1.  The set-up and values used to calculate four $K_a$ values from data within the buffer region.

2.  The set-up and values used to calculate four $K_a$ values from data taken between the buffer region and the equivalence point.

Date _____  Signature _____

# Investigation 25

## INTRODUCTION

Aqueous solutions can be neutral, acidic or basic.  An aqueous solution is neutral when its hydronium ion concentration is $1 \times 10^{-7}$ mol $L^{-1}$ (pH = 7), acidic when the hydronium ion concentration is larger than $1 \times 10^{-7}$ mol $L^{-1}$ (pH < 7), and basic when the hydronium ion concentration is smaller than $1 \times 10^{-7}$ mol $L^{-1}$ (pH > 7).  The hydroxide ion ($OH^-$) concentration in a neutral solution is equal to the hydronium ion solution; in a basic solution it is larger than the hydronium ion concentration.  In an aqueous solution the product of the concentrations of the hydronium ion and the hydroxide ion is constant and has the value $1 \times 10^{-14}$ ($K_w$) at 25° in dilute solutions.

Many chemical reactions performed by chemists in the laboratory, occurring in the abiotic environment, or proceeding in the cells of organisms require that the aqueous medium have a hydronium or hydroxide ion concentration within a rather narrow range.  Human blood, for example, normally has a pH between 7.3 and 7.5.  Should the pH of the blood change by as little as 0.5 pH units, death usually results.  If one drop (0.05 mL) of 6 M hydrochloric acid is added to one liter of pure distilled water, the pH of the water will change by 3.6 pH units from pH 7.0 to pH 3.4.  If blood had no more resistance to pH change than distilled water, a little acid entering the blood stream (for instance, after drinking a glass of fruit juice) would quickly produce a fatal reaction.  Fortunately, blood is protected against pH changes.  Addition of a drop of 6 M hydrochloric acid to one liter of blood causes a pH change less than 0.05 units.  Human blood and many other biological, largely aqueous fluids are protected against serious pH changes by buffers. These liquids are said to be buffered.

Buffer action is not only needed in the acidic region but also in the basic region.  The pH of ocean water is usually in the range of 7.9 to 8.3.  Ocean water is buffered by reactions of aluminosilicates.  This slightly basic pH favors absorption of carbon dioxide from the air, regulates the carbon dioxide concentration in the atmosphere, and may reduce the greenhouse effect.

Buffers are substances that keep the pH of an aqueous solution fairly constant by reacting with hydronium ions from added acids or hydroxide ions from added bases.  Most biological buffers are rather complex.  The chemist uses simple buffers consisting of a weak acid plus its salt that is formed when it is reacted with a strong base (e.g., $CH_3COOH/CH_3COONa$) for buffer action in the acidic region or a weak base plus its salt that is formed when it is reacted with a strong acid (e.g., $NH_3/NH_4Cl$) for buffer action in the basic region.

This experiment deals with the ammonia/ammonium chloride buffer and its buffer capacity.

## CONCEPTS OF THE EXPERIMENT

The $NH_3/NH_4Cl$ buffer functions in the basic region.  When small amounts of acid (e.g., HCl) are added to a buffer solution, the pH changes very little because the hydronium ions from the acid react with the base (equation 1).  When a strong base is added (e.g., NaOH), the hydroxide ions

$$NH_3 + H^+ \rightarrow NH_4^+ \qquad\qquad (1)$$

will react with the ammonium ion to form water and ammonia (equation 2).  Because the buffer

$$NH_4^+ + OH^- \rightarrow NH_3 + H_2O \tag{2}$$

solution contains $NH_3$ as well as $NH_4^+$, the solution is protected from pH changes upon addition of bases as well as acids. Two questions have to be answered: How can a buffer solution be prepared that buffers at a specified pH? How much acid or base can such a buffer solution tolerate without excessive change of pH?

In a solution prepared by adding ammonium chloride, the salt of the strong acid HCl and the weak base ammonia, to an aqueous solution of ammonia, the following equilibria must be obeyed (equations 3, 4). The two constants $K_b$ and $K_a$ are related by the expression $K_a \times K_b = K_w$.

$$NH_3 + H_2O \rightleftharpoons NH_4^+ + OH^- \qquad K_b = \frac{\left[NH_4^+\right]\left[OH^-\right]}{\left[NH_3\right]} \tag{3}$$

$$NH_4^+ + H_2O \rightleftharpoons NH_3 + H_3O^+ \qquad K_a = \frac{\left[NH_3\right]\left[H^+\right]}{\left[NH_4^+\right]} \tag{4}$$

If we take the equation for $K_a$, rearrange the terms, take the logarithm, and introduce $pH = -\log[H^+]$ and $pK_a = -\log K_a$, equation 5 is obtained. Exactly the same equation is obtained from

$$[H^+] = K_a\frac{\left[NH_4^+\right]}{\left[NH_3\right]} \rightarrow \log[H^+] = \log K_a + \log\frac{\left[NH_4^+\right]}{\left[NH_3\right]} \rightarrow -\log[H^+] = -\log K_a - \log\frac{\left[NH_4^+\right]}{\left[NH_3\right]} \rightarrow$$

$$pH = pK_a - \log\frac{\left[NH_4^+\right]}{\left[NH_3\right]} \tag{5}$$

the expression (3) with consideration that $[H^+][OH^-] = K_w$ and $K_a K_b = K_w$. Equation 5 tells that the pH of an $NH_3/NH_4Cl$ buffer solution is determined by the concentration ratio $[NH_4^+]/[NH_3]$ and the equilibrium constant $K_a$. When the ratio $[NH_4^+]/[NH_3]$ is one, the pH of the buffer solution is given by the value for $pK_a$. The acid dissociation constant for the ammonium ion is 5.6 x $10^{-10}$ and $pK_a$ is 9.26. The pH of the buffer solution with $[NH_4^+] = [NH_3]$ is, therefore, 9.26 at any concentration. The pH of the buffer solution can be changed by varying the ratio of $[NH_4^+]/[NH_3]$. If this ratio is larger than 1 ($[NH_4^+]>[NH_3]$), the logarithm of this ratio will be positive, $-\log$ will be negative, and the pH will be lower than 9.26. Conversely, when $[NH_4^+]<[NH_3]$, the logarithm of the ratio will be negative, $-\log$ will be positive, and the pH larger than 9.26. For practical work, ratios of $[NH_4^+]/[NH_3]$ of 10 to 0.1 can be employed. The range determines the pH limits within which the ammonia/ammonium chloride buffer can be used as $pK_a \pm 1$ (between 8.26 and 10.26). Outside this region, the buffering action is very weak. If a solution must be buffered at another pH, then a salt/base pair with a $pK_a$ close to the desired pH must be chosen.

Equation 5, known as the Henderson-Hasselbach equation, is the fundamental relation describing buffer systems. An $NH_3/NH_4Cl$ buffer is prepared by adding a certain quantity of ammonium chloride to a solution of ammonia with an "analytical" concentration of $[NH_3]_a$ to produce a solution with an "analytical" concentration of $[NH_4Cl]_a$ or $[NH_4^+]_a$. These "analytical" concentrations are calculated from the grams of $NH_3$ and $NH_4Cl$ present in one liter of solution and the appropriate molecular masses neglecting the interactions of the ammonium ion with water (equation 4) and of ammonia with water (equation 3). A rigorous treatment must consider these interactions. Ammonia reacts to a very slight extent to form ammonium and hydroxide ions. This reaction, responsible for the basic character of ammonia solution, diminishes the ammonia concentration and increases the concentration of ammonium ion. The ammonium ions dissociate to a very slight extent into ammonia and hydronium ion increasing the ammonia concentration and decreasing the concentration of ammonium ions. At equilibrium the "analytical" concentrations of $NH_3$ and $NH_4^+$ will be different from the equilibrium concentrations. Because $K_a$ ($5.6 \times 10^{-10}$) and $K_b$ ($1.8 \times 10^{-5}$) are small, the differences between analytical and equilibrium concentrations are also small. Therefore, the analytical concentrations are used with the Henderson-Hasselbach equation. The error introduced through this considerable simplification is negligible with buffer systems in the pH range 3 to 11. Outside this range a much more complex equation that uses the equilibrium concentrations must be employed.

Every buffer system has a buffer capacity which measures the ability to "neutralize" added acid or base and keep the pH of the buffered solution almost constant. Any addition of acid or base to a buffered solution will somewhat change the pH; however, the buffer minimizes these changes and renders them frequently negligible.

In the $NH_3/NH_4Cl$ buffer, the hydrogen (hydronium) ions from the added acid are "neutralized" by the reaction with $NH_3$ to produce $NH_4^+$ ions. This reaction decreases the ammonia concentration, increases the concentration of ammonium ions, increases the ratio $[NH_4^+]/[NH_3]$, and – because of the logarithmic function – very slightly decreases the pH. How much acid can be added without serious change in pH will depend on the concentration of the $NH_3$, the species that "neutralizes" the acid. The buffer capacity BC is rigorously defined as the moles of added acid ($\Delta A$) required to change the pH ($\Delta pH$) by one unit (equation 6). The buffer capacities of solutions

$$BC = \frac{\Delta A \, (\text{moles})}{\Delta pH} = \frac{dA}{dpH} \text{ (differential form)} \qquad (6)$$

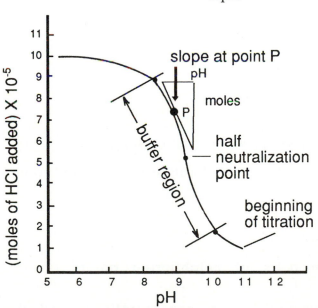

Figure 1: Titration curve for the titration of 5 mL 0.20 M $NH_3$ with 0.2 M HCl at a constant total volume of 10 mL.

containing $NH_3$ and $NH_4Cl$ at different ratios can best be deduced from a titration curve obtained by adding hydrochloric acid to a solution of ammonia and measuring the pH after each addition. In such a titration the total volume of the solution increases and a correction for dilution would have to be applied. This correction can be avoided when solutions of ammonia and hydrochloric acid are mixed and water is added to a constant total volume. When the measured pH (x-axis) of these solutions is plotted versus the moles of hydrochloric acid added (y-axis), the slope of the resulting curve at each point gives the buffering capacity (Fig. 1). The buffer capacity at the endpoint is about zero because the number of moles of acid added is very small and the pH change is large. The buffer capacity at the beginning of the titration is similarly small for the same reasons. The slope and the buffer capacity are largest at the half-neutralization point and decrease on either side of the half-neutralization point toward the beginning of the curve and toward the endpoint. Ideally, the slopes at points on such a curve are obtained as the first derivative of the curve "moles acid" versus "pH". However, a mathematical expression for this curve is not easily obtained. Therefore, the slopes are determined in this experiment by approximation. A series of solutions are prepared from 0.2 M ammonia, 0.2 M hydrochloric acid, and water. The pH's of these solutions are measured. The curve between two neighboring points is assumed to be linear. The slope (the buffering capacity) is calculated from the differences between the moles of acid added and the two pH values ($\Delta$moles/$\Delta$pH). The calculated buffer capacity is ascribed to a solution with a pH exactly between the two pH values used [(pH$^1$ + pH$^2$)/2] and between the moles of acid added [(moles$^1$ + moles$^2$)/2]. The following example using part of the titration curve should clarify this procedure:

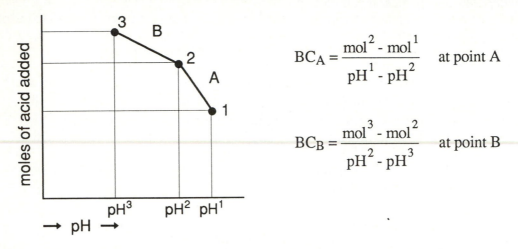

$$BC_A = \frac{mol^2 - mol^1}{pH^1 - pH^2} \quad \text{at point A}$$

$$BC_B = \frac{mol^3 - mol^2}{pH^2 - pH^3} \quad \text{at point B}$$

The buffer capacity has the units $mol(pH)^{-1}$. The concentrations of $NH_4^+$ and $NH_3$ at each point for which the buffer capacity was obtained (A, B, . . . ) must be calculated from the initial concentration of ammonia, the number of moles of acid added, and the total volume of the solution. The ammonium ion concentration is calculated from the moles of acid added with the consideration of the neutralization reaction on equation 1.

## ACTIVITIES

- Complete the PreLab Exercises before coming to the laboratory.

- Prepare solutions from 0.2 M ammonia, 0.2 M hydrochloric acid, and water required to construct a titration curve.

- Measure the pH of these solutions.

- Plot your data (moles acid added versus pH).

- Calculate the buffer capacities from pairs of neighboring points on the curve.

- Calculate ammonia concentrations, ammonium ion concentrations, and the ratios $[NH_4^+]/[NH_3]$ for each point for which buffer capacities were obtained.

- Plot the buffer capacities versus moles of acid added.

- Use the plots to select buffers of a certain buffer capacity and pH value.

## SAFETY

*Wear approved eye protection and exercise proper care in working with acids and bases. If any solution is spilled, clean it up immediately by first sponging up the solution, then rinsing the sponge and the area with water. In case of skin contact with reagents, wash quickly and thoroughly with water.*

## PROCEDURES

In this experiment you will work with one or more partners as assigned by your instructor. Every member of the team has to keep complete records and submit a separate Report Form. Plan the division of work carefully.

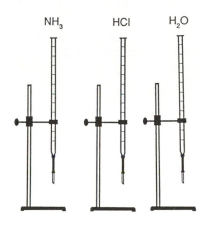

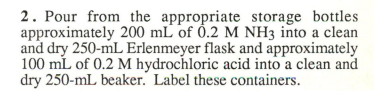

**1.** Obtain three 50-mL burets. Check the burets for cleanliness and proper delivery of liquid using distilled water. If necessary, clean the burets (Appendix D). Mount the burets on ringstands and label them "NH₃", "HCl", and "H₂O".

**2.** Pour from the appropriate storage bottles approximately 200 mL of 0.2 M $NH_3$ into a clean and dry 250-mL Erlenmeyer flask and approximately 100 mL of 0.2 M hydrochloric acid into a clean and dry 250-mL beaker. Label these containers.

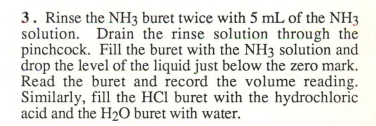

**3.** Rinse the NH₃ buret twice with 5 mL of the NH₃ solution. Drain the rinse solution through the pinchcock. Fill the buret with the NH₃ solution and drop the level of the liquid just below the zero mark. Read the buret and record the volume reading. Similarly, fill the HCl buret with the hydrochloric acid and the H₂O buret with water.

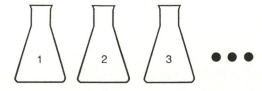

**4.** Collect five (or more) clean and dry containers (beakers, Erlenmeyer flasks) that can hold 50 or more mL. If necessary, clean and dry the containers.

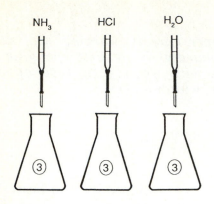

**5.** Prepare the solutions specified in the table on the Report Form in batches of five or more by draining the required volumes of $NH_3$ solution, HCl solution, and water into an appropriately marked container. Shake the 20 mL of solution in the container for proper mixing.

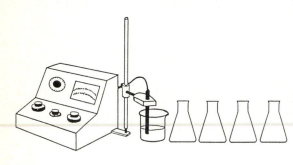

**6.** When a series of solutions is ready, measure the pH of these solutions with a pH meter. If necessary, calibrate the pH meter according to instructions provided by your instructor. Follow the directions made available to you for the proper use of the pH meter (also see Investigation 24). Record the measured pH values.

**7.** Discard the solutions once you have recorded the pH values. Clean and dry the containers. Prepare the next set of solutions and measure their pH's. Repeat this procedure until all of the solutions have been prepared and their pH values determined.

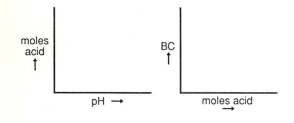

**8.** Prepare a plot "moles of acid added" versus "pH". Calculate the buffer capacities (14 values) from each pair of neighboring points, $[NH_4^+]$ and $[NH_3]$ for the midpoint between each neighboring pair of pH values and the ratios $[NH_4^+]/[[NH_3]$. Plot the buffering capacities versus the moles of acid added.

**9.** Complete the Report Form.

# PRELAB EXERCISES

## INVESTIGATION 25: THE $NH_3/NH_4Cl$ BUFFER SYSTEM

Name:_____

Instructor:_____    ID No.:_____

Course/Section:_____    Date:_____

1. Define buffer capacity in a sentence and by an equation.

2. Give examples for:
   A weak acid and a salt of this weak acid with a strong base.

   A weak base and a salt of this weak base with a strong acid.

   A salt from a strong acid and a strong base.

   A salt from a weak acid and a weak base.

3. A buffer solution with a $[NH_4^+]/[[NH_3]$ ratio of 1 is diluted 10-fold. Does the pH of the solution change on dilution? Use equation 5 to find the answer to this question.

4. A solution was prepared by mixing 4.00 mL 0.2 M $NH_3$, 1.00 mL of 0.2 M HCl, and 5.00 mL of water. Calculate the concentration of $NH_3$ and $NH_4^+$ in the final solution.

5. A solution prepared by adding "a" moles of acid to an ammonia solution had a pH of 8.0. When "b" (b<a) moles of acid were added to another sample of this ammonia solution, the pH was 8.2. Calculate the buffer capacity.

# REPORT FORM

## INVESTIGATION 25:  THE NH3/NH4Cl BUFFER SYSTEM

Name: _____    ID No.: _____

Instructor: _____    Course/Section:_____

Partner's Name (if applicable): _____    Date: _____

Show calculations on attached pages.

1.  Complete the table

| Solution number | mL 0.2M $NH_3$ | mL 0.2M HCl | mL $H_2O$ | $[NH_3]$ | $[NH_4^+]$ | pH | $\Delta pH$ | mol HCl added | $\dfrac{\Delta mol\ HCl}{\Delta pH}$ |
|---|---|---|---|---|---|---|---|---|---|
| 1 | 10 | 0 | 10.0 | | | | – | 0 | 0 |
| 2 | 10 | 0.5 | 9.5 | | | | | | |
| 3 | 10 | 1.0 | 9.0 | | | | | | |
| 4 | 10 | 2.0 | 8.0 | | | | | | |
| 5 | 10 | 4.0 | 6.0 | | | | | | |
| 6 | 10 | 4.5 | 5.5 | | | | | | |
| 7 | 10 | 5.0 | 5.0 | | | | | | |
| 8 | 10 | 5.5 | 4.5 | | | | | | |
| 9 | 10 | 6.0 | 4.0 | | | | | | |
| 10 | 10 | 7.0 | 3.0 | | | | | | |
| 11 | 10 | 8.0 | 2.0 | | | | | | |
| 12 | 10 | 8.5 | 1.5 | | | | | | |
| 13 | 10 | 9.0 | 1.0 | | | | | | |
| 14 | 10 | 9.5 | 0.5 | | | | | | |
| 15 | 10 | 10 | 0 | | | | | | |

2.  Prepare a plot of "moles of acid added" versus "pH".  Attach the graph.

3.  Prepare a plot of "buffer capacity" versus "moles of acid added" using the numbers corresponding to the midpoint between the pairs of neighboring experimental values that were used in the calculations.  For each point entered into the "BC - moles acid" graph calculate the ratio $[NH_4^+]/[NH_3]$.  Write these values next to the pertinent points in this graph.  Attach the graph.

4.  At which $[NH_4^+]/[NH_3]$ ratio is the buffer capacity at a maximum?

5.  You need to prepare an ammonia/ammonium chloride buffer with a pH of 8.5. Use the results of this experiment to write a recipe for the preparation of 500 mL of such a buffer. Find the buffer capacity for this buffer.

6.  Aniline, $C_6H_5NH_2$, has a $K_b$ of $3.8 \times 10^{-10}$ and benzoic acid has an acid dissociation constant $K_a = 6.6 \times 10^{-5}$. Which substances are needed to make a buffer with aniline, with benzoic acid? In which pH region could these buffers be used?

Date _____ Signature _____

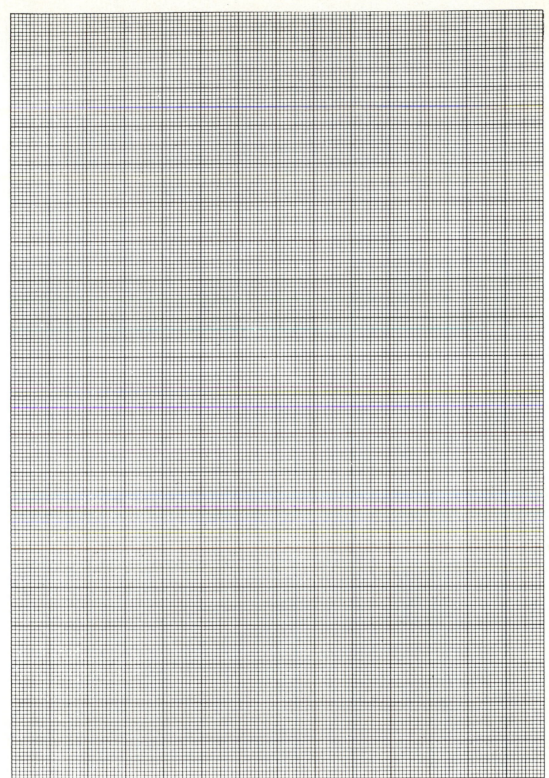

Moles of Acid Added

pH

Buffer Capacity

Moles of Acid Added

# Investigation 26

## ENTHALPY OF DISSOCIATION OF ACETIC ACID

## INTRODUCTION

Chemical reactions transforming starting materials into products either liberate heat (exothermic reactions) or absorb heat (endothermic reactions). When a reaction is carried out under constant pressure – for instance, in an open beaker at barometric pressure – the heat associated with such a reaction is known as enthalpy. The symbol $\Delta H$ is used for enthalpy. $\Delta H$ values are generally tabulated for the formation of one mole of product under one atmosphere pressure at 25°C. Enthalpies obtained under these conditions are known as standard molar enthalpies. Enthalpies must be determined experimentally by performing reactions under well defined conditions in a calorimeter. Enthalpies are very useful thermodynamic data: they are needed in the design of furnaces and gas-fired boilers, in the development of industrial chemical processes, and in the prediction of the spontaneity of chemical reactions. Enthalpy alone does not determine whether or not a mixture of reagents will spontaneously react; entropy also has to be considered. With a large number of experimentally determined enthalpies available, one can calculate enthalpies of many other reactions that cannot be measured – for a variety of reasons – in a calorimeter. The basis for these calculations is the experience that a perpetual motion machine – a gadget capable of generating energy from nothing – cannot be built. This firm principle associated with the first law of thermodynamics has never been violated in spite of numerous past and current attempts. This principle applied to chemical reactions is known as Hess' Law. In this experiment you will use Hess' Law to calculate the enthalpy of dissociation of acetic acid from measured enthalpies of neutralization of hydrochloric acid and acetic acid with sodium hydroxide.

## CONCEPTS OF THE EXPERIMENT

When one liter of 1.0 M hydrochloric acid reacts with one liter of 1.0 M sodium hydroxide solution, sodium chloride is formed in a neutralization reaction that is associated with a molar enthalpy of neutralization, ${}^{S}\Delta H_{M}^{N}$ (equation 1). Because hydrochloric acid is a strong acid and

$$H^{+}(aq) + Cl^{-}(aq) + Na^{+}(aq) + OH^{-}(aq) \rightarrow H_2O + Na^{+}(aq) + Cl^{-}(aq) \qquad (1)$$

sodium hydroxide is a strong base, these electrolytes are completely dissociated into anions and cations in aqueous solution. The ions are hydrated (surrounded by water molecules). This fact is expressed in equation 1 by the parenthetical (aq) following the formula of each ion. Sodium chloride, the salt produced in the neutralization reaction, is also a strong electrolyte and, therefore, completely dissociated in aqueous solution. The states of $Cl^{-}$ and $Na^{+}$ are the same before and after the reaction; these ions do not participate in the reaction. The enthalpy associated with this reaction must then come from the combination of $H^{+}$ and $OH^{-}$ to form undissociated $H_2O$. Experiments have shown that ${}^{S}\Delta H_{M}^{N}$ has almost the same value for all neutralization reactions between strong acids and bases.

When acetic acid, a weak acid ($pk_a$ 4.75), is similarly reacted with sodium hydroxide, neutralization takes place also with liberation of heat ${}^{W}\Delta H_{M}^{N}$ (equation 2). In contrast to the solution of hydrochloric acid, a 1.0 M solution of acetic acid has almost all of the acetic acid

$$CH_3COOH(aq) + Na^+ + OH^- \rightarrow CH_3COO^-(aq) + Na^+(aq) + H_2O \qquad (2)$$

molecules in undissociated form as indicated by the formula $CH_3COOH$ (aq) in equation 2. The enthalpy associated with this reaction of a weak acid, $^W\Delta H_M^N$, is smaller than for the reaction of a strong acid, $^S\Delta H_M^N$, because some energy is used to tear $H^+(aq)$ from $CH_3COOH$. The neutralization of acetic acid can be separated on paper into two steps: the dissociation of acetic acid (equation 3) followed by the neutralization of the generated $H^+(aq)$ by $OH^-(aq)$. The second step is the same as in the neutralization of a strong acid with a strong base (equation 1) and must have the same enthalpy, $^S\Delta H_M^N$. The dissociation of the acetic acid is associated with the molar enthalpy of dissociation, $^W\Delta H_M^D$.

$$CH_3COOH(aq) \rightarrow CH_3COO^-(aq) + H^+(aq) \qquad (3)$$

The stepwise reaction and the direct neutralization define a cyclic process.

$$CH_3COO^-(aq) + H^+(aq)$$

$$^W\Delta H_M^D \nearrow \qquad \qquad ^S\Delta H_M^N \searrow \quad + OH^-$$

$$+ OH^-$$

$$CH_3COOH \xrightarrow{\quad\quad\quad\quad} CH_3COO^-(aq) + H_2O$$

$$^W\Delta H_M^N$$

The starting materials and the products are the same regardless of the path taken. Hess' Law demands in this situation that the enthalpies associated with the two paths must be equal (equation 4). Two of these enthalpies $\left( ^S\Delta H_M^N, {}^W\Delta H_M^N \right)$ can be determined by experiment and the third enthalpy can then be calculated. If equation 4 does not hold, a perpetual motion machine could be constructed. If, for instance, the direct neutralization of acetic acid (equation 2) were to produce

$$^W\Delta H_M^D + {}^S\Delta H_M^N = {}^W\Delta H_M^N \qquad (4)$$

more heat per mole than would be needed to reverse the neutralization by the two-step path, heat would be left over after the cycle had regenerated the starting material. Repeating this cyclic reaction over and over again would generate an unlimited amount of heat without energy input. This heat produced from "nothing" could be used to generate steam, drive a steam turbine, and make electricity that could power an electric car. Because nobody has yet succeeded in such a "heat-from-nothing" endeavor, Hess' Law (equation 4) still holds and can be used with confidence in this experiment.

To distinguish an exothermic reaction from an endothermic reaction, the $\Delta H$ values must have a sign. The sign is determined by looking at the reaction from the participants' (the chemicals', the systems') point of view. In an exothermic reaction heat is given up by the system to the surroundings; heat is lost by the substances; a loss has negative aspects and the value of $\Delta H$ for an exothermic reaction carries a minus sign. In an endothermic reaction heat flows from the surroundings into the system; heat is gained by the substances; a gain has positive aspects and the value of $\Delta H$ for an endothermic reaction carries a plus sign.

The enthalpies are determined by performing reactions in a calorimeter. In the ideal case, a calorimeter is a device that completely insulates the reacting solutions from the surroundings and forces all the heat liberated in the reaction to raise the temperature of the aqueous solution. The temperature before the reaction and the temperature after the reaction is completed can be measured and the temperature rise calculated. This calculation (equation 5) is based on the First Law of Thermodynamics: the heat generated by the reaction must be equal to the heat gained by the solution. If the volume of the solution is known, the density of 1.02 g mL$^{-1}$ must be used to calculate the mass.

$$\Delta H = m_s C_s(T_{BR} - T_{AR}) \tag{5}$$

$\Delta H$:    enthalpy of the reaction performed with known quantities of reagents
$m_s$:    mass of solution in calorimeter
$C_s$:    specific heat capacity of the solution (3.97 J g$^{-1}$ deg$^{-1}$)
$T_{BR}$:    temperature of solution before reaction
$T_{AR}$:    temperature of solution after instantaneous reaction

Unfortunately, this ideal situation cannot be realized. The calorimeter does not completely prevent the loss of heat to the surroundings. In addition, some of the heat is lost by warming the calorimeter. These heat losses are taken into account through a calorimeter constant "K". The calorimeter constant is the heat taken up by the calorimeter and lost to the surroundings per degree temperature rise. This constant with the units J deg$^{-1}$ is determined experimentally by mixing a known mass $\left(m_W^R\right)$ of room temperature water ($T_R$) with a known mass $\left(m_W^B\right)$ of boiling water ($T_B$) and determining the temperature of the resulting mixture ($T_M$). The heat lost by the boiling water must be equal to the sum of the heat gained by the room-temperature water, the heat gained by the calorimeter, and the heat gained by the surroundings (equation 6). This equation can be solved for the calorimeter constant "K" (equation 7). The ideal equation 5 must now be modified

$$m_W^B C_W(T_B - T_M) = m_W^R C_W(T_M - T_R) + K(T_M - T_R) \tag{6}$$

$C_W$: specific heat of water (4.18 J g$^{-1}$ deg$^{-1}$)

$$K = \frac{m_W^B C_W(T_B - T_M) - m_W^R C_W(T_M - T_R)}{(T_M - T_R)} \tag{7}$$

by introducing the calorimeter constant. The heat set free by the reaction must now be equal to the heat gained by the solution plus the heat taken up by the calorimeter and the surroundings (equation 8). All the quantities on the right hand side of this equation are known at the end of the experiment and the enthalpy of the reaction can be calculated.

$$\Delta H = m_s C_s(T_{BR} - T_{AR}) + K(T_{BR} - T_{AR}) \tag{8}$$

Two more details need to be considered. First, the boiling point of water increases with increasing atmospheric pressure. The temperature of the boiling water, $T_B$, must be taken from a graph of boiling temperature versus pressure. The pressure in the laboratory is read from a barometer. Secondly, it is not possible to directly measure the temperature that would be reached upon instantaneous mixing of hot and cold water or instantaneous liberation of the enthalpy of a reaction. These processes and the response of the measuring devices take time during which heat is lost to the surroundings. Therefore, this "instantaneous" temperature is found by extrapolation.

The temperature of the water or the solution at room temperature is measured four times at 40-second intervals. Immediately after the fourth reading the hot water or the base is added, the

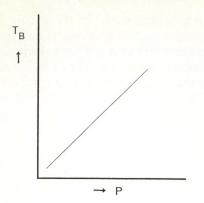

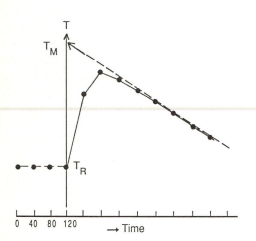

solutions are mixed and the temperatures read again at the same intervals. The measured temperatures are plotted versus time. The linear range in the temperature-time curve is extrapolated to the temperature axis. The "instantaneous" temperature, $T_M$, is obtained from the intersection of the straight extrapolation line with the temperature axis.

Equation 4 is given in terms of one mole of product. In these experiments, much smaller amounts of acids and bases are used (50 mL of 1.0 M solutions). The number of moles actually used must be calculated from the volumes and the concentrations. All other equations apply to these smaller quantities. The calculated enthalpies for the smaller quantities must then be converted to enthalpies for one mole.

In these experiments triple-beam balances, thermistors, thermometers, and graduated cylinders are employed. Each of these measuring instruments gives results associated with a certain precision expressed as a relative standard deviation. Each number obtained in this experiment by direct measurement must be reported with its relative standard deviation. These relative standard deviations can be obtained from earlier experiments, by repeating measurements, or from your instructor. The final result, the molar enthalpy of dissociation of acetic acid, cannot be more precise than the value with the largest relative standard deviation (Appendix H).

# ACTIVITIES

- Complete the PreLab Exercises before coming to the laboratory.

- Prepare a thermistor-computer system or a thermometer-water system for temperature/time measurements.

- Construct a calorimeter from two styrofoam cups.

- Find the calorimeter constant by mixing room-temperature and boiling water in the calorimeter and determining the temperature of the water mixture.

- Mix equimolar quantities of hydrochloric acid and sodium hydroxide solution and determine the "instantaneous" temperature at the time of mixing.

- Repeat this experiment with solutions of acetic acid and sodium hydroxide.

- Calculate the molar enthalpy of the dissociation of acetic acid.

- Complete the Report Form.

## SAFETY

*Wear approved eye protection. Exercise great care in handling acids and bases. Avoid spilling these solutions. Should a spill occur, soak up the spill with a sponge and rinse the spill area with a sponge full of tap water. Protect your hands with gloves during the clean-up operation.*

## PROCEDURES

For this experiment form a team with one or more students as directed by your instructor. Plan the division of tasks among the team members for efficient use of equipment, timely collection of data, and avoidance of jams at the balances. Every team member must submit a completed Report Form. The temperature data needed in this experiment can be collected either with a thermometer and a watch showing seconds or with a thermistor (a temperature-sensing device) connected to a computer in which a program for the collection of temperatures at various time intervals resides. Your instructor may ask certain teams to carry out the experiment with the help of the computer/thermistor, and assign others to use the watch/thermometer. If time permits, the experiment should be repeated with the other group's method. The directions in bold type are for the computer method. The directions in italics are for the thermometer/watch method. The directions printed in normal type are needed for both methods.

## A. PREPARATION OF THE EQUIPMENT

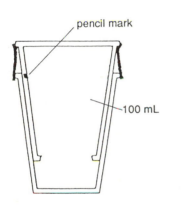

pencil mark

100 mL

**A-1.** Construct a calorimeter by cutting the bottom from a styrofoam cup. Push the bottomless cup firmly into an intact cup. Weigh the cup assembly on the triple-beam balance to ±0.01 g. Record the mass and the relative standard deviation associated with this mass. With a graduated cylinder measure 100 mL of tap water into the cup assembly. Mark the liquid level inside the cup with a pencil line. Empty the calorimeter and dry it with a towel.

**A-2.** With a 100-mL graduated cylinder measure 200 mL of distilled water into a clean 250-mL beaker. Set the beaker aside to allow the water to reach room temperature.

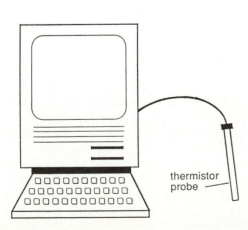

thermistor probe

**A-3. Turn on the computer that has a thermistor connected to one of the input ports. Check that the number on the disk on which the program for this experiment is stored matches the number on the computer. If the numbers are the same, insert the disk into DISK DRIVE 1. (The program being described is "General Lab Interfacing" from Project SERAPHIM, Chem. Dept., Univ. of Wisc.) The computer will boot when the disk is inserted. When the booting operation is completed, press RETURN. Repeat this procedure until you see on the monitor the option to select "General**

Laboratory Interfacings" or "Alternate Thermistor Program". Select "General Laboratory Interfacing" by pressing the direction keys (← or →) until "General Laboratory Interfacing" is highlighted.

Then press RETURN.

A-4. General information is now displayed on the monitor. Disregard this information. Press any key to bring the menu titled "General Laboratory Interfacing" to the screen. Select from the menu the "Thermistor" program by entering the number corresponding to the "Thermistor" program.

Another menu with the title "Thermistor Project Seraphim" appears on the screen. Press return twice and then select "Continuous Temperature Sensing" by entering the number corresponding to "Continuous Temperature Sensing" and pressing RETURN.

A-5. The thermistor in contact with laboratory air should cause the room temperature to be displayed on the monitor. Compare the thermistor temperature with the temperature read from a thermometer. If these temperatures are not the same within ±2°C, check with your instructor. Hold the thermistor between your fingers. The temperature on the screen should be between 36 and 37°C. If the thermistor has passed these tests, press Q to return to the menu "Thermistor Project Seraphim". If the thermistor did not pass the tests, contact your instructor.

## B. DETERMINATION OF THE CALORIMETER CONSTANT

B-1. With a 100-mL graduated cylinder measure 60 mL of the distilled water from the set aside 250-mL beaker into the cup assembly. Weigh the cup assembly containing the water on the triple-beam balance to ±0.01 g. Record the mass.

B-2. Pour approximately 80 mL of distilled water into a 125-mL Erlenmeyer flask. Heat the flask with a burner to bring the water to a gentle boil. Keep the water boiling gently.

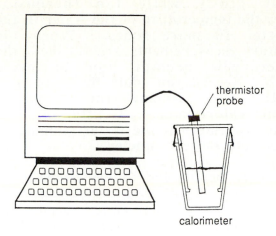

thermistor probe

calorimeter

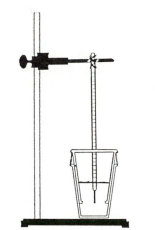

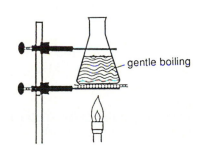

gentle boiling

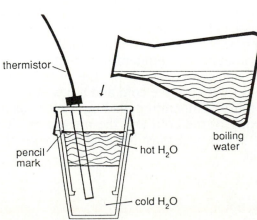

thermistor

pencil mark

hot H$_2$O

cold H$_2$O

boiling water

**B-3.** Place the thermistor in the water in the cup assembly. From the menu "Thermistor Project Seraphim" select "Sample Temperatures" by entering the number corresponding to "Sample Temperatures" and pressing RETURN.

The computer will prompt you to enter the time interval in seconds at which you would like to have the temperature determined. Enter 40 and press RETURN. The next screen will provide the option to select "Tabular Display" or "Graphic Display". Select "Tabular Display" by entering the number corresponding to "Tabular Display".

The next screen will ask you to "Enter the Number of Samples Desired". Enter 14 and press RETURN. The next screen will prompt you to "Press Enter When Ready to Start Sampling".

**B-3'.** *Use a string to tie a thermometer (0-120°C range, 0.1° divisions) to a ring clamp fixed to a ringstand at an appropriate height. Place the calorimeter on the base of the ringstand and adjust the ring clamp to have the bulb of the thermometer completely immersed in the water. One team member keeps time by calling out at 40 second intervals. Another team member reads the thermometer whenever time is called and announces the temperature reading. Time and temperatures are recorded for three readings.*

**B-4.** The thermistor is in the calorimeter water ready to measure its temperature every 40 seconds. Protect your thumb and middle finger with short pieces of lengthwise-cut rubber tubing. Make sure that the water in the 125-mL Erlenmeyer flask is boiling gently. Then press RETURN and allow three temperature readings to accumulate. As soon as the third reading appears on the screen, take the 125-mL Erlenmeyer flask with the boiling water by its neck. When the fourth temperature appears on the screen, quickly pour as much boiling water (~40 mL) from the Erlenmeyer flask into the calorimeter as needed to bring the water level to the pencil mark. Swirl the water in the cup assembly gently while the thermistor measures the temperature of the water in the calorimeter 10 more times. When the temperature measurements are finished, the computer will prompt you to press RETURN. Record your data first and then press RETURN. Remove the thermistor from the calorimeter. Gently tap the thermistor to allow all the water clinging to it to drop into the calorimeter.

**B-4'.** *Shortly before the fourth temperature reading, take the 125-mL Erlenmeyer flask with the gently boiling water (fingers protected from the heat with pieces of rubber tubing) by its neck. Immediately after the fourth temperature reading, pour as much boiling water (~40 mL) from the Erlenmeyer flask into the calorimeter as needed to bring the water level to the pencil mark. Move the cup or the ringstand base in a slow circular motion to mix the water. Read and announce the temperature every 40 seconds until 10 more readings have been obtained. Then remove the thermometer from the calorimeter. Record the time/temperature pairs.*

**B-5.** The prompts "Enter S to Save the Data" and "Press RETURN to Return to Main Menu" will appear on the screen. Press RETURN. There is no need to save the data if you have recorded them in your notebook.

**B-6.** Take the calorimeter with the water to the triple-beam balance and weigh it to ±0.01 g. Record the mass.

**B-7.** Empty the water from the calorimeter into the sink and dry the calorimeter with a towel.

**B-8.** The calorimeter constant can be calculated from the masses of the room-temperature and boiling water, the temperature of the room-temperature water in the calorimeter, the temperature of the boiling water, and the temperature of the water after instantaneous mixing of the cold and hot water. The time of the fourth temperature reading is taken as the time of mixing of the cold and the hot water. The temperature at this time must be obtained by graphic extrapolation.

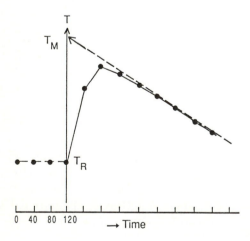

# C. ENTHALAPY OF NEUTRALIZATION OF 1.0 M HYDROCHLORIC ACID BY 1.0 M SODIUM HYDROXIDE SOLUTION.

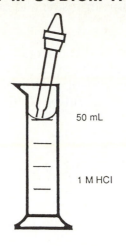

50 mL

1 M HCl

**C-1.** Take a clean and dry graduated cylinder (50-mL or 100-mL) to the 1.0 M hydrochloric acid storage bottle. Cautiously pour the acid into the graduated cylinder until the liquid level is just above the 50-mL mark. With a clean dropper remove some of the acid until the meniscus touches the 50-mL mark. Place the excess acid in the dropper into the designated waste container. Pour the 50 mL of acid into the cup assembly. Make sure that all the liquid is transferred.

**C-2.** Into another clean and dry graduated cylinder pour 50 mL of 1.0 M sodium hydroxide solution as described above. The hydrochloric acid and the sodium hydroxide solution must have the same temperature. This condition is fulfilled when the storage bottles have been in the laboratory for several hours.

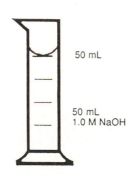

50 mL acid

**C-3.** Prepare the computer (Section A), place the thermistor into the hydrochloric acid in the calorimeter, and collect three temperatures of the acid at 40-second intervals (Procedure B-4). Immediately after the fourth reading, pour the 50 mL of sodium hydroxide solution into the calorimeter, gently swirl the solution, and allow the computer to collect ten additional temperatures.

50 mL

50 mL
1.0 M NaOH

**C-3'.** *Suspend a thermometer in the acid in the calorimeter, take three temperature readings at 40-second intervals, and immediately after the fourth reading, add the base. Swirl the solution gently and take 10 more temperature readings.*

**C-4.** When the computer has accumulated all the required temperature readings, record the data in your notebook.

**C-5.** Empty the salt solution from the calorimeter into the sink. Rinse the calorimeter with distilled water. Dry the calorimeter inside and outside with a towel.

**C-6.** Use the discussion in the section "Concepts of the Experiment" to guide you in calculating the molar enthalpy of neutralization of hydrochloric acid by sodium hydroxide. Pay attention to the precision of your results.

## D. ENTHALPY OF NEUTRALIZATION OF 1.0 M ACETIC ACID BY 1.0 M SODIUM HYDROXIDE SOLUTION

**D-1.** Repeat Procedures C-1 through C-5 using 50 mL of 1.0 M acetic acid in the place of the HCl solution.

**D-2.** Use the discussion in the section "Concepts of the Experiment" to guide you in calculating the molar enthalpy of neutralization of acetic acid by sodium hydroxide and then the molar enthalpy of dissociation of acetic acid.

# PRELAB EXERCISES

**INVESTIGATION 26: ENTHALPY OF DISSOCIATION OF ACETIC ACID**

Name:_____

Instructor:_____       ID No.:_____

Course/Section:_____       Date:_____

1. Write net ionic equations for the neutralization of a strong acid with a strong base, a weak acid with a strong base, and a weak base with a strong acid.

2. Define Hess' Law and the First Law of Thermodynamics.

3. The following data were collected in a determination of a calorimeter constant: 30.0 g water at 25.0° mixed with 30.0 g of water at 75.0°, extrapolated instantaneous temperature at mixing was 48.0°. Calculate the calorimeter constant K.

(over)

4. For an exothermic reaction carried out in a calorimetric experiment the following temperature data were collected at 1-minute intervals with the exothermic reaction beginning at the fourth temperature reading: 22.0°, 22.1°, 21.9°, 22.0°, 26.7°, 29.1°, 30.5°, 30.4°, 30.2°, 30.0°, 29.8°, 29.6°, 29.4°, 29.3°. Find the temperature expected for instantaneous reaction.

5. How many moles of hydronium ions were converted to water when 40.0 mL of 1.0 M hydrochloric acid were neutralized by sodium hydroxide?

6. The enthalpies can be measured for all reactions but one (dashed arrow) in the following cycle:

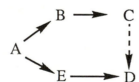

Calculate the molar enthalpy of reaction $\Delta H_{CD}$, for the step C→D in terms of the experimentally determinable enthalpies of the other steps.

# REPORT FORM

## INVESTIGATION 26: ENTHALPY OF DISSOCIATION OF ACETIC ACID

Instructor:_____     Name: _____

Course/Section:_____     ID No.: _____

Partner(s)name(s):_____     Date:_____

---

A.  Test readings on the thermistor-computer system:
    Thermistor in air:      _____°C     Thermometer in air (room):    _____°C
    Thermistor between fingers:    _____°C
                 Report all data with relative standard deviations.

B.  Determination of the Calorimeter Constant "K"
    Barometric Pressure:  _____             Boiling Point of Water:  _____
    Mass of Calorimeter:  _____ ± ___ g     Temperature of Room Temp. Water
    Mass of Calorimeter and                    at time of mixing:  _____°C
       Room Temp. Water:_____ ± ___ g       Extrapolated temperature of
    Mass of Calorimeter + Room Temp.           mixture:  _____°C
       + Hot Water:  _____ ± ___ g          $\Delta T_{Boiling}$: _____°C  $\Delta T_{Room\ Temp.}$: _____°C
    Mass of Room Temp. Water:  _____ ± ___ g  Mass of Boiling Water:  _____ ± ___ g

    Time/Temperature Data: 0 s ____°C; 40 s ____°C; 80 s ____°C; 120 s ____°C;
       160 s ____°C; 200 s ____°C; 240 s ____°C; 280 s ____°C; 320 s ____°C; 360 s ____°C;
       400 s ____°C; 440 s ___°C; 480 s ____°C; 520 s ____°C; 560 s ____°C.

    Heat lost by boiling water: _____ ± __ J    Heat gained by room temp. water: ___ ± __ J
    Heat lost to calorimeter and surroundings: _____ ± ___ J
    Calorimeter Constant "K": _____ ± ___ J deg$^{-1}$
    Attach calculations and time/temperature graph.

C.  Enthalpy of Reaction for HCl(aq) + NaOH (aq)
    Volume of 1.0 M HCl: ____ ± ___ mL      Mass of solution:____ ± ___ g;
    Moles of HCl: ____ ± ___ mol            Volume of 1.0 M NaOH: ____ ± ___ mL
    Mass of solution:____ ± ___ g           Moles of NaOH: ____ ± ___ mol
    Total volume of solution: _____ ± __ mL Total mass of solution: _____ ± ___ g
    Temperature of HCl solution at time of mixing: _____ ± ___ °C
    Extrapolated Temperature at time of mixing for instantaneous neutralization:_____ ± ___ °C

    Time/Temperature Data: 0 s ____°C; 40 s ____°C; 80 s ____°C; 120 s ____°C;
       160 s ____°C; 200 s ____°C; 240 s ____°C; 280 s ____°C; 320 s ____°C; 360 s ____°C;
       400 s ____°C; 440 s ___°C; 480 s ____°C; 520 s ____°C; 560 s ____°C.

    Heat gained by NaCl solution: _____ ± __ J    Heat gained by cal. & surroundings: ___ ± __ J
    Heat liberated by the reaction: _____ ± ___ J

    $\Delta H$ formation of one mole of $H_2O$: ____ ± ___ J mol$^{-1}$
    Attach calculations and time/temperature graph.

D. Enthalpy for $CH_3COOH(aq) + NaOH\ (aq)$

　Volume of 1.0 M $CH_3COOH$: ___ ± ___ mL  Mass of solution: ___ ± ___ g

　Moles of $CH_3COOH$: _____mol　　　　Volume of 1.0 M NaOH: _____ ± ___ mL

　Mass of solution: _____ ± ____ g　　　　Moles of NaOH: _____mol

　Total volume of solution: _____ ± ___ mL　Total mass of solution: _____ ± ___ g

　Temperature of $CH_3COOH$ solution at time of mixing: _____ ± ___ °C

　Extrapolated temperature at time of mixing for instantaneous neutralization: _____ ± ___ °C

　Time/Temperature Data: 0 s ____°C; 40 s ____°C; 80 s ____°C; 120 s ____°C; 160 s ____°C; 200 s ____°C; 240 s ____°C; 280 s ____°C; 320 s ____°C; 360 s ____°C; 400 s ____°C; 440 s ___°C; 480 s ____°C; 520 s ____°C; 560 s ____°C.

　Heat gained by $CH_3COONa$ solution: _____ ± ___ J

　Heat gained by calorimeter and surroundings: _____ ± ___ J

　Heat liberated by reaction: _____ ± ___ J

　Heat liberated by the neutralization of one mole of $CH_3COOH$: _____ ± ___ J

E. Molar Enthalpy of Dissociation of $CH_3COOH$

　Molar Enthalpy for $H^+ + OH^-$: _____ ± ____ J mol$^{-1}$

　Molar Enthalpy for $CH_3COOH + OH^-$: _____ ± _____ J mol$^{-1}$

　Molar Enthalpy for $CH_3COOH \rightarrow CH_3COO^- + H^+$: _____ ± ____ J mol$^{-1}$

　Is the dissociation of acetic acid exothermic or endothermic?

ATTACH YOUR GRAPHS!

Date_____　　Signature_____

# Investigation 27

## ELECTROCHEMICAL GALVANIC CELLS

## INTRODUCTION

Electricity pervades industrial societies. The flick of a switch lights up homes, streets, and entire cities. Electrical energy drives motors that make it possible to start automobiles with little effort, run air conditioning systems, keep the inside of refrigerators cool, drive tools, and power almost pollution-free vehicles. The ingenious use of electricity brought us electronic calculators and computing machines ranging from personal computers to powerful main-frame computers with huge memories and lightening operating speed. In this list of electrical appliances the low-tech flashlight should not be forgotten because it draws its electricity from the oldest sources of direct electric current.

A controllable source of direct electric current (DC) has only been available for the past two hundred years. Luigi Galvani, after whom the galvanic cells are named, observed in 1791 the contraction of the muscles from frog legs when the two ends of the tissues were in contact with two different metals. In this experiment the frog tissue served as the aqueous solution that conducts electricity and as an indicator (through the contraction) of flowing electricity. The presence of two different metals in such arrangements was soon recognized as a necessary condition for the electric phenomena to appear. Allesandro Volta, after whom the unit of the electric potential (volt) is named, began to experiment with stacks consisting of alternating disks of metals such as silver and zinc that were separated by disks of absorbent paper saturated with a salt solution. These Voltaic stacks delivered a constant direct current when wires were connected to the two ends of such a stack. These Voltaic stacks are the forerunners of the batteries that are now used in flashlights, cars, watches, and calculators.

An important characteristic of batteries is the electric potential difference measured in volts between the electrodes of a galvanic cell. This potential difference is largely determined by the nature of the metals that serve as electrodes. Physicists and chemists constructed galvanic cells (batteries) and measured the potential differences characteristic of a large number of metal combinations. From such measurements a list of standard electrode potentials was developed. The potential of the standard hydrogen electrode was arbitrarily assigned a value of zero volt. The standard electrode potential is measured when a metal electrode is dipped into an aqueous solution containing one mole of a salt of this metal per liter and is connected via a voltmeter to a hydrogen electrode (a platinum wire surrounded by bubbles of hydrogen gas at one atmosphere pressure) dipped into an aqueous solution with a hydronium ion concentration of one mole per liter. A list of such standard electrode potentials allows the standard cell potential of any galvanic cell to be calculated. This list is also useful in estimating the reduction and oxidation characteristics of elements and their compounds.

This experiment was designed to provide the opportunity to gain experience in electrochemistry by constructing several galvanic cells, measuring their potentials, generating a short list of electrode potentials, and applying a potential measurement to the determination of the formation constant of the copper complex $[Cu(NH_3)_4]^{2+}$.

## CONCEPTS OF THE EXPERIMENT

When a strip of copper is placed into a 0.1 M aqueous solution of $CuSO_4$, neither the solution nor the copper strip changes visually. However, changes do take place on an atomic level. In a very simplified way, copper metal consists of $Cu^{2+}$ ions embedded in a sea of electrons. The piece of copper is electrically neutral because for each positively charged $Cu^{2+}$ there are two negatively charged electrons present. In contact with water, the copper ions show a certain tendency to leave the metal and move into the aqueous phase. This "dissolution" of a copper ion is promoted by the energetically favorable formation of a shell of water molecules around the copper ion. When copper ions leave the electrode, the electrode acquires a negative charge because the negatively

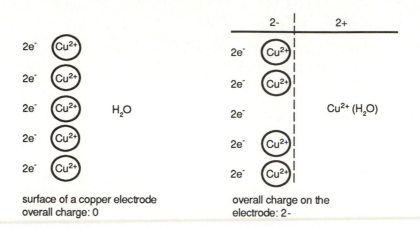

charged electrons formerly associated with the $Cu^{2+}$ ions remain with the electrode. As the $Cu^{2+}$ ions leave the electrode, a potential difference is created between the electrode and the surrounding solution: the solution becomes positively charged and the electrode negatively charged. This potential difference prevents additional $Cu^{2+}$ ions from migrating into the aqueous phase. A dynamic equilibrium is established that leaves the copper electrode negatively charged.

The same process occurs when a zinc strip is placed into a 0.1 M solution of zinc nitrate. However, experiments have shown that zinc ions in the electrode have a larger tendency to transfer into the aqueous phase than copper ions. For this reason under the same experimental conditions the zinc electrode will have more electrons that are not paired with cations than the copper electrode. The zinc electrode will be negative with respect to the copper electrode.

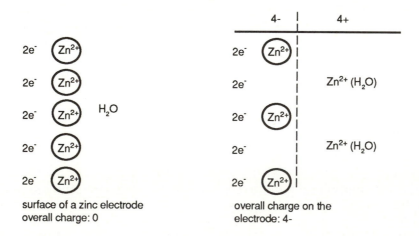

When a copper strip dipping into a 0.1 M $CuSO_4$ solution (the copper half-cell) is connected with a wire to the zinc strip dipping into a 0.1 M $Zn(NO_3)_2$ solution (the zinc half-cell), electrons will flow from the zinc electrode where an electron excess exists to the copper electrode where an electron shortage exists relative to the zinc electrode. The electrons arriving at the copper electrode will combine with a $Cu^{2+}$ ion from the aqueous phase and make it join the copper electrode. Because the zinc electrode has become temporarily less negative, the reduced potential difference between the zinc electrode and the zinc nitrate solution allows additional zinc ions to leave the electrode and increase the negative charge on the electrode. These electrons move to the copper electrode and convert another copper ion to a copper atom. As the galvanic cell operates, copper ions from the solution are plated onto the copper electrode, the copper electrode gains mass, and the concentration of $Cu^{2+}$ ions in the solution decreases. Zinc ions leave the zinc electrode, the zinc electrode loses mass, and the concentration of $Zn^{2+}$ ions in solution increases. These processes are summarized by equations 1 and 2 for the half-cell reactions and by equation 3 for the overall cell reaction.

$$\text{Cu half cell: } Cu^{2+}(aq) + 2e^- \longrightarrow Cu\ (s) \qquad (1)$$
$$\text{Zn half cell: } Zn(s) \longrightarrow Zn^{2+}(aq) + 2e^- \qquad (2)$$
$$\text{Cell: } Cu^{2+}(aq) + Zn(s) \longrightarrow Cu(s) + Zn^{2+}(aq) \qquad (3)$$

As the $Cu^{2+}$ ions are reduced by the electrons coming from the zinc electrode, they leave the solution and join the electrode. However, they do not take the negatively charged sulfate ions with them. The sulfate ions without their positive $Cu^{2+}$ counter ions would create a negative potential that impedes and finally stops the flow of electrons from the zinc to the copper electrode. In the zinc half-cell, zinc ions move from the electrode into the solution but do not bring with them negative counter ions, for instance, nitrate ions. The solution will become positively charged and the developing potential difference will stop additional zinc ions from leaving the electrode and electrons from moving to the copper electrode.

To prevent the galvanic cell from shutting down, a salt bridge supplies anions to the zinc half-cell and cations to the copper half-cell. The electric circuit is now closed and the cell can operate

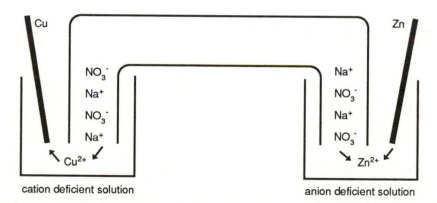

cation deficient solution                    anion deficient solution

until the zinc electrode is completely dissolved or all the copper ions in solution have plated onto the electrode. As the cell operates, the concentration of the zinc nitrate solution increases and the concentration of the copper sulfate solution decreases. These concentration changes influence the cell potential. Therefore, the cell potential is measured by placing a voltmeter into the external connection between the two electrodes before appreciable concentration changes have occurred. The complete experimental arrangement consists of a copper half-cell, a half-cell with another metal electrode, a salt bridge, and a voltmeter appropriately connected to the electrodes.

In the copper-zinc cell, the electrons flow from the Zn electrode to the copper electrode: the zinc electrode is the negative electrode (the anode), the copper electrode the positive electrode (the

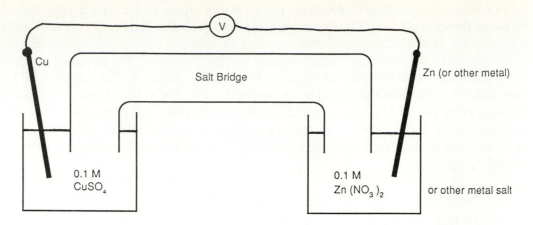

cathode).   At the anode zinc is oxidized to $Zn^{2+}$; at the cathode $Cu^{2+}$ is reduced to elemental copper.

The voltmeter can be connected to the electrodes in two ways:  red test lead to Cu, black test lead to Zn; or red test lead to Zn, black test lead to Cu.  Only one arrangement will allow the cell voltage to be read.  Note whether a useful reading is obtained with the red or the black lead connected to the negative zinc electrode.  Then check which test lead is connected to the positive terminal of the voltmeter.  Consistent application of the proper connection scheme makes it possible to identify the positive and negative electrodes in a galvanic cell.

The cell potentials are influenced by the concentrations of the salt solutions in the two half-cells.  When both of these concentrations are 0.1 molar, the measured potential at 25° is the standard potential.  The Nernst equation (equation 4) allows the cell potential for other concentrations to be calculated from the standard cell potential.  Written for the copper-zinc cell, equation 4 becomes transformed into equation 5.

$$E_{cell} = E^o_{cell} - \frac{0.059}{n} \log \frac{[M^{n+}]anode}{[M^{n+}]cathode}$$

(4)

$$E_{cell} = 1.10 - \frac{0.059}{2} \log \frac{[Zn^{2+}]}{[Cu^{2+}]}$$

(5)

Because the $Zn^{2+}$ and the $Cu^{2+}$ concentrations are both 0.1 molar, the ratio $[Zn^{2+}]/[Cu^{2+}]$ is 1, the log of 1 is zero, and $E_{cell}$ should be equal to 1.10 volt.  The cell voltage measured in this experiment will probably be much smaller.  The cause of this discrepancy lies in experimental difficulties with the preparation of the electrodes and the proper functioning of the salt bridge.

The potential of a half-cell, a metal electrode dipping into a solution of one of its salts, cannot be measured.  A galvanic cell consisting of two half cells has a potential that can be measured and can be calculated as the difference of the potentials of the two half-cells.  Such a difference is independent of the choice of the reference half-cell to which the potential of zero is assigned.  The hydrogen half-cell has been chosen as the reference half-cell by international agreement.  Because the copper half-cell is used in all of the galvanic cells in this experiment, a half-cell potential of zero is initially assigned to the copper half-cell.  The measured cell potential is attributed entirely to the other half-cell.  By agreement, a negative half-cell potential is assigned to the half-cell reaction written as a reduction reaction when the electrons flow from the half-cell electrode to the copper electrode; a positive half-cell potential when the electrons flow from the copper electrode to the other electrode. As an example, the following galvanic cells for which experimental cell potentials and the polarities of the electrodes are given are considered.

$$\overset{-}{\text{Co}} \mid \text{CoSO}_4, 0.1 \text{ M} \parallel \text{CuSO}_4, 0.1 \text{ M} \mid \overset{+}{\text{Cu}} \qquad \text{cell potential: } 0.62 \text{ V}$$

$$\overset{-}{\text{Cu}} \mid \text{CuSO}_4, 0.1 \text{ M} \parallel \text{AgNO}_3, 0.1 \text{ M} \mid \overset{+}{\text{Ag}} \qquad \text{cell potential: } 0.46 \text{ V}$$

The half-cell reactions written as reductions are:

$$Cu^{2+} + 2e^- \longrightarrow Cu \qquad E = 0: \text{ by definition for this experiment}$$
$$Co^{2+} + 2e^- \longrightarrow Co \qquad E = -0.62 \text{ V: Co electrode is negative}$$
$$Ag^+ + e^- \longrightarrow Ag \qquad E = +0.46 \text{ V: Ag electrode is positive}$$

The galvanic cell built from the Ag and Co half cells is calculated as the difference between the potentials of the Ag half-cell and the Co half-cell. The cell potential must always be positive.

$$E_{cell}^{Ag/Co} = E_{cell}^{Ag} - E_{cell}^{Co} = +0.46 - (-0.62) = 1.08 \text{ V}$$

A table of reduction potentials for these three elements arranged in increasing order with consideration of the signs can now be established (reference $Cu^{2+} + 2e^- \rightleftharpoons Cu \quad E = 0$). A conversion to the scale with the $H_2/H^+$ electrode as the zero point can be accomplished by adding the potential of the $Cu/Cu^{2+} - H_2/H^+$ cell (+0.34 V) to the Cu-based values.

| Half-cell reaction | Potential: Cu = 0 | Potential: $H_2 = 0$ |
|---|---|---|
| $Co^{2+} + 2e^- \rightleftharpoons Co$ | E = −0.62 V | −0.28 V |
| $Cu^{2+} + 2e^- \rightleftharpoons Cu$ | E = 0.0 V | +0.34 V |
| $Ag^+ + e^- \rightleftharpoons Ag$ | E = +0.46 V | +0.80 V |

Such tables can be used to calculate cell potentials, estimate the reducing and oxidizing power of elements and their ions, and calculate equilibrium constants. The cell potentials are independent of the reference point.

Galvanic cells can also be constructed from two-half cells with the same metal as electrodes but different concentrations of a metal salt. An example for such a concentration cell is:

$$Cu \mid CuSO_4, 0.1 \text{ M} \parallel CuSO_4, 1.0 \text{ M} \mid Cu$$

The cell potential of a concentration cell can be calculated with a modified Nernst equation (equation 6). For a Cu concentration cell with n = 2, $[M^{n+}]_{dil}/[M^{n+}]_{conc}$ is 0.1, the log of the ratio

$$E_{conc. \, cell} = -\frac{0.059}{n} \log \frac{[M^{n+}]_{dilute}}{[M^{n+}]_{concentrated}} \qquad (6)$$

is −1, and $E_{cell}$ approximately +0.03 V. Because the concentration ratios that are achievable are rather limited, the cell potentials of concentration cells are rather small. However, when the metal ion in one of the half-cells is reacted with a ligand to form a coordination compound that hardly dissociates and, thus, keeps the free metal ion concentration very low and constant, reasonable cell potentials that can be easily measured are achievable.

$Cu^{2+}$ ions form a very stable tetraammine complex (equation 7). The formation constant K for this complex is defined by equation 8. If the formation constant K is very large, most of the

$$Cu^{2+} + 4NH_3 \rightleftharpoons [Cu(NH_3)_4]^{2+} \tag{7}$$

$$K = \frac{[Cu(NH_3)_4{}^{2+}]}{[Cu^{2+}][NH_3]^4} \tag{8}$$

$Cu^{2+}$ ions are present in the form of the complex, $[Cu^{2+}]_{dil}$, is very small; the concentration ratio is small; the negative logarithm of this ratio is large; and the cell potential appreciable. The potential of a cell consisting of a Cu strip in a 0.1 M $CuSO_4$ solution and a Cu strip in a copper tetraammine solution can be measured ($E_{conc.\ cell}$). All quantities in equation 6 are now known with the exception of the free copper ion concentration, $[Cu^{2+}]_{dil}$, in the copper tetraammine solution. This concentration can be calculated from the value of the cell potential and used to estimate the formation constant for the copper complex.

The copper tetraammine solution is prepared by adding 1.0 mL of concentrated ammonia (14.4 M) to 24.0 mL of the 0.1 M $CuSO_4$ solution. From these data the ammonia concentration and the concentration of $[Cu(NH_3)_4]^{2+}$ at equilibrium can be calculated on the basis of the stoichiometry given in equation 7. One can safely assume that very little free $Cu^{2+}$ will be present at equilibrium and its concentration is obtained from the potential measurement. These values introduced into equation 8 produce a value for the formation constant of the copper tetraammine complex.

## ACTIVITIES

- Complete the PreLab Exercises before coming to the laboratory.
- Prepare 0.1 M solutions from solid metal salts.
- Prepare 0.1 M solutions by diluting more concentrated solutions.
- Construct galvanic cells.
- Measure the potentials of these galvanic cells.
- Prepare a solution of tetraamminecopper(II) sulfate.
- Construct a copper concentration cell.
- Measure the potential of this concentration cell.
- Calculate the formation constant for the copper tetraammine complex.
- Prepare a list of half-cell potentials.
- Complete the Report Form.

## SAFETY

*Wear approved eye protection in the laboratory. Work with concentrated ammonia in a well-ventilated hood. Clean spills immediately with a sponge and water. Dispose of solutions and chemicals as directed by your instructor.*

# PROCEDURES

## A. PREPARATION OF SOLUTIONS

**A-1**. Two hundred and fifty milliliters of each of the following solutions are needed for this experiment.

| | |
|---|---|
| 0.10 M $CuSO_4$ | 0.10 M $FeSO_4$ |
| 0.10 M $SnCl_2$ | 0.10 M $Zn(NO_3)_2$ |

These solutions are to be prepared by dissolving the required quantities of the solid salts [$CuSO_4 \cdot 5H_2O$, $FeSO_4 \cdot 7H_2O$, $SnCl_2 \cdot 2H_2O$, $Zn(NO_3)_2 \cdot 6H_2O$] in distilled water or by appropriate dilution of more concentrated solutions of these salts with distilled water. Your instructor will inform you which solutions are to be prepared from the solid salts and which by dilution, and distribute these tasks among groups of students. The solutions will be shared.
*[See Investigation 21 for directions for preparation of a $SnCl_2$ solution.]*

### Preparation of Solutions from Salts

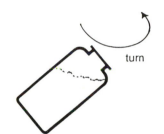

**A-2**. Check the storage bottle from which you will take the salt. Turn the bottle and observe the salt inside. The salt should be in powder form and flow freely when the bottle is turned. If the salt has "caked" and does not flow freely, shake the bottle, tap it gently against the palm of your hand, or - as a last resort - break up the cake with a clean spatula. Consult your instructor should you have problems with obtaining a free-flowing powder.

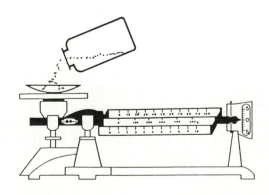

**A-3**. When your salt flows freely in the bottle, weigh a clean and dry watchglass on the triple-beam balance to 0.1 g. Record the mass of the watchglass. Advance the mass riders on the balance by the mass of salt required for the solution to be prepared. Open the bottle, bring it just above the watchglass, carefully roll and tap the container to transfer salt to the watchglass until the balance returns to equilibrium. Masses within 0.2 g of the required amount are acceptable. Use the 1 g rider to restore the balance to exact equilibrium. Record the mass.

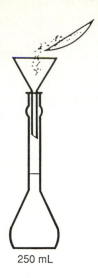

250 mL

250 mL mark

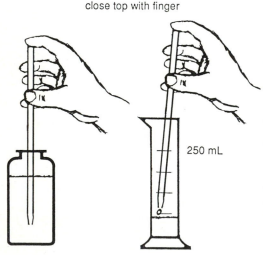

close top with finger

250 mL

Practice the dropwise release
in the storage bottle.

**A-4.** Place a funnel into a clean, but not necessarily dry, 250-mL volumetric flask. Hold the watchglass with the salt over the funnel, tilt the watchglass and tap it with your finger to cause the salt to slide into the funnel. With the watchglass still over the funnel, use a gentle stream of distilled water to rinse the last crystals of salt from the watchglass into the funnel and the flask. Remove the funnel. Fill the flask half full with distilled water. Swirl the flask until the salt is completely dissolved. Fill the flask with distilled water to the mark. The last milliliter must be added dropwise until the bottom of the meniscus touches the calibration mark. Mix the solution by inverting the stoppered flask several times. Label the flask with the formula of the salt and the concentration.

## Preparation of Solutions by Dilution

**A-5.** The concentrated salt solutions provided will be marked with their concentrations expressed in moles per liter. From this concentration calculate the volume (in mL) needed to prepare 250 mL of the 0.1 M solution.

**A-6.** Take a clean and dry 250-mL graduated cylinder to the storage bottle containing the concentrated solution. Carefully pour into this cylinder a few milliliters less than the calculated volume of the concentrated solution. Place a glass tube with one end drawn out to a tip into the storage bottle holding the concentrated solution. Close the top with one of your fingers. Bring the glass tube above the liquid in the 250-mL graduated cylinder and release the solution dropwise by slightly lifting your finger until the meniscus of the solution rests on the required graduation mark. Place the glass tube with the unused solution back into the storage bottle. Stir the solution in the graduated cylinder with a clean and dry glass rod until the solution is homogeneous. Label the cylinder with the formula of the salt and the concentration.

## B.  SETUP OF THE VOLTMETER

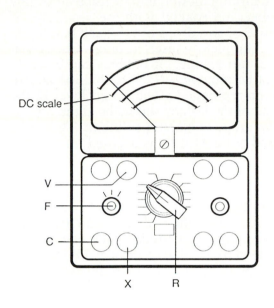

DC scale

V

F

C

X    R

**B-1**. A voltmeter is needed in this experiment to measure potential differences between the metallic electrodes. Your instructor will provide information about the proper use of the voltmeter available in your laboratory. The instructions given here refer to the Simpson voltmeter.

**B-2**. On the Simpson voltmeter
- Set the Function Switch F to "+DC".
- Set the Range Switch R to "2.5".
- Plug the black test lead into the jack C marked "Common –".
- Plug the red test lead into the jack marked "+1 V".

**B-3**. In this configuration the voltmeter will measure voltages between 0 and 1 volt. To read this range, locate the black scale marked "D.C." at both ends. Under this scale are three ranges of numbers: 0-250, 0-50, and 0-10. Read the 0-10 numbers and divide the reading by ten to obtain the voltage value.

**B-4**. Should the needle peg out at 10 (voltage larger than 1V), plug the red test lead into the jack X marked "+" on the meter. The meter now measures voltages between 0 and 2.5 volt. Read the 0-250 scale and divide the reading by 100 to obtain the voltage.

## C.  PREPARATION OF GALVANIC CELLS AND MEASUREMENT OF THEIR POTENTIALS

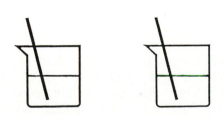

**C-1**. Take two clean and dry 100-mL beakers. Into one of these beakers pour approximately 40 mL of the 0.1 M $CuSO_4$ solution. Into the other beaker pour 40 mL of the 0.1 M $Zn(NO_3)_2$ solution. Obtain a strip of copper and a strip of zinc. It is advisable to clean one end of each strip with sandpaper. Set the copper strip into the copper sulfate solution and the zinc strip into the zinc nitrate solution. The cleaned ends should be submerged.

**C-2**. To construct the salt bridge, obtain a glass U-tube. From cotton balls roll two cotton plugs each approximately 1 cm long and with a diameter slightly larger than the inside diameter of the U-tube.

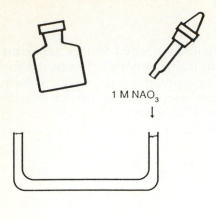

1 M NaNO$_3$

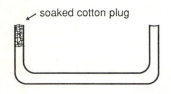

soaked cotton plug

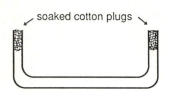

soaked cotton plugs

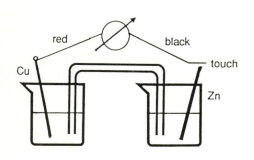

red black

Cu touch

Zn

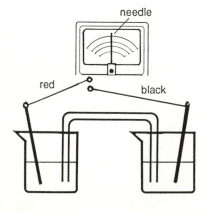

needle

red black

**C-3.** With a dropper fill the U-tube completely with a 1 M NaNO$_3$ solution. Plug one end of the filled U-tube with one of the cotton plugs. "Screw" the plug slowly with a turning motion into the tube and allow the plug to become soaked with the solution. If needed, add one or two drops of the NaNO$_3$ solution on top of the completely inserted plug. Check that no air bubbles are visible. If bubbles can be seen, the plug must be removed, NaNO$_3$ solution added, and the U-tube replugged. The liquid level in the unplugged end of the U-tube should be at the rim. If not, add a few drops of NaNO$_3$ solution. Then plug the second end of the U-tube as detailed for the first end.

**C-4.** Connect the solution in the two beakers with the salt bridge. Clip the red test lead from the voltmeter to the copper electrode, and touch the clip from the test lead to the zinc electrode. Observe the needle on the voltmeter. Does the needle move along the scale to the right or past the zero point to the left?

**C-5.** Reverse the connections to the metallic electrodes. Connect the clip from the red test lead to the Zn electrode; touch the Cu electrode briefly with the clip from the black test lead. Observe the needle of the voltmeter. Select the connection that moves the needle upscale to the right. The electrode connected to the red test lead causing a movement of the needle to the right is the positive electrode.

**C-6.** When the needle has stabilized, read the voltage developed by the cell. You might have to wait several minutes until the needle has become stable. Record the voltage of this cell and identify the negative electrode (electrode with electron excess in comparison with the other electrode) and the positive electrode.

**C-7.** Disconnect the voltmeter and replace the zinc nitrate beaker with an empty 100-mL beaker that will support the salt bridge. Remove the zinc electrode from the beaker, rinse the electrode with distilled water, and return the electrode to the storage container. Pour the zinc nitrate solution into a waste container. Rinse the beaker with distilled water. Dry the inside of the beaker with a towel.

**C-8**. Construct the following cells using the appropriate solutions and metallic electrodes.

- Fe I 0.1 M FeSO$_4$ II 0.1 M Cu(NO$_3$)$_2$ I Cu
- Sn I 0.1 M SnCl$_2$ II 0.1 M Cu(NO$_3$)$_2$ I Cu
- Zn I 0.1 M Zn(NO$_3$)$_2$ II 0.1 M Cu(NO$_3$)$_2$ I Cu

An iron nail coated with tin is used as the Sn electrode. Measure the cell potentials and determine the polarity of the electrodes (Procedures C-4 through C-7). Reverse the notations above for those cells that do not have the electrode written on the left as their anode (the negative electrode).

**C-9**. Arrange the elements used in this experiment in the order of increasing reduction potential with respect to copper. Calculate the reduction potentials with respect to the standard hydrogen electrode.

# D. THE POTENTIAL OF A Cu CONCENTRATION CELL AND THE FORMATION CONSTANT OF [Cu(NH$_3$)$_4$]$^{2+}$

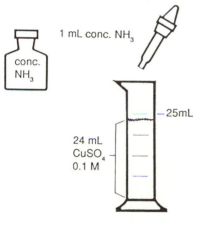

**D-1**. Fill a 25-mL clean and dry graduated cylinder with 24 mL of the 0.1 M CuSO$_4$ solution. Use the procedures described in A-6. Take the graduated cylinder with the 24 mL of CuSO$_4$ solution to the hood where a dropping bottle filled with concentrated ammonia is stored. Add the concentrated ammonia (14.4 M) dropwise (approximately 20 drops) until the meniscus of the liquid in the graduated cylinder rests on the 25 mL mark. With a clean and dry stirring rod thoroughly mix the solution that will become dark blue.

**D-2**. Pour the dark blue solution into a clean and dry 100-mL beaker. Place a cleaned strip of copper into the blue solution. Connect the solution via the salt bridge with the 0.1 M CuSO$_4$ solution in the second beaker. Connect the voltmeter to the two copper electrodes and read the potential. Identify the polarities of the electrodes (Procedures C-4, C-5 and C-6). This is experiment 4 on the Report Form

**D-3**. Discard the solutions into appropriate waste containers. Rinse the electrodes and return them to the storage containers. Store the voltmeter as directed by your instructor.

**D-4**. From the measured cell potential calculate the $Cu^{2+}$ concentration in the $Cu^{2+}/NH_3$ solution. With this concentration and the help of the discussion in the section "Concepts of the Experiment" calculate the formation constant of the complex $[Cu(NH_3)_4]^{2+}$.

# PRELAB EXERCISES

## INVESTIGATION 27: ELECTROCHEMICAL GALVANIC CELLS

Name:_____

Instructor:_____

ID No.:_____

Course/Section:_____

Date:_____

1. Define:

   Galvanic cell

   Standard cell potential

   Cell potential

   Half-cell

   Concentration cell

   Concentration

2. Calculate the masses of $CuSO_4 \cdot 5H_2O$, $FeSO_4 \cdot 7H_2O$, $SnCl_2 \cdot 2H_2O$, and $Zn(NO_3)_2 \cdot 6H_2O$ needed to prepare 250 mL of a 0.10 M solution. (Also record your answers on the Report Form.)

(over)

3. Calculate the volume of a 0.22 M solution of $SnCl_2$ needed to make 250 mL of a 0.10 M solution.

4. Concentrated ammonia (1.0 mL, 14.4 M) is added to a solution of $M^+$ (0.10 M, 19.0 mL). Assume that the volumes are additive. The metal ion forms a complex $[M(NH_3)_2]^+$. The concentration of free (uncomplexed) $M^+$ obtained from potential measurements in a galvanic cell is $1.0 \times 10^{-5}$ mol $L^{-1}$. Calculate the concentration of ammonia and $[M(NH_3)_2]^+$. Estimate the formation constant for the complex.

5. A galvanic cell consisted of $M \mid M^+$ 0.1 M $\parallel [M(NH_3)_2]^+ \mid M$ with the $[M(NH_3)_2]^+$ solution having been prepared as described in problem 4. The cell had a cell potential of 0.24 V. Calculate the concentration of uncomplexed $M^+$ in the ammonia-containing solution.

6. A galvanic cell $Pb \mid Pb^{2+}$ 1.0 M $\parallel Cu^{2+}$ 1.0 M $\mid Cu$ is allowed to operate and deliver a direct current.

   Which electrode is positive?

   Which negative?

   In which direction do the electrons flow?

   Which solution will become more concentrated?

   Which electrode will gain mass?

# REPORT FORM

## INVESTIGATION 27:  ELECTROCHEMICAL CELLS

Instructor: _____     Name: _____

Course/section: _____     ID No.: _____

Partner's Name:_____     Date:_____

## PREPARATIONS

Masses of salts needed to prepare 250 mL 0.10 M solutions:

$CuSO_4 \cdot 5H_2O$: _____ g          $FeSO_4 \cdot 7H_2O$: _____ g

$SnCl_2 \cdot 2H_2O$: _____ g          $Zn(NO_3)_2 \cdot 6H_2O$: _____ g

Volumes of solutions needed to prepare 250 mL 0.10M solutions:

$CuSO_4 \cdot 5H_2O$: _____ mL of _____ M soln.     $FeSO_4 \cdot 7H_2O$: _____ mL of _____ M soln.

$SnCl_2 \cdot 2H_2O$: _____ mL of _____ M soln.     $Zn(NO_3)_2 \cdot 6H_2O$: _____ mL of _____ M soln.

Attach copies of laboratory notebook pages showing data for the weighings and dilutions.

## CELL POTENTIALS

| Exp. | Galvanic Cell | Electrodes +(Cathode)   – (Anode) | | Cell Potential (V) |
|------|---------------|-----------|-----------|--------------------|
| 1 | | | | |
| 2 | | | | |
| 3 | | | | |
| 4 | | | | |

## Table of Reduction Potentials in Order of Increasing Potential

| Non-Copper half-cell reaction written as oxidation | Half-cell potentials based on Cu half-cell = 0 | Half-cell potentials based on $H_2/H^+$ half-cell = 0 (Subtract 0.337V) |
|----|----|----|
| 1. | | |
| 2. | | |
| 3. | | |

## Cell Potentials Calculated For The Three Galvanic Cells Using The Half-Cell Potentials Listed In Your Textbook.

| galvanic cell | electrode | | Cell potentials based upon textbook values | |
|---|---|---|---|---|
| | +(cathode) | -(anode) | $H_2/H^+$ half-cell = 0 | Cu half-cell = 0 |
| 1. | | | | |
| 2. | | | | |
| 3. | | | | |

## ANALYSIS OF DATA

On a separate sheet of paper compare the above table of textbook values with the data in the previous two tables and calculate and compare, using the two sources, the standard potentials for the following cells.

$Fe/Fe^{2+}//Sn^{2+}/Sn$

$Zn/Zn^{2+}//Fe^{2+}/Fe$

$Zn/Zn^{2+}//Sn^{2+}/Sn$

## FORMATION CONSTANT OF $[Cu(NH_3)_4]^{2+}$

Concentration of $Cu^{2+}$ in a solution prepared from 24 mL 0.1 M $Cu^{2+}$ and 1.0 mL of water:

_____ M

Concentration of $NH_3$ in a solution prepared from 1.0 mL 14.4 M $NH_3$ and 24 mL of water:

_____ M

$E_{conc. cell}$ for Cu I $CuSO_4$ 0.1 M II $Cu^{2+}$ from $Cu(NH_3)_4^{2+}$ I Cu: _____ V

Concentration of $Cu^{2+}$ in the dark blue solution: _____ M

Concentration of $NH_3$ not bound to $Cu^{2+}$: _____ M

Concentration of $[Cu(NH_3)_4]^{2+}$: _____ M

Formation Constant for $[Cu(NH_3)_4]^{2+}$: _____

.

Date _____ Signature _____

# Investigation 28

## INTRODUCTION

Transition elements and their ions have a great tendency to react with molecules and anions possessing at least one lone electron pair to yield coordination compounds. A brief, general discussion of coordination compounds can be found in the introduction to Investigation 15.

Cobalt, the element with atomic number 27, forms – as a member of the first transition series – many well-studied coordination compounds. In these complexes Co(0), Co(II) or Co(III) is surrounded by four or six ligands in tetrahedral or octahedral arrangements. Examples of such cobalt complexes are:

$K_4[Co(CN)_4]$
potassium
tetracyanocobaltate(0)

$Na_2[CoCl_4]$
sodium
tetrachlorocobaltate(II)

$[Co(NH_3)_6]Cl_3$
hexaamminecobalt(III)
chloride

$Co_2(CO)_8$
octacarbonyldicobalt(0)

$[Co(H_2O)_6]SO_4$
hexaaquocobalt(II)
sulfate

$NH_4[Co(SO_3)_2(NH_3)_4]$
ammonium
tetraamminedisulfitocobaltate(III)

Cobalt complexes have been extensively studied as carriers of molecular oxygen. With the choice of proper ligands, Co(II) complexes take on molecular oxygen from the air by incorporating $O_2$ as one of the ligands. By changing the reaction conditions, the Co(II)-$O_2$ complex can be made to release the bound oxygen. Such a process could be useful as a low-energy way of separating oxygen from nitrogen in air.

Cobalt complexes also have biological importance. In 1926, the consumption of large amounts of liver was discovered as a cure for pernicious anemia. Pernicious anemia is a serious illness characterized by a marked and progressive decrease of the number and an increase in the size of red blood cells and by pallor, weakness, and gastrointestinal and nervous disturbances. Twenty-two years later the substance responsible for curing pernicious anemia was isolated from liver tissue. This substance, now known as cyanocobalamin or vitamin $B_{12}$, was recognized to be a cobalt complex. Its structure was determined in 1957. R. B. Woodward (Harvard University) and A. Eschenmoser (Swiss Federal Institute of Technology, Zürich) announced the synthesis of vitamin $B_{12}$. Ninety separate reactions, 100 co-workers and eleven years were needed to synthesize this "complex" molecule (see structure below).

Methylcobalamin (R = $CH_3$ in the structure below) is synthesized by a variety of microorganisms, but not by plants and animals. Methylcobalamin has been implicated in the methylation of heavy metals to their toxic methyl derivatives. An example of such a conversion is the transformation of inorganic mercury to methylmercury halides and dimethyl mercury. Therefore, the reactions of methylcobalamin with inorganic compounds are important for an understanding of the cycling of heavy metals in the environment. These reactions are studied in many laboratories. Because methylcobalamin and other Vitamin $B_{12}$ derivatives are difficult to synthesize, not easily separated from organisms, and expensive, model compounds that are easily prepared in the laboratory and behave in a manner similar to the "natural" complex molecules are frequently used. Such a model compound is chloro(pyridine)bis(dimethylglyoximato)cobalt(III), the compound you will prepare in this experiment.

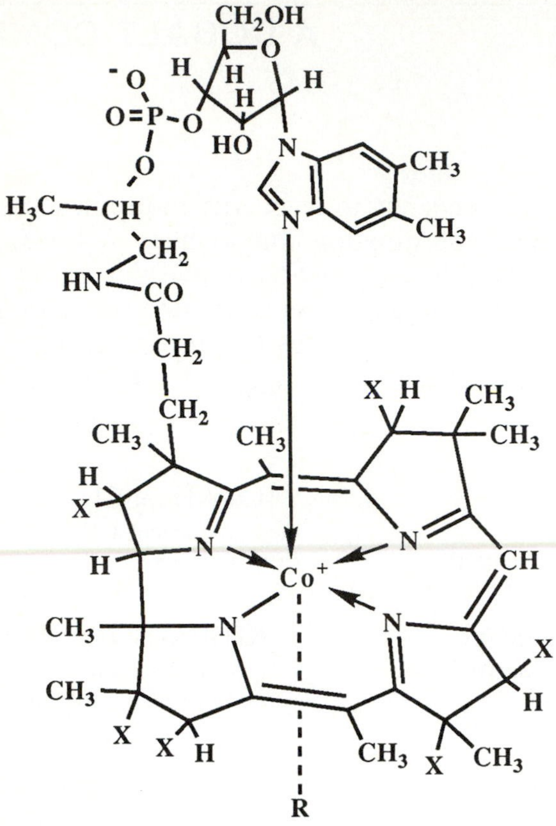

R = CN  Cyanocobalamin
R = CH₃  Methylcobalamin
X = –CH₂CONH₂

## CONCEPTS OF THE EXPERIMENT

The starting materials – cobalt dichloride hexahydrate, pyridine and 2,3–dioxobutane dioxime, known as dimethylglyoxime – are easily accessible chemicals. The reaction (equation 1) is carried out in ethanolic solution in which the dimethylglyoxime is more soluble than in water. In the first

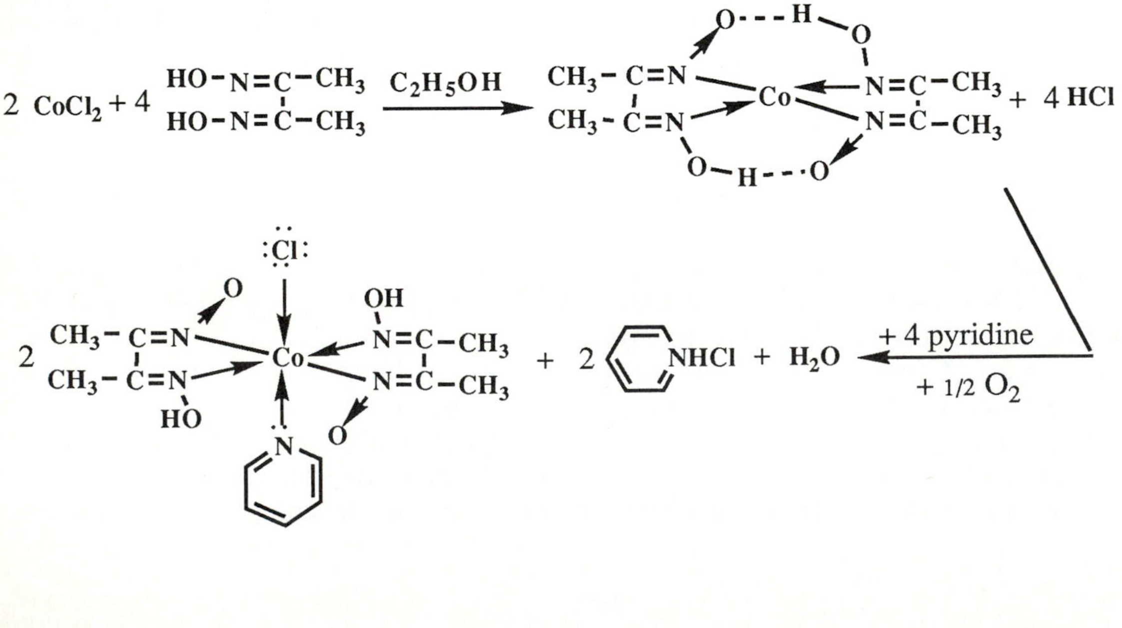

$$(1)$$

step bis(dimethylglyoximato)cobalt(II) is formed by replacement of $H_2O$, $Cl^-$, and ethanol bound to Co(II) by the anion of dimethylglyoxime that is derived from a tautomeric form of dimethylglyoxime (equation 2). A small excess of dimethylglyoxime will improve the yield of the

$$CH_3 - C = N - \ddot{O}H \\ CH_3 - C = N - OH \rightleftharpoons \begin{matrix} H \\ CH_3 - C = N \rightarrow O \\ CH_3 - C = N \diagdown OH \end{matrix} \rightleftharpoons \begin{matrix} CH_3 - C = N \diagup^O \\ \quad \ddot{} ^- \\ CH_3 - C = N \diagdown OH \end{matrix} + H^+ \qquad (2)$$

complex. When pyridine is added and air is bubbled through the solution, cobalt(II) is oxidized to cobalt(III). The final product is formed by coordination of one pyridine molecule and one chloride anion to the cobalt(III) center. This complex is insoluble in ethanol at room temperature, precipitates, and is easily isolated by filtration. Formally, the cobalt carries three positive charges. You started with $Co^{2+}$ in $CoCl_2$ and removed one electron from the cobalt during the oxidation with air. The charge of the resulting $Co^{3+}$ is neutralized by the negative charges on the chloride anion and the two dimethylglyoximate anions. Thus, the complex is electrically neutral.


## ACTIVITIES

- **Complete the PreLab Exercises before coming to the laboratory and have your instructor check the amount of reagents you plan to use.**

- **Prepare and isolate the cobalt complex.**

- **Weigh your product.**

- **Calculate your yield.**


## *SAFETY*

*Be very careful with open flames and hotplates to avoid igniting the ethanol used. In case of contact with any of the chemicals used, wash immediately and thoroughly with soap and water. Be careful in cutting, polishing, and inserting the glass tubing.*


## PROCEDURES

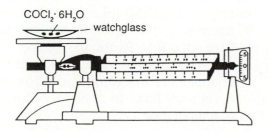

COCl$_2$·6H$_2$O — watchglass

1. Weigh a clean, dry watchglass to the nearest 0.01 g on the triple-beam balance. Add to the watchglass the approximate amount of $CoCl_2 \cdot 6H_2O$ needed for the preparation of 0.60 g of the complex. Record the mass used (to ± 0.01 g) in your notebook. Transfer the $CoCl_2 \cdot 6H_2O$ quantitatively into a clean 100-mL beaker.

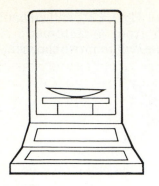

**2.** Using a clean, dry watchglass and the analytical balance, weigh out the mass of dimethylglyoxime approximately corresponding to 110 percent of the mass required to convert the amount of $CoCl_2 \cdot 6H_2O$ you are using to the complex. Have your instructor check the amount of dimethylglyoxime before proceeding.

Record the mass used and completely transfer the compound into the beaker containing the $CoCl_2 \cdot 6H_2O$.

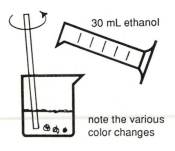

30 mL ethanol

note the various color changes

**3.** Slowly add 30 mL of 95-percent ethanol to the mixture of cobalt chloride and dimethylglyoxime. Record in your notebook the color changes observed during the initial stages of this procedure.

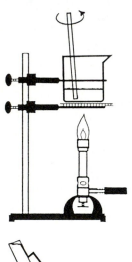

**4.** Place the 100-mL beaker containing the alcoholic mixture on a ringstand. Bring the mixture to a very gentle boil. *Exercise caution not to ignite the alcohol vapors.* Some of the solid material in the mixture may not dissolve. Discontinue boiling after 5 min. **Warning:** *In case of fire, turn off the burner, remove the stirring rod with tongs, and smother the fire with a damp towel or sponge.*

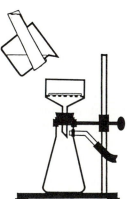

**5.** While the solution is still hot, filter it through a Buchner funnel into a 250-mL vacuum flask. (Review, if necessary, the section on vacuum filtration in Appendix F.) Rinse the beaker and Buchner funnel with 3 to 4 mL of 95-percent ethanol. Disconnect the aspirator and allow the solution in the vacuum flask to cool to within 10 degrees of room temperature.

**6.** Take the vacuum flask containing the filtered ethanolic solution to the fume hood. *Keep the pyridine container and your flask inside the fume hood* while you add the calculated number of drops of pyridine. Swirl the flask gently to mix the contents. Record any observation of color change and other indications of a reaction.

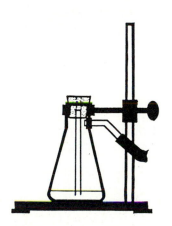

**7.** Obtain a one-hole rubber stopper that fits the top of the vacuum flask. Cut a piece of glass tubing just long enough to reach from about halfway into the stopper to about 1 cm above the bottom of the flask. Fire polish both ends of the tubing, allow the tubing to cool, and *carefully* insert one end of the tubing about halfway into the one-hole stopper (from the bottom of the stopper). Insert the stopper and tubing into the vacuum flask. Check that the tubing extends well below the surface of the liquid, but does not touch the bottom of the flask. Connect the aspirator and adjust the aspirator to draw a gentle stream of air bubbles through the solution. Bubble air through the solution for 20 min. Record any observation of color change and crystal formation.

**8.** Disconnect the aspirator and transfer the reaction mixture completely into a clean 250-mL beaker. Use a small amount of distilled water from your squeeze bottle to wash the last crystals from the flask into the beaker.

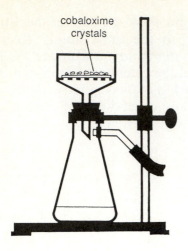

cobaloxime crystals

**9.** Filter the mixture using vacuum filtration. Wash the crystals with 10 mL of distilled water and then with 10 mL of 95-percent ethanol. Draw air through the crystals for 5 min.

**10.** Weigh the crystals on a sheet of smooth paper. Record the mass of the paper and the mass of the paper plus crystals to the nearest 0.01 g. Record a description of the appearance of the crystals.

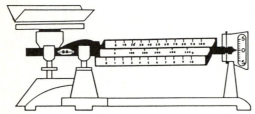

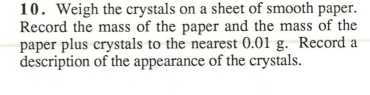

Show your product to your instructor who will direct you to the appropriate storage container. Ask your instructor to mark the product "grade" space on your Report Form.

**11.** On the basis of the mass of $CoCl_2 \cdot 6H_2O$ used and the mass of product obtained, calculate the percentage yield for your preparation. Use your calculations and notebook records to complete the Report Form.

# PRELAB EXERCISES

## INVESTIGATION 28:  A COBALT COMPLEX

Name:_____

Instructor:_____          ID No.:_____

Course/Section:_____          Date:_____

---

1.  Write and balance the equation using molecular formulas for the reaction described below.

    Cobalt(II) chloride hexahydrate plus pyridine, dimethylglyoxime and molecular oxygen yield chloro(pyridine)bis(dimethylglyoximato)cobalt(III) plus pyridinium chloride and water.

2.  Calculate the mass of $CoCl_2 \cdot 6H_2O$ required for the preparation of 0.60 g of the complex. Assume a 70-percent yield in this experiment and consider $CoCl_2 \cdot 6H_2O$ to be the limiting reactant.

3.  The mass of $CoCl_2 \cdot 6H_2O$ calculated in problem 2 above should be reacted with 110% of the stoichiometrically required amount of dimethylglyoxime.  How many grams of dimethylglyoxime are needed?

(over)

4.  A 10 percent excess of pyridine above the stoichiometrically required amount for the mass of $CoCl_2 \cdot 6H_2O$ from problem 2 should be added to the reaction mixture. How many grams of pyridine will be needed? Assuming that one drop of liquid pyridine weighs 0.04 g, calculate the number of drops of pyridine to be added.

5.  List the precautions you should observe when handling ethanol and pyridine.

6.  If you had started with 4.11 g of $CoCl_2 \cdot 6H_2O$ as the limiting reactant and had obtained 4.11 g of complex, what would have been the percent yield?

# REPORT FORM

## INVESTIGATION 28:  A COBALT COMPLEX

Name: _____     ID No.: _____

Instructor: _____     Course/Section: _____

Partner's Name (if applicable): _____     Date: _____

---

Show all calculations on attached pages.

Cobalt(II) chloride hexahydrate: _____ g mole$^{-1}$     Dimethylglyoxime: _____ g mole$^{-1}$

Pyridine: _____ g mole$^{-1}$

Chloro(pyridine)bis(dimethylglyoximato)cobalt(III): _____ g mole$^{-1}$

### Reactants used

| $CoCl_2 \cdot 6H_2O$ | Dimethylglyoxime | Pyridine |
|---|---|---|
| Watchglass + compound: _____ g | Watchglass + compound: _____ g | _____ drops, corresponding to |
| Watchglass: _____ g | Watchglass: _____ g | approximately: _____ g |
| $CoCl_2 \cdot 6H_2O$: _____ g | Dimethylglyoxime: _____ g | |

### Product

Paper + complex: _____ g

Paper: _____ g

Theoretical yield (based on $CoCl_2 \cdot 6H_2O$ used): _____ g

Complex: _____ g     Percentage yield: _____ %
in your experiment

Product quality   ❏ good   ❏ average   ❏ poor     Instructor's initials _____.

### Observations during Procedure 3

## Observations during Procedure 6

## Observations during Procedure 7

Date _____ Signature _____

# Investigation 29

## CALCIUM IN NATURAL MATERIALS

## INTRODUCTION

Calcium is the fifth most abundant element in the earth's crust, occurring in a wide variety of minerals such as *gypsum* ($CaSO_4 \cdot 2H_2O$), *calcite* ($CaCO_3$), *fluorite* ($CaF_2$), *perovskite* ($CaTiO_3$, a useful titanium ore), *scheelite (CaWO_4*, a tungsten ore), and the widely distributed *phosphate rock* [$Ca_3(PO_4)_2$]. Calcium is also a component of many of the more complex minerals. A number of these minerals are commercially important. Gypsum is used in making wallboard (sheetrock) and plaster of paris. Phosphate rock is a valuable source of various phosphate fertilizers. Calcium carbonate, as *calcite, limestone,* or seashells can be decomposed by heating to form lime:

$$CaCO_3(s) \longrightarrow CaO(s) + CO_2(g)$$

Oyster shells are about 95 percent $CaCO_3$ and their abundance in seashore areas led to the development of an important seaside industry, the recovery of magnesium from seawater. Lime, produced by heating crushed oyster shells, is added to seawater to precipitate valuable magnesium hydroxide:

$$\underset{\text{as oyster shell}}{CaCO_3} \overset{\text{heat}}{\longrightarrow} \underset{\text{lime}}{CaO + CO_2}$$

$$CaO(s) + H_2O \longrightarrow Ca^{2+}(aq) + 2OH^-(aq)$$

$$Ca^{2+}(aq) + 2OH^-(aq) + Mg^{2+}(aq) + 2Cl^-(aq) \longrightarrow \underset{\text{precipitate}}{Mg(OH)_2(s)} + Ca^{2+}(aq) + 2Cl^-(aq)$$
$$\underset{\text{in seawater}}{}$$

Oyster shell chips are also sold as a food supplement for chickens because the added dietary $Ca^{2+}$ results in improved strength of eggshells.

To assure that the correct amount of calcium oxide is added to seawater and that the chicken food contains the required amount of calcium carbonate, the raw materials must be assayed for the calcium concentration. This experiment demonstrates the determination of calcium by complexometric titration with ethylenediaminetetraacetate (EDTA).

## CONCEPTS OF THE EXPERIMENT

Ethylenediaminetetraacetic acid in the form of its disodium salt is quite soluble in water. Such solutions are suitable to quantitatively determine metal ions by titration at appropriate pH values.

$$HOC-CH_2 \quad CH_2-CO^-Na^+$$
$$Na^+ \ ^-OC-CH_2 \diagdown NCH_2CH_2N \diagup CH_2-COH$$
EDTA

disodium dihydrogen
ethylenediaminetetraacetate

Calcium can be determined by titration with EDTA. When a solution of EDTA is added to a solution of $Ca^{2+}$, a calcium-EDTA complex is formed in which the $Ca^{2+}$ ion is surrounded octahedrally by four oxygen atoms from the carboxylate groups and two nitrogen atoms. The complex ion carries two negative charges (equation 1).

$$HOOCCH_2 \diagdown NCH_2CH_2N \diagup CH_2COONa$$
$$NaOOCCH_2 \diagup \qquad \diagdown CH_2COOH$$

$$+ \ Ca^{2+} + 2\ Cl^-$$

$\longrightarrow$

$$+\ 2\ Na^+$$
$$+\ 2\ Cl^-$$
$$+\ 2\ H^+$$

(1)

ethylenediaminetetraacetatocalcium(II)

During this reaction two moles of hydronium ion are formed per mole of Ca-EDTA complex. Therefore, the solution will become acidic during the titration. The Ca-EDTA complex is less stable in acidic than in basic solution. The addition of a pH-10 ammonia/ammonium chloride buffer neutralizes the hydronium ions and keeps the solution basic during the titration.

To aid you in recognizing the endpoint of the titration, the point at which the volume of EDTA solution added is just sufficient to complex all the $Ca^{2+}$ ions in the solution according to equation 1, a metal indicator is added to the calcium solution. Eriochrome Black T is used for the determination of $Ca^{2+}$. This indicator gives a blue solution in basic medium in the absence of calcium ions. When calcium ions are present, the indicator forms a complex with $Ca^{2+}$ that has a red color. The $Ca^{2+}$ indicator complex is less stable than the $Ca^{2+}$-EDTA complex. When EDTA solution is added to the $Ca^{2+}$ solution containing a few drops of Eriochrome Black T solution, the EDTA reacts first with the $Ca^{2+}$ ions not complexed by the indicator. When all the free $Ca^{2+}$ ions have been converted to the EDTA complex, additional EDTA will take away $Ca^{2+}$ ions from the

$pK_1 = 6.3$

$pK_2 = 11.5$

(K's for phenolic OH groups)

Eriochrome Black T (NaH$_2$R)

$$NaH_2R \quad \rightleftharpoons \quad [NaHR]^- \quad + Ca^{2+} \longrightarrow [NaCaR] + H^+$$
| below pH 6 | between pH 7 and 11 | pH 10 |
| red | blue | red |

$Ca^{2+}$ indicator complex liberating the indicator. Before the endpoint is reached, some of the indicator molecules will still be complexed with $Ca^{2+}$. The solution will have the blue color of the indicator without $Ca^{2+}$ and the red color of the $Ca^{2+}$-indicator complex. Therefore, the solution is the blue-red mixed color of purple just before the endpoint. The endpoint is reached when all the $Ca^{2+}$ ions have been extracted by EDTA from Eriochrome Black T and the solution has a pure blue color.

The concentration and the mass of $Ca^{2+}$ in the titrated solution can be calculated from the volume of EDTA solution required to reach the endpoint, the molarity of the EDTA solution, and the volume of the $Ca^{2+}$ solution titrated. The stoichiometry for the reaction between $Ca^{2+}$ and EDTA is given by equation 1.

## ACTIVITIES

- Complete the PreLab Exercises before coming to the laboratory.

- Prepare a standard $Ca^{2+}$ solution of 0.01 molarity.

- Prepare an EDTA solution (approximately 0.01 M).

- Standardize the EDTA solution with the standard $Ca^{2+}$ solution.

- Prepare a $Ca^{2+}$ solution from a natural material consisting largely of calcium carbonate.

- Titrate the unknown $Ca^{2+}$ solution with EDTA.

- Calculate the percent calcium in your unknown.

- Complete the Report Form.

## SAFETY

*Wear approved eye protection while working in the laboratory. Handle corrosive acids and bases with great care. Avoid spills.*

*If spills occur, clean them with plenty of water and a wet sponge. Should acids or bases come in contact with your skin, flush the affected area with copious amounts of water. Then contact your instructor.*

## PROCEDURES

---

### A. PREPARATION OF THE STANDARD $Ca^{2+}$ SOLUTION

**A-1.** Place a 10 cm x 10 cm weighing paper on the pan of an analytical balance and determine its mass to the nearest 0.1 mg. Then weigh out the mass of reagent-grade calcium carbonate required for the preparation of 100 mL of a 0.01 molar solution of $Ca^{2+}$. It is not necessary to weigh out the exact amount. Any amount within ±5 mg of the calculated quantity is acceptable. Record all the masses.

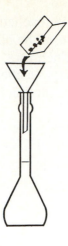

**A-2.** Transfer the calcium carbonate into a clean – but not necessarily dry – 100-mL volumetric flask. Tap the creased weighing paper held over the funnel on the volumetric flask to make all the $CaCO_3$ crystals slide off the paper. With a stream of distilled water from your squeeze bottle, flush the calcium carbonate from the funnel into the flask and from the sides of the flask to its bottom. Add 10 mL of distilled water and five drops of 6 M HCl. Swirl the flask to ensure intimate contact and reaction between the calcium carbonate and the hydrochloric acid. If any particles remain after a few minutes of swirling, add two or three more drops of 6 M HCl and continue swirling until all solid has dissolved. To the clear solution add distilled water to the 100-mL mark. Stopper and invert the flask several times to ensure homogeneity. Label the flask "Standard $Ca^{2+}$ solution".

## B. PREPARATION OF THE EDTA SOLUTION

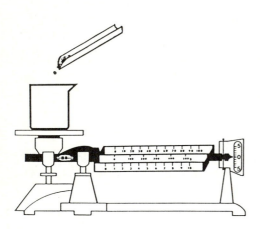

**B-1.** On the triple-beam balance weigh a clean and dry 500-mL beaker to the nearest 0.1 g. Advance the balance weights by the amount calculated for the EDTA salt and add approximately this mass of the salt into the beaker *being careful not to add too much*. Record the mass data.

**B-2.** Remove the beaker from the balance and add 200 mL of distilled water. Stir the mixture until all solute has dissolved, then add two crystals of $MgCl_2$ and two pellets of NaOH (*corrosive*). Continue stirring until the solution is homogeneous. Label the beaker "EDTA solution". Cover the beaker with a sheet of paper.

# C. STANDARDIZATION OF THE EDTA SOLUTION

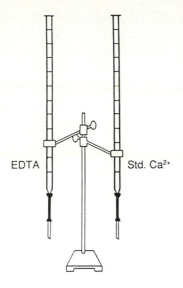

**C-1.** Obtain and, if necessary, clean (Appendix D) two 50-mL burets. Mount these securely on a ringstand. Rinse and fill the left buret with the 0.01 M "EDTA solution" and attach a proper label to the buret. Then rinse and fill the other buret with the "Standard $Ca^{2+}$ solution" and attach a proper label to the buret. Drain just enough liquid from each buret into a "waste" flask to expel any air from the buret tips and to bring the liquid levels to just below the zero marks.

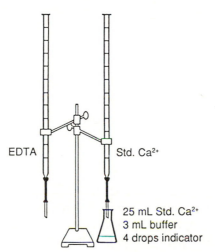

**C-2.** Record the buret reading in the Standard $Ca^{2+}$-buret, then drain about 25 mL of the Standard $Ca^{2+}$ solution into a clean 125-mL Erlenmeyer flask. Record the final buret reading. Add to the flask 3 mL of a pH 10 buffer solution and three drops of Eriochrome Black T indicator.

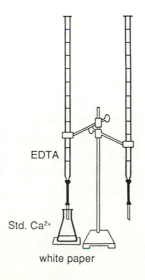

white paper

**C-3.** Set the flask containing the "Standard $Ca^{2+}$ solution" on a sheet of white paper under the buret containing the EDTA solution. Record the initial volume in this buret. Then begin the titration slowing to a dropwise addition near the endpoint. Swirl the flask during the titration. Near the endpoint the indicator will change from red to purple. At the endpoint, the indicator color will be a pure blue with no tint of red. Record the final buret reading.

**C-4**. Repeat Procedures C-2 and C-3 with volumes of the standard $Ca^{2+}$ solution that consume 35 to 40 mL of the EDTA solution. Calculate the molarity of the EDTA solution for each titration. Average the molarities and calculate the relative standard deviation. If the relative standard deviation is more than two percent, consult your instructor for suitable improvements to your titration technique.

# D. DETERMINATION OF CALCIUM IN A NATURAL MATERIAL

**D-1**. Obtain a sample of a natural material consisting largely of $CaCO_3$ (limestone, oyster shell, stalagmite, coral) and the approximate percentage of $CaCO_3$ in the sample from your instructor. Place a piece of the sample the size of a dime in a clean, dry mortar. Use the pestle to grind the sample to a fine powder.

**D-2**. Calculate the mass of your sample needed (with consideration of the approximate percentage of $CaCO_3$ in the sample) to prepare 100 mL of a 0.01 M $Ca^{2+}$ solution. On the analytical balance weigh a creased piece of smooth paper (10 cm x 10 cm) to the nearest 0.1 mg. Add to the paper the approximate mass of the powdered sample needed to prepare a 0.01 M $Ca^{2+}$ solution. Masses within 10 mg of the calculated mass are acceptable. Record all masses in your notebook.

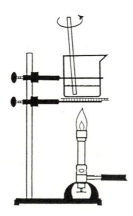

**D-3**. Transfer the weighed powder *completely* into a clean 100-mL beaker. Add 15 mL of distilled water and 30 drops of 6 M HCl. Warm the mixture *very gently* while stirring constantly. When the reaction has stopped and no $Ca^{2+}$- containing solid remains, discontinue heating.

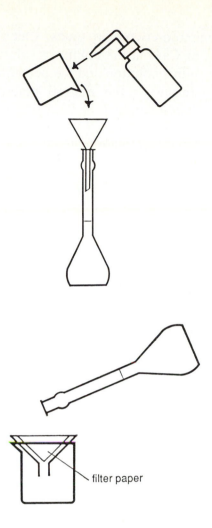

filter paper

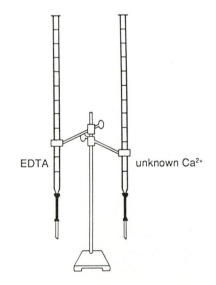

EDTA          unknown Ca²⁺

**D-4**. Allow the mixture to cool to room temperature, then transfer it completely into a clean 100-mL volumetric flask, rinse the beaker several times with small portions of distilled water and pour these rinses into the volumetric flask. Drop two pellets of NaOH (Caution: *corrosive*) into the flask and add distilled water to the 100-mL mark. Stopper the flask and invert it several times to mix the contents.

**D-5**. Fold an 11-cm diameter Whatmann No. 1 filter paper into quarters. Open one of the "pockets" formed and fit the filter paper into a powder funnel supported over a clean, dry 250-mL beaker. Filter the $Ca^{2+}$ mixture from the volumetric flask into the beaker. Label the beaker as "Unknown $Ca^{2+}$ solution", cover the beaker with a piece of paper, and set it aside for later use. (Rinse the volumetric flask with distilled water, discarding the rinses.)

**D-6**. Rinse the buret used for the "Standard $Ca^{2+}$ solution" with distilled water. Then rinse and fill this buret with your "Unknown $Ca^{2+}$ solution" and re-label the buret properly. If necessary, refill the buret containing the "EDTA solution". Drain enough liquid from both burets into a "waste" flask to expel air from the buret tips and to drop the liquid levels below the zero marks. Record the buret readings.

**D-7**. Drain approximately 30 mL of the unknown $Ca^{2+}$ solution into a clean 125-mL Erlenmeyer flask. Record the buret readings to the nearest 0.1 mL. Add to the flask 3 mL of a pH 10 buffer solution and 3 drops of Eriochrome Black T indicator. Titrate the $Ca^{2+}$ solution to a pure blue endpoint. Record the buret reading.

**D-8**. Repeat the titration two more times. Calculate the concentration of $Ca^{2+}$ in the unknown solution, the mass of $Ca^{2+}$ in this solution and the percentage of $Ca^{2+}$ in the sample of the natural material. Perform this calculation for each of your titrations. Calculate the average percentage of calcium in your solid sample, the standard deviation, and the relative standard deviation.

# PRELAB EXERCISES

## INVESTIGATION 29:  CALCIUM IN NATURAL MATERIALS

Name: _____

Instructor: _____     ID No.: _____

Course/Section:_____     Date:_____

1. Calculate the molecular mass of disodium dihydrogen ethylenediaminetetraacetate dihydrate.

2. Calculate the mass of pure (reagent grade) $CaCO_3$ required for the preparation of 100 mL of a 0.01 M $Ca^{2+}$ solution.

3. Calculate the mass of disodium dihydrogen ethylenetetraminetetraacetate dihydrate ($Na_2C_{10}H_{14}N_2O_8 \cdot 2H_2O$) required for the preparation of 200 mL of a 0.01 M solution of EDTA.

(over)

4.  Calculate the mass of an impure limestone (containing 80% by mass of $CaCO_3$) required to make 100 mL of a 0.01 M $Ca^{2+}$ solution.

5.  Write a balanced equation for the reaction of EDTA with $Ca^{2+}$.

# REPORT FORM

## INVESTIGATION 29: CALCIUM IN NATURAL MATERIALS

Name: _____ ID No.: _____

Instructor: _____ Course/Section: _____

Partner's Name (if applicable): _____ Date: _____

---

Show calculations on attached pages.

### MASS DATA

Natural Material                Pure CaCO3                EDTA Salt

  sample + paper: _____ g    sample + paper: _____ g    EDTA + beaker: _____ g

        paper: _____ g        paper: _____ g        beaker: _____ g

(Powdered)
natural material: _____ g    Pure CaCO3: _____ g    EDTA salt: _____ g

### TITRATION DATA

#### Standardization

| VOLUME | Standard Ca$^{2+}$ | | | EDTA | | |
|---|---|---|---|---|---|---|
|  | Trial 1 | Trial 2 | Trial 3 | Trial 1 | Trial 2 | Trial 3 |
| final volume |  |  |  |  |  |  |
| initial volume |  |  |  |  |  |  |
| volume used |  |  |  |  |  |  |

### Molarity of EDTA Solution

Trial 1 _____ M          Average: _____

Trial 2 _____ M          Standard Deviation: _____

Trial 3 _____ M          Relative Standard Deviation: _____

## Unknown

| VOLUME | Unknown $Ca^{2+}$ | | | EDTA | | |
|---|---|---|---|---|---|---|
| | Trial 1 | Trial 2 | Trial 3 | Trial 1 | Trial 2 | Trial 3 |
| final volume | | | | | | |
| initial volume | | | | | | |
| volume used | | | | | | |

## Molarity of Unknown $Ca^{2+}$ Solution

Trial 1 _____ M        Average:_____ M

Trial 2 _____ M        Standard Deviation:_____

Trial 3 _____ M        Relative Standard Deviation:_____

Mass of $Ca^{2+}$ in 100 mL of Solution

Trial 1 _____ g        Average:_____

Trial 2 _____ g        Standard Deviation:_____

Trial 3 _____ g        Relative Standard Deviation:_____

Average mass of $Ca^{2+}$ in sample:_____ g

Mass of sample:_____ g

Percent Ca in sample:_____

Average Percent Ca in sample:_____

Average Percent $CaCO_3$ in sample:_____

Date _____ Signature _____

# Investigation 30

## SPECTROPHOTOMETRIC DETERMINATION OF GLUCOSE

### INTRODUCTION

A major product of photosynthesis is the compound glucose, $C_6H_{12}O_6$, a simple sugar (monosaccharide). Glucose is produced by plants from carbon dioxide and water and is an essential organic chemical for animals and man. The oxidation of glucose to carbon dioxide and water in the cells of organisms supplies the energy needed by muscles and for the maintenance of body temperature. The glucose concentration in human blood is maintained at approximately 100 mg per 100 mL by the regulatory action of the liver. Excess glucose is converted in the liver to glycogen, a polymeric carbohydrate. Glycogen stored in the liver is depolymerized to glucose when glucose is needed by the body. Large excesses of glucose are converted to fat that is stored in adipose tissue. Eating too many sweets makes a fat person.

When the body's mechanism for the control of the glucose concentration in blood malfunctions, serious medical problems may result. In hyperglycemia, a condition resulting from improper regulation of the conversion of stored glycogen to glucose, the concentration of glucose in the blood is well above normal levels. A deficiency of the production of insulin affects the cellular combustion of glucose resulting in a condition known as diabetes. In hyperglycemia and diabetes the glucose concentrations in the blood are elevated. At approximately 180 mg glucose/100 mL blood or serum, excess glucose passes through the kidneys and is excreted in the urine. At glucose concentrations above 400 mg/100 mL, the patient may fall into a coma and may die if the concentration is not quickly lowered.

Tests for the presence of the concentration of glucose in urine and/or the blood are used to check for diseases associated with glucose metabolism. Before the development of precise and accurate chemical methods for the determination of glucose, doctors resorted to tasting the urine of patients; sweet urine suggested trouble. This experiment introduces a photometric method for the determination of glucose which is a great improvement over the urine-tasting methodology. In clinical laboratories glucose is now determined by enzymatic methods.

### CONCEPTS OF THE EXPERIMENT

Glucose does not absorb visible light; therefore, photometric measurements in the visible region of the electromagnetic spectrum cannot be used for the determination of glucose. An indirect method must be used. Glucose is a reducing monosaccharide. It contains an aldehyde functional group that can be oxidized by a suitable oxidizing agent to a carboxylic acid group. Such an oxidizing agent is copper sulfate dissolved in an alkaline solution of sodium tartrate, Benedict's Reagent. A simplified equation for the oxidation of glucose by Benedict's Reagent is

$$
\begin{array}{l}
\text{H} \\
\text{C=O} \\
\text{H·C-OH} \\
\text{HO·C-H} \\
\text{H–C-OH} \\
\text{H–C·OH} \\
\text{CH}_2\text{OH}
\end{array}
+ 2\,Cu^{2+} + 4\,OH^- \longrightarrow
\begin{array}{l}
\text{H} \\
\text{O} \\
\text{C=O} \\
\text{H·C-OH} \\
\text{HO·C-H} \\
\text{H–C-OH} \\
\text{H–C·OH} \\
\text{CH}_2\text{OH}
\end{array}
+ Cu_2O + 2\,H_2O \qquad (1)
$$

given in equation 1. The $Cu^{2+}$ as a reactant is present as the blue citrate or tartrate complex and is converted to the insoluble Cu(I) oxide during the reaction. As the reaction proceeds, copper(I) oxide will precipitate until all the glucose is oxidized. The concentration of the blue $Cu^{2+}$-tartrate complex decreases as copper(I) oxide is formed. When the reaction is complete, the transmittance (absorbance) of the reaction mixture can be measured with a spectrophotometer and the glucose concentration calculated from the measured absorbance. The higher the glucose concentration, the more copper(I) oxide is formed, and the less intense is the color of the remaining $Cu^{2+}$-tartrate solution.

A spectrophotometer is the instrument with which the intensity of the color can be measured. Such instruments vary in complexity, performance, ease of operation, and cost, but have the following components in common (Figure 1):

•the radiation source: a lamp that emits light consisting of a range of wavelengths needed for an experiment. A tungsten lamp is most often used for photometric measurements in the visible region of the spectrum.

•a dispersing device: filters, prisms, or gratings may be used to separate light into its constituent colors (wavelengths). These devices allow certain small wavelength-regions to be selected (a small band of wavelengths in the red region centered at 735 nm for this experiment).

•a sample chamber: a compartment within the instrument into which the solution kept in a cuvet is placed in such a manner that the light of selected wavelength passes through a defined thickness of the solution.

•a cuvet: test tubes or special cells with two optical surfaces (that are never to be touched), uniform wall thickness, and exactly defined inner diameter (path length) are used to hold the solution. A path length of 1.00 cm is frequently used. High quality cuvets are made from quartz and are available in matched pairs.

•a radiation detector: radiation detectors [such as photoelectron multiplier (PM) tubes] produce an electric current that is proportional to the amount of radiation reaching them per unit time. The electric current is then amplified and drives a pen on a recorder or deflects a needle on a calibrated scale.

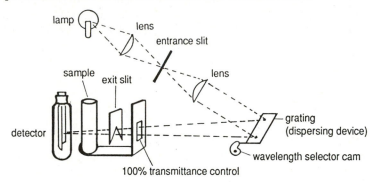

Figure 1:  Principal Parts of a Spectrophotometer

A spectrophotometer has control knobs to adjust zero and 100% transmittance, a lever to block the light beam from reaching the detector, a wavelength selector knob, and a meter or electronic display showing the transmittance, absorbance, or both (Figure 2).

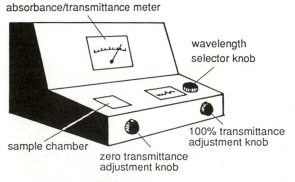

Figure 2:  Controls and Display on a Spectrophotometer

When light of appropriate wavelength and intensity "$I_o$" passes through a light-absorbing solution, some of the light is absorbed, and the light leaving the solution has a lower intensity "$I$" than the entering light. The following mathematical relation connects these intensities with the pathlength, d, and the concentration, C, of the absorbing species (equation 2).

$$-\log\frac{I}{I_o} = \varepsilon dC \tag{2}$$

ε: molar absorptivity; d = path length; C = concentration of absorbing species

The ratio $I/I_o$ is called transmittance, usually expressed in percent (100 $I/I_o$). The negative logarithm of the ratio $I/I_o$ is called absorbance. Spectrophotometers may display the reduction of light intensity as percent transmittance, as absorbance, or both. If only one of these quantities is displayed, equation 3 may be used to calculate the other.

$$A = -\log\frac{\% \text{ transmittance}}{100} \tag{3}$$

The molar absorptivity, $\varepsilon$, in equation 2 is the absorbance caused by a 1.00-cm layer of a 1.0 molar solution of the absorbing species. The molar absorptivity is a characteristic of the absorbing molecule in a certain solvent at a specified wavelength (usually the wavelength at an absorption maximum). The molar absorptivity changes with wavelength and may change with the nature of the solvent.

For the determination of the concentration of an absorbing species, the molar absorptivity does not have to be known when a calibration curve is constructed with standard solutions of the absorbing substance for which the concentrations are known. The path length is kept constant. Under these conditions, equation 2 reduces to equation 4.

$$A = k \cdot C \tag{4}$$

The plot of the measured absorbances versus the concentrations is the calibration curve. From the measured absorbance of an unknown, its concentration can be found with the help of the calibration curve. In most spectrophotometric determinations the absorbance increases with increasing concentration of the absorbing species. In the indirect determination of glucose the situation is not quite so simple.

The solutions of glucose used for the establishment of the calibration curve are obtained by taking volumes of a standard stock solution of glucose (100 mg glucose per 100 mL solution) in the range from 0.5 mL to 4.0 mL and diluting them with distilled water to 6.0 mL. Then 1.00 mL of the Benedict reagent is added. Three different glucose concentrations are associated with each solution:

- the glucose concentration in the undiluted aliquot from the stock solution (always 100 mg/100 mL).

- the glucose concentration in the aliquot diluted with distilled water to 6.0 mL.

- the glucose concentration in the solution after addition of 1.00 mL Benedict's Reagent. Each solution now has a total volume of 7.0 mL.

The glucose concentrations in the aliquots diluted to 6.0 mL shall be used for the construction of the calibration curve because for the determination of glucose in urine, 6.0 mL urine would be used under the conditions of this experiment. From the measured absorbance the glucose concentration in the urine sample could be found directly from the calibration curve without further calculation.

The calibration curve for the determination of glucose is obtained when the measured absorbances are plotted against the glucose concentrations in the calibration standards. Because the concentration of the blue $Cu^{2+}$-tartrate complex decreases with increasing glucose concentration, the calibration curve has a negative slope and an intercept at the absorbance axis that corresponds to the absorbance of 1.00 mL of Benedict's Reagent diluted with distilled water to 7.0 mL. The best straight line is drawn through the experimental points. The concentrations of the unknowns are obtained from the measured absorbances as shown in Figure 3.

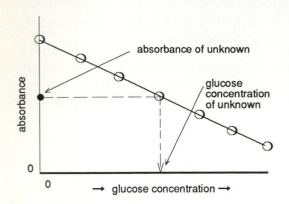

Alternatively, a least-squares equation in the form $A = kC + A_0$ could be obtained with a calculator programmed for a least-squares treatment of data. The program produces the least-squares slope k (negative in this case) and the least-squares intercept $A_0$. The concentrations of the unknown can then be calculated using the absorbances measured for the unknown solutions.

Figure 3.  Calibration Curve

## ACTIVITIES

- Complete the PreLab Exercises before coming to the laboratory.

- Prepare a standard solution of glucose.

- Prepare seven solutions of known glucose concentrations.

- Prepare two solutions of unknown glucose concentrations.

- Determine the absorbances of the seven known and the two unknown solutions after they have reacted with 1.00 mL of Benedict's solution.

- Calculate the concentration of glucose in the unknown solutions before and after dilution with the added water and Benedict's solution.

## SAFETY

*You are working in a laboratory in which chemical reagents are set out; wear approved eye protection.*

# PROCEDURES

## A.  PREPARATION OF SOLUTIONS

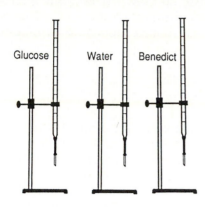

Glucose    Water    Benedict

**A-1.** Select a partner for this experiment or work in groups as designated by your instructor.  Set up three clean burets.  If necessary clean the burets (Appendix D). Label the burets "Glucose", "Water", "Benedict".

**A-2.** Weigh a creased weighing paper on the analytical balance to the nearest 0.1 mg.  Then weigh out a 90 to 100-mg sample of glucose.  Record all masses to the nearest 0.1 mg.

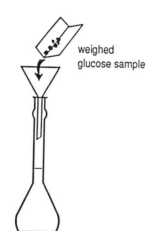

weighed
glucose sample

**A-3.** Place approximately 50 mL of distilled water into a clean 100-mL volumetric flask.  Transfer the glucose sample quantitatively into the volumetric flask.  Rinse any glucose adhering to the funnel or the inside of the neck of the flask into the solution with a stream of distilled water from your squeeze bottle.  Then fill the flask to the mark with distilled water.  Stopper the flask tightly and mix the solution by inverting the flask several times.

waste

**A-4.** Rinse the "Glucose" buret twice with 5-mL portions of the glucose solution.  Then fill this buret to just above the 20-mL mark with the glucose solution.  Similarly, rinse and completely fill the "Water" buret with distilled water, and the "Benedict" buret to just above the 20-mL mark with the Benedict reagent provided by your instructor. Drain enough liquid from each buret into a waste container to expel air from the buret tip.

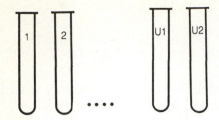

**A-5.** Clean and dry seven test tubes of the appropriate size (volume at least 7 mL). Label the dry test tubes near the top with a grease pencil with numbers 1 through 7. Obtain two test tubes of "unknown" glucose solutions from your instructor. Label these two test tubes with codes assigned by your instructor and record the volumes of the two unknowns.

| Tube No. | mL glucose | mL $H_2O$ |
|----------|-----------|-----------|
| 1 | 0.0 | |
| 2 | 0.5 | |
| 3 | 1.0 | |
| 4 | 2.0 | |
| 5 | 2.5 | |
| 6 | 3.0 | |
| 7 | 4.0 | |

**A-6.** From the glucose buret drain the volumes of glucose solution given in the Table into the test tubes. Record all buret readings to the nearest 0.01 mL. The volumes added do not have to be exactly 0.50, 1.00, or 4.00 mL. Volumes close to these values such as 0.55, 1.03, or 4.12 mL are acceptable and allow faster work.

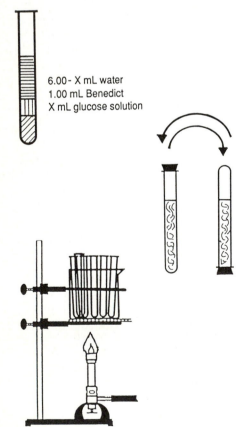

6.00 - X mL water
1.00 mL Benedict
X mL glucose solution

**A-7.** Add to each of the nine test tubes 1.00 mL of Benedict's reagent. This volume must not deviate from 1.00 mL by more than $\pm$ 0.02 mL. Then add a sufficient volume of distilled water to each test tube to reach a total volume of 7.00 mL in each tube. The volumes of distilled water added must not deviate from the calculated volume by more than 0.03 mL. Record all volume readings to the nearest 0.01 mL in your notebook.

**A-8.** Stopper each test tube tightly with a clean, dry stopper. Invert each test tube several times to mix the contents thoroughly. Pour 150 mL water into a 400-mL beaker. Label and unstopper each test tube. Place all nine unstoppered test tubes into the beaker of water. Heat the beaker with a Bunsen flame and bring the water to a gentle boil. Keep the water boiling gently for 30 minutes, then turn off the burner, place the tubes into a test tube rack, and allow them to cool until they can be handled comfortably.

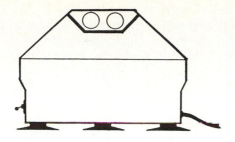

**A-9**. Centrifuge the test tubes for two minutes. Check Appendix E for proper centrifuge techniques. Place the centrifuged tubes into the test tube rack. Avoid stirring up the precipitate at the bottom of the tubes.

## B.  PREPARATION OF THE SPECTROPHOTOMETER

**B-1**. From your instructor obtain information about the spectrophotometer available in your laboratory. Follow the instructions to prepare the photometer for absorbance measurements.  The instructions given here are for the Bausch & Lomb Spectronic 20.

**B-2**. Obtain a cuvet from your instructor.  Turn on the spectrophotometer assigned to your group and allow it to warm up for a few minutes.  Set the wavelength with the wavelength selector knob to 735 nm.

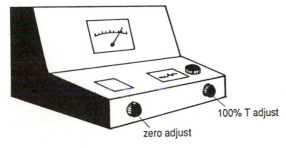

100% T adjust

zero adjust

**B-3**. With the sample chamber empty and closed and the light beam blocked, adjust the "zero transmittance" knob until the meter indicates zero transmittance (100% absorbance). Rinse the cuvet with distilled water and then fill the cuvet with distilled water.  When handling the cuvet, do not touch the optical surfaces.  Make sure that these surfaces are clean.  If they need cleaning because of fingerprints or liquid droplets, consult your instructor.  Insert the cuvet filled with distilled water into the sample chamber with the mark on the cuvet positioned toward the front of the spectrophotometer. Adjust the "100% transmittance" knob until the meter shows 100% transmittance (zero absorbance). Repeat the zero-transmittance and 100% transmittance adjustments until the zero and 100% transmittance readings remain unchanged.

## C.  TRANSMITTANCE (ABSORBANCE) MEASUREMENTS

to aspirator

**C-1**.   Connect a rubber hose of suitable length and diameter to a water aspirator.  Into the hose stick a short piece of a plastic tube with one end drawn to a tip.  With the aspirator turned on, place the end of the plastic tube into the cuvet and remove all of the distilled water.  Carefully – without stirring up the precipitate – rinse the cuvet with a few drops of the solution from test tube No. 1. Completely draw this rinse solution from the cuvet. Then fill the cuvet with the remaining solution from test tube No. 1. Try not to transfer any of the solid. If bubbles adhere to the inside surfaces of the cuvet, shake the cuvet gently to dislodge the bubbles. Insert the cuvet into the sample chamber (mark on the cuvet to the front) and read and record the transmittance (absorbance) value.

**C-2**. Remove the cuvet, draw the solution from the cuvet, rinse it with a few drops of solution from test tube No. 2, and fill the cuvet with the solution from tube No. 2. Measure the transmittance (absorbance). Repeat this procedure until the transmittances (absorbances) of the solutions in all nine test tubes have been measured.

**C-3**. Clean the cuvet, place the spectrophotometer on standby according to directions from your instructor, and clean the test tubes and other glassware.

**C-4**. Calculate the concentration of glucose in your standard solution in units of mg of glucose per milliliter. Then calculate the mg of glucose in 6 mL of the solution (mL standard glucose solution plus mL of distilled water) in each test tube. From these concentrations for test tubes No. 1 through 7 find the mg of glucose in 100 mL of such solutions [(mg glucose in 6 mL) x (100/6)]. Plot the absorbance of each of these solutions versus the glucose concentrations expressed in mg glucose per 100 mL solution. Draw the best straight line through the calibration points. Your instructor may require you to calculate the least-squares intercept and slope using a calculator with a least-squares (straight line) program. Write an equation for the best straight line that describes the calibration curve.

**C-5**. Calculate the concentrations of your two unknown glucose solutions from the measured absorbances using the calibration graph and the least-squares calibration equation. The concentrations of your unknowns are to be reported as the concentrations before and after any water or Benedict's solution was added. Complete the Report Form.

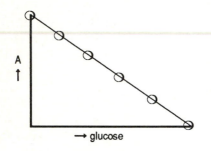

→ glucose

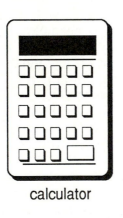

calculator

# PRELAB EXERCISES

## INVESTIGATION 30: SPECTROPHOTOMETRIC DETERMINATION OF GLUCOSE

Name:_____

Instructor:_____     ID No.:_____

Course/Section:_____     Date:_____

1. Calculate the molecular mass of glucose.

2. An aliquot (1.00 mL) of a glucose solution containing 95.1 mg glucose in 100 mL solution is diluted with distilled water to 6.00 mL Calculate the concentration of glucose in this dilute solution in mg glucose per 100 mL solution and in moles per glucose in 1.0 liter of solution.

3. The calibration equation for a spectrophotometric determination was found to be:

$$A = -0.32C + 0.72$$
A: Absorbance     C: Concentration (mg/100 mL)

An unknown solution had an absorbance of 0.20. Calculate the concentration of the unknown solution.

4. Calculate the equation for the linear calibration curve from the following set of absorbance/concentration (mg/100 mL) data: 0.80/0.0. 0.55/20, 0.16/50, 0.69/10, 0.41/30, 0.07/60, 0.30/40. (A least-squares program, such as that on many hand-held calculators, for the equation of a straight line is the preferred method for solving this type of problem. However, if you are unable to use a program on a calculator, you may determine the slope, intercept, and equation by graphical means.)

(over)

345

5. For the experimental conditions described in question 4 the absorbance of an unknown is 0.35. Find the concentration of the unknown. Describe the method you used to obtain your answer.

6. A solution had an absorbance of 0.12. Calculate the transmittance and the percent transmittance.

7. A solution has a percent transmittance of 50. Calculate the transmittance and the absorbance.

8. A solution was determined to contain 10.0 mg of glucose per mL. The solution had been prepared by mixing 1.52 mL of an unknown glucose solution, 1.00 mL of Benedict's solution, and 4.09 mL of water. Calculate the concentration of the original glucose solution.

# REPORT FORM

## INVESTIGATION 30: SPECTROPHOTOMETRIC DETERMINATION OF GLUCOSE

Name: _____     ID No.: _____

Instructor: _____     Course/Section:_____

Partner's Name (if applicable): _____     Date:_____

---

Show all calculations on attached pages.

### Glucose Standard

Glucose + Weighing paper: _____ g     Concentration of the glucose solution:

Weighing paper: _____ g     _____ moles/Liter

Glucose: _____ g     _____ mg glucose/mL

### Data for Glucose Solutions

| Quantity | Tube Number | | | | | | | | |
|---|---|---|---|---|---|---|---|---|---|
|  | 1 | 2 | 3 | 4 | 5 | 6 | 7 | UK1 | UK2 |
| Glucose Solution (mL) | 0.0 | | | | | | | | |
| Benedict (mL) | 1.0_ | 1.0_ | 1.0_ | 1.0_ | 1.0_ | 1.0_ | 1.0_ | 1.0_ | 1.0_ |
| Water (mL) | | | | | | | | | |
| Total Solution Volume (mL) | | | | | | | | | |
| Transmittance | | | | | | | | | |
| % Transmittance | | | | | | | | | |
| Absorbance | | | | | | | | | |
| Total Solution Volume minus Benedict (mL) | | | | | | | | | |
| Glucose Concentration mg/100 mL (dil. sol.) | | | | | | | | | |

347

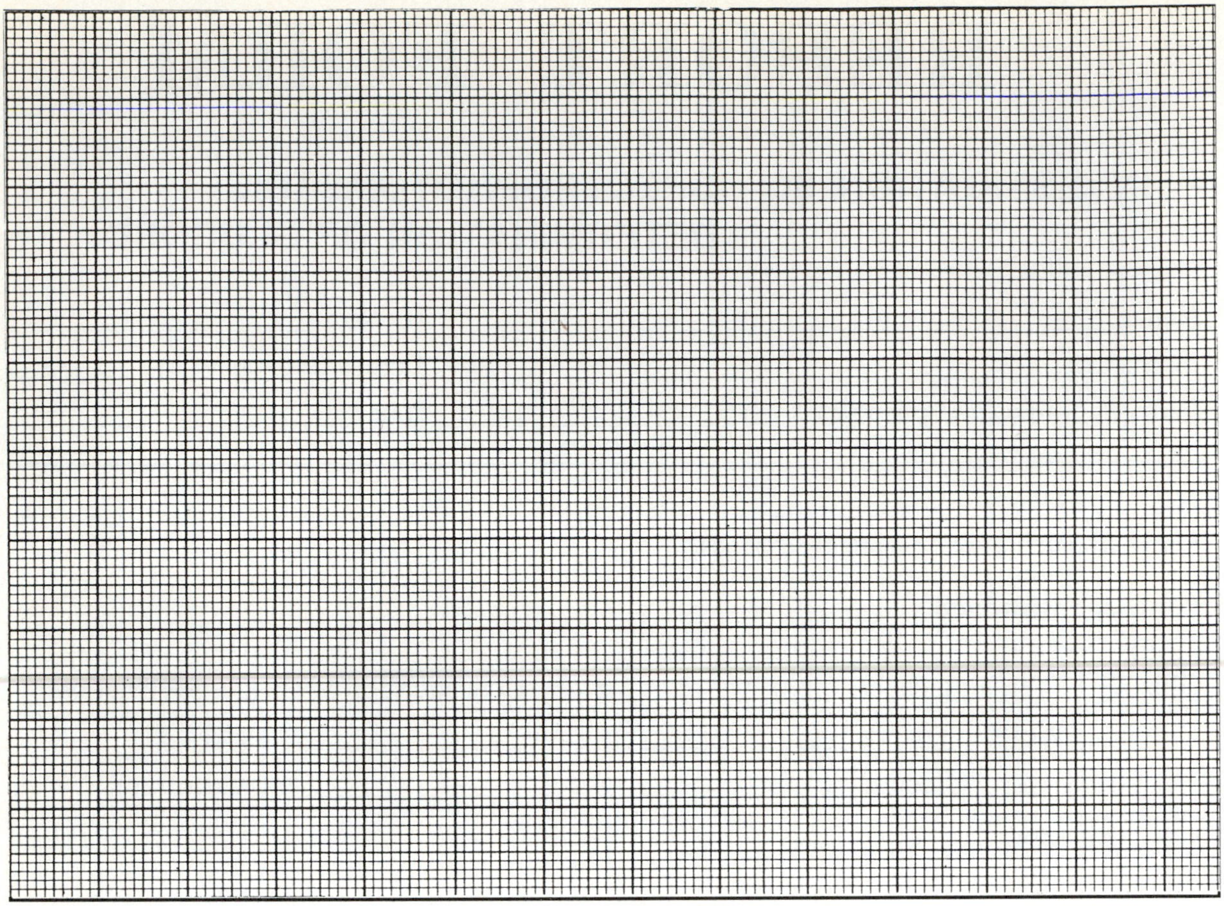

Absorbance ↑

0

0           → mg glucose/100 mL →

Calibration Plot for the Determination of Glucose

Equation (determined by a least-squares program ___ or by graphical means ___):

     Slope k: _____           Intercept $A_0$: _____

          Equation: _____

Unknown Solutions

| Unknown No. | Concentration of Glucose | | | |
| --- | --- | --- | --- | --- |
| | From Calibration Curve | | From Least-Squares Equation | |
| | mg/100 mL | Molarity | mg/100 mL | Molarity |
| 1 | | | | |
| 2 | | | | |

Concentration of Unknown Solutions before Water or Benedict's Solution was added:

         Unknown # ____ = _____mg / mL        Unknown # ____ = _____mg / mL
         (Attach your calculations)

             Date _____    Signature _____

# Investigation 31

## COMPOSITION OF AN Fe(III)-SCN COMPLEX

### INTRODUCTION

Transition metals and even main group elements form coordination compounds often called complexes. When complexes are prepared from a transition metal compound and suitable ligands (see Investigations 15 and 28 for examples), the composition of the complex must be ascertained. This task does not pose any problems, at least in principle, when a complex can be isolated and purified. Elemental analyses performed by commercial laboratories, magnetic measurements, information from ultraviolet, visible, and infrared spectra, data obtained by nuclear magnetic resonance spectroscopy, and single-crystal X-ray crystallography may unequivocally establish the chemical nature and valence state of the central metal atom, the type of ligands, and their mode of bonding. However, many complexes cannot be isolated. They exist only in solution. Attempts to isolate them might change their composition. The isolated complex might not have the same composition as the complex in solution. Our inability to isolate complexes does not make these compounds less important than their isolatable relatives. Many metal complexes that have a major role in the cycling of elements in nature cannot be isolated. Other complexes existing only in solution are used by analytical chemists to detect the presence of a metal ion. One such complex is the deep red coordination compound formed when $Fe^{3+}$ and $SCN^-$ (thiocyanate) ions react with each other. Fortunately, many techniques are now known that allow the determination of the composition of such non-isolatable complexes.

This experiment provides the opportunity to become acquainted with such a method, Job's method of continuous variation, by applying it to the determination of the $Fe^{3+}-SCN^-$ ratio in the dark-red iron-thiocyanate complex.

### CONCEPTS OF THE EXPERIMENT

Solutions of an iron(III) salt react with solutions of potassium thiocyanate to produce a soluble, deep-red iron(III)-thiocyanate coordination compound (equation 1).

$$Fe^{3+} + nSCN^- \rightleftharpoons Fe(SCN)_n^{(3-n)+} \qquad (1)$$

The iron-thiocyanate ratio in this complex can be determined by Job's method of continuous variation. The formation of the deep-red complex is governed by the equilibrium expression

$$K_f = \frac{[Fe(SCN)_n^{(3-n)+}]}{[Fe^{3+}][SCN^-]^n} \qquad (2)$$

(equation 2). $K_f$ is the formation constant for the complex. All concentrations in equation 2 – expressed in moles per liter – are the concentrations at equilibrium. To find the composition of the complex [the "n" in $Fe(SCN)_n$], the concentration of the complex at equilibrium is measured spectrophotometrically in a series of solutions containing $Fe^{3+}$ and $SCN^-$.

The solutions of $Fe^{3+}$ and $SCN^-$ to be mixed must have the same concentration (~0.00160 M). Volumes of these solutions are mixed according to Table 1. When the solutions are prepared in this way, the sum of the initial concentrations of $Fe^{3+}$ and $SCN^-$ before any reaction takes place is the same ($1.60 \times 10^{-3}$ M) in all the solutions (equation 3).

$$[Fe^{3+}]_i + [SCN^-]_i = C \qquad\qquad (3)$$

To simplify the mathematics, $K_f$ (equation 2) can be assumed to be infinite under some circumstances. When $K_f$ is infinite, either $[Fe^{3+}]$ or $[SCN^-]$ in equation 2 can be considered to be zero because the ion that is the limiting reagent ($Fe^{3+}$ or $SCN^-$) will be nearly completely converted to the iron-thiocyanate complex.

To show how Job's method leads to the composition of a complex, the $Fe^{3+}-SCN^-$ system will be discussed with the assumption (that does not correspond to reality) of $Fe(SCN)_3$ as the formula for the complex.

When $[Fe^{3+}]_i = 0$ and $[SCN^-]_i = 0.00160$, no complex can be formed and $[Fe(SCN)_3] = 0$. When $[Fe^{3+}]_i = 0.27 \times 10^{-3}$ M and $[SCN^-]_i = 1.33 \times 10^{-3}$ M, $Fe^{3+}$ is the limiting reagent. To form $Fe(SCN)_3$ from all the iron present, $[SCN^-]_i$ must be at least three times $[Fe^{3+}]_i$ or $0.81 \times 10^{-3}$ M. Because $[SCN^-]_i = 1.33 \times 10^{-3}$, the concentration is sufficient to form $Fe(SCN)_3$, the concentration of which will be $0.27 \times 10^{-3}$ M. The results of similar calculations for the remaining five solutions are listed in Table 2.

Table 1. Iron(III) -Thiocyanate solutions

| Reagent | solution (test tube) No. | | | | | | | | |
|---|---|---|---|---|---|---|---|---|---|
| | 1 | 2 | 3 | 4 | 5 | 6 | 7 | 8 | 9 |
| mL $Fe^{3+}$ | 0 | 1 | 2 | 2.5 | 3 | 3.5 | 4 | 5 | 6 |
| mL $SCN^-$ | 6 | 5 | 4 | 3.5 | 3 | 2.5 | 2 | 1 | 0 |
| total mL | 6 | 6 | 6 | 6 | 6 | 6 | 6 | 6 | 6 |

Table 2. Data calculated for a complex with an assumed formula of $Fe(SCN)_3$*

| Reagent | solution (test tube) No. | | | | | | |
|---|---|---|---|---|---|---|---|
| | 1 | 2 | 3 | 5 | 7 | 8 | 9 |
| $[Fe^{3+}]_i$ | 0.0 | 0.27 | 0.53 | 0.8 | 1.07 | 1.33 | 1.6 |
| $[SCN^-]_i$ | 1.6 | 1.33 | 1.07 | 0.8 | 0.53 | 0.27 | 0.0 |
| {$[Fe^{3+}]_i + [SCN^-]_i$} | 1.6 | 1.6 | 1.6 | 1.6 | 1.6 | 1.6 | 1.6 |
| $[Fe^{3+}]_{eq}$** | 0.00 | 0.00 | 0.17 | 0.53 | 0.89 | 1.24 | 1.6 |
| $[SCN^-]_{eq}$** | 1.6 | 0.52 | 0.00 | 0.00 | 0.00 | 0.00 | 0.0 |
| $[Fe(SCN)_3]_{eq}$** | 0.00 | 0.27 | 0.36 | 0.27 | 0.18 | 0.09 | 0.0 |

\* All concentrations were multiplied by 1000. For example, 0.00027 M is listed as 0.27(0.00027 x 1000).

\*\* These values were calculated under the assumption $K_f = \infty$ and $Fe(SCN)_3$ as the formula for the complex.

Inspection of the last entries in Table 2 reveals that [Fe(SCN)$_3$] increases, reaches a maximum corresponding to a solution with initial concentrations for $Fe^{3+}$ and $SCN^-$ between the concentrations given for solution No. 2 and solution No. 3, and then decreases to zero. The exact location of the maximum can be determined by plotting [Fe(SCN)$_3$]$_{eq}$ versus [Fe$^{3+}$]$_i$ (Fig. 1) and extending the solid lines until they intersect. The initial concentrations of $Fe^{3+}$ and $SCN^-$ that would give this maximal Fe(SCN)$_3$ concentration can be read from the [Fe$^{3+}$]$_i$/[SCN$^-$]$_i$ axis as [Fe$^{3+}$]$_i$ = 0.0004 M and [SCN$^-$]$_i$ = 0.0012 M. In a solution initially containing $Fe^{3+}$ at 0.0004 moles/liter and $SCN^-$ at 0.0012 moles/liter, all $Fe^{3+}$ and $SCN^-$ is used up to form the complex because [Fe$^{3+}$]$_i$/[SCN$^-$]$_i$ = 1/3 as required by the stoichiometry of the complex. For the general stoichiometry Fe(SCN)$_n$, the ratio of [Fe$^{3+}$]$_i$/[SCN$^-$]$_i$ at the maximum of the curve will be equal to 1/n, from which n can be obtained.

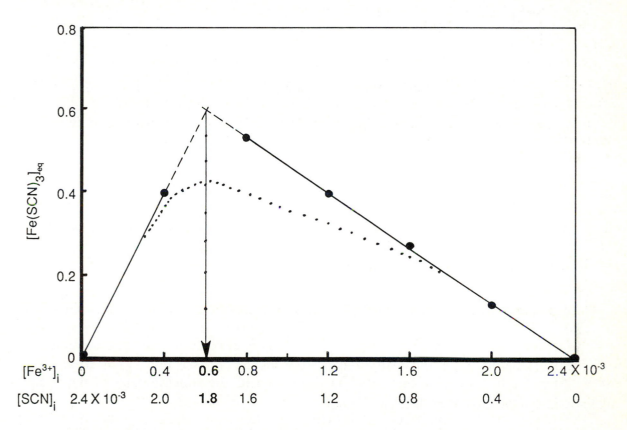

Fig. 1.  Job plot for the $Fe^{3+}$–$SCN^-$ system under the assumption that the formula for the complex is Fe(SCN)$_3$. The concentrations used in this plot are listed in Table 2.

A detailed and much more involved mathematical analysis proves that the location of the maximum in a Job's plot depends only on the stoichiometry of the complex and not on the value of $K_f$. As $K_f$ departs from the assumed value of infinity toward smaller values, the maximum broadens and the plot may take the shape of the dotted line in Fig. 1. With such a shape the location of the maximum (the point on the curve through which the tangent is parallel to the [Fe$^{3+}$]$_i$ axis) is graphically not determinable as precisely as in the case $K_f = \infty$.

To produce a Job's plot, the concentration of the complex at equilibrium must be determined in the solutions. In the $Fe^{3+}$-$SCN^-$ system, the concentration of the complex is conveniently obtained by spectrophotometry. The Fe-SCN complex absorbs light of 450 nanometers ($1nm = 1 \times 10^{-9}$m). The reagents and the solvent (water) are transparent at this wavelength. When light of 450 nm with an incident intensity $I_0$ is passed through a solution of the complex, some of the light is absorbed by the complex. The light after passage through the solution has an intensity I ($I < I_0$). The absorbance A of the solution is defined by equation 4. If no light is absorbed, $I = I_0$, $I/I_0 = 1$,

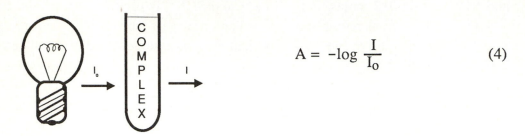

$$A = -\log \frac{I}{I_0} \qquad (4)$$

and $A = 0$. If one tenth of the incident light is absorbed, nine tenths pass through, $I = 0.9I_0$, $I/I_0 = 0.9$, and $A = 0.046$. When nine tenths of the incident light is absorbed, $I = 0.1I_0$, $I/I_0 = 0.1$, and $A = 1.0$. The absorbance is related to the concentration of the complex in solution by Beer's Law (equation 5).

$$A = \varepsilon \times d \times [Fe(SCN)_n^{(3-n)+}] \qquad (5)$$

    d:   length of the lightpath within the solution (cm)

    $\varepsilon$:   molar absorption coefficient (L mol$^{-1}$cm$^{-1}$); the value for $\varepsilon$ is given by the absorbance of a 1.0 molar solution of 1 cm pathlength

    $[Fe(SCN)_n^{(3-n)+}]$:   concentration of the complex (mol L$^{-1}$)

In this experiment the pathlength is kept constant. Because $\varepsilon$ is a constant characteristic for the iron-thiocyanate complex, the absorbance measured with a spectrophotometer is proportional to the concentration of the complex. A plot of the absorbances determined for the various solutions listed in Table 1 versus the initial concentrations of $Fe^{3+}$ and $SCN^-$ (see Fig. 1) will produce a Job's curve. After the maximum of this curve has been located, the stoichiometry of the complex is calculated from the ratio $[Fe^{3+}]_i^{max}/[SCN^-]_i^{max}$.

The broader the maximum in the Job's curve, the more uncertain the exact location of the maximum becomes. The experimental ratio $[Fe^{3+}]_i/[SCN^-]_i$ must, therefore, be rounded up or down. If the broad maximum in Fig. 1 would have given a ratio of $0.58/1.82 = 0.32 = 10/31$, one could have concluded that the formula for the complex is $[Fe_{10}(SCN)_{31}]^-$. Although such high molecular mass complexes (called clusters) are known, they are not very common. Complexes in solution generally have a simple stoichiometry such as 1:1, 1:2, 1:3, 1:4, 1:6. Therefore, it is justified to round the experimental ratio to a ratio corresponding to a stoichiometry expressible in small numbers. $Fe(SCN)_4$ requires $1/4 = 0.25$; $Fe(SCN)_3$ $1/3 = 0.33$, and $Fe(SCN)_2$ $1/2 = 0.5$. The experimental ratio of 0.32 is closest to 0.33 (difference only 0.01) and rounding up from 0.32 to 0.33 is justified. The complex would then be thought to have the composition of $Fe(SCN)_3$.

As the number of ligands per metal ion increases, the fractions expressing these stoichiometries in the Job's plot are not very different. A complex $ML_6$ has a ratio of 0.16, and $ML_7$ has a ratio of 0.14. This difference is the same as the experimental error in the example above. Job's method will not be able to differentiate between $ML_6$ and $ML_7$ under these conditions.

Job's method as described here is only applicable to systems in which the reagents do not absorb light at the wavelength the complex absorbs. Difficulties will also be encountered when more than one complex forms and the complexes have similar absorption spectra. Fortunately, the $Fe^{3+}$-$SCN^-$ system investigated in this experiment is not troubled by these limitations.

To obtain a value for the formation constant $K_f$ (equation 2), the equilibrium concentration of the complex, the stoichiometry of the complex, and the equilibrium concentrations of uncomplexed $Fe^{3+}$ and $SCN^-$ must be known. The stoichiometry for the complex is known from the Job's plot. The concentration of the complex is obtainable from the measured absorbance provided $\varepsilon$ and d are known. An approximate value for $\varepsilon$ can be calculated using the absorbances measured for solutions No. 2 and 8 (Table 1). In these mixtures either $Fe^{3+}$ or $SCN^-$ is present in concentrations several times higher than required by stoichiometry. Under the assumption that the concentration of the complex is either equal to $[Fe^{3+}]_i$ or $[SCN^-]_i/n$, the molar absorption coefficient $\varepsilon$ can be calculated (equation 6). The value of d can be obtained by measuring the inner diameter of the test tube.

$$\varepsilon = \frac{A_{(2 \text{ or } 8)}}{[Fe(SCN)_n^{(3-n)+}] \times d_{(2 \text{ or } 8)}} \qquad (6)$$

With values of $\varepsilon$ and d known, and A measured for the solutions, the concentration of the complex in a solution can be calculated (equation 7). With $[Fe(SCN)_n]^{3-n}$ known, the equilibrium

$$[Fe(SCN)_n^{(3-n)+}] = \frac{A}{\varepsilon x d} \qquad (7)$$

concentrations of $[Fe^{3+}]$ and $[SCN^-]$ are given by equations 8 and 9.

$$[Fe^{3+}]_{eq} = [Fe^{3+}]_i - [Fe(SCN)_n^{(3-n)+}] \qquad (8)$$

$$[SCN^-]_{eq} = [SCN^-]_i - n[Fe(SCN)_n^{(3-n)+}] \qquad (9)$$

The expression for $K_f$, from which an approximate value for $K_f$ can be calculated, is given by equation 10.

$$K_f = \frac{A}{\varepsilon d ([Fe^{3+}]_i - [Fe(SCN)_n^{(3-n)+}])([SCN^-]_i - n[Fe(SCN)_n^{(3-n)+}])^n} \qquad (10)$$

or

$$K_f = \frac{[complex]_{eq}}{[Fe^{3+}]_{eq} ([SCN^-]_{eq})^n}$$

## ACTIVITIES

- Become familiar with the principles of Job's method.
- Prepare nine solutions in which the sum of the iron and thiocyanate concentrations is constant.
- Measure the absorbances of these solutions.
- Prepare a Job's plot.
- Determine the composition of the Fe-SCN complex.
- Calculate the formation constant for this complex.

## SAFETY

*Wear approved eye protection at all times. Dispose of chemicals as directed by your instructor. All chemical spills are to be cleaned up at once as directed by your instructor.*

## PROCEDURES

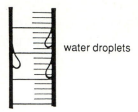

water droplets

**1.** Obtain two burets. Fill the burets with distilled water and drain them through the stopcock. Examine the inside of the buret. Should water drops clinging to the inside wall be visible, clean the buret. Label one buret "Fe", the other "SCN".

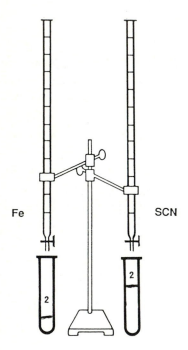

Fe         SCN

**2.** Take two appropriately labeled, clean and dry 250-mL Erlenmeyer flasks to your instructor to obtain 50 mL each of 0.0016 M $Fe(NO_3)_3$ in 0.1 M $HNO_3$ and 0.0016 M KSCN in 0.1 M $HNO_3$ solutions. Rinse the "Fe" buret with the $Fe^{3+}$ solution, the "SCN" buret with the $SCN^-$ solution. Fill each buret to just above the 25-ml mark with its designated solution. Discharge any air bubbles that might be trapped in the buret tip.

**3.** Obtain nine small, clean and dry matching test tubes with a volume of 8 to 10 mL. Label these tubes "1", "2", ..."9" on the top part of each tube.

**Note:** *Do not touch the middle sections of these tubes with your fingers. Grease or oil on the tube surface will affect the absorbance measurements.*

Add to each test tube the volume of $Fe^{3+}$ and $SCN^-$ solutions given in Table 1. Record the volumes in your notebook. Mix the contents of each tube well by rocking the tube and vigorously tapping it several times.

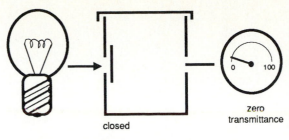

closed

zero
transmittance

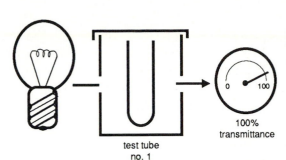

test tube
no. 1

100%
transmittance

**4.** Turn on the spectrophotometer, allow it to warm up, and set the wavelength to 450 nm. With the empty sample chamber closed, adjust the "zero transmittance" knob until the dial reads zero transmittance. Place test tube No. 1 into the sample chamber and adjust the "100% transmittance" knob until the dial reads 100% transmittance (absorbance = 0).

Repeat the zero-transmittance and 100% transmittance adjustments until the dial remains at zero (chamber closed, empty) and at 100 (test tube No. 1 in chamber).

**5.** Measure the absorbances of the solutions in test tubes No. 2 through 9. Record all absorbance values in your notebook. Measure the inner diameters of your test tubes.

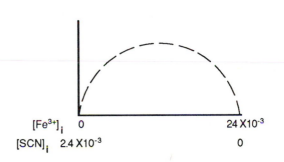

| | | |
|---|---|---|
| $[Fe^{3+}]_i$ | 0 | $24 \times 10^{-3}$ |
| $[SCN]_i$ | $2.4 \times 10^{-3}$ | 0 |

**6.** Plot "absorbance" versus initial concentration of $Fe^{3+}$ and $SCN^-$. Connect the points by a smooth line. French curves will be useful in this task. Locate the maximum and the concentration of $[Fe^{3+}]_i$ and $[SCN^-]_i$ corresponding to this maximum. From the ratio of these concentrations determine "n" in $Fe(SCN)_n^{(3-n)+}$. You may have to round up or round down your concentration ratio to obtain a reasonable composition for the complex.

**7.** If time permits, make additional solutions and measure their absorbances to better define the maximum in the plot.

**8.** Use the procedure discussed in the last part of the section "Concepts of the Experiment" and the expression for $K_f$ derived there (equation 10) and your experimental value for n to calculate an approximate value for the formation constant.

# PRELAB EXERCISES

## INVESTIGATION 31: COMPOSITION OF AN Fe(III)-SCN COMPLEX

Name:________________________________

Instructor:________________________________ ID No.:________________________________

Course/Section:____________________________ Date:________________________________

---

1. Define:

   Absorbance

   Coordination compound

   Complex (Does your textbook distinguish between coordination compound and complex?)

   Nanometer

   Spectrophotometer

2. Write an equation expressing Beer's Law and define all terms in the equation.

(over)

3. What are the units along the horizontal axis (abscissa) of a Job's plot?

4. Which range of wavelengths, expressed in nanometers, covers the visible range of the electromagnetic spectrum? (Hint: Check your textbook for assistance with this question.)

# REPORT FORM

## INVESTIGATION 31: COMPOSITION OF AN Fe(III)-SCN COMPLEX

Name: _____    ID No.: _____

Instructor: _____    Course/Section: _____

Partner's Name (if applicable): _____    Date:_____

---

Concentration of the $Fe^{3+}$ solution (3 significant figures):    _____ M

Concentration of the $SCN^-$ solution (3 significant figures):    _____ M

Diameter of the test tube: _____ cm

| Tube No. | 1 | 2 | 3 | 4 | 5 | 6 | 7 | 8 | 9 |
|---|---|---|---|---|---|---|---|---|---|
| mL $Fe^{3+}$ | 0.00 | | | | | | | | |
| mL $SCN^-$ | | | | | | | | | 0.00 |
| Absorbance | | | | | | | | | |
| $[Fe^{3+}]_i$ | 0.00 | | | | | | | | |
| $[SCN^-]_i$ | | | | | | | | | 0.00 |

Attach a plot of absorbance versus $([Fe^{3+}]_i + [SCN^-]_i)$.  [See Figure 1].

Based upon the plot of absorbance versus $([Fe^{3+}]_i + [SCN^-]_i)$ the formula of the complex is:

_____.

| Tube No. | 1 | 2 | 3 | 4 | 5 | 6 | 7 | 8 | 9 |
|---|---|---|---|---|---|---|---|---|---|
| Absorbance | | | | | | | | | |
| $[\text{complex}]_{eq}$ | 0.00 | | | | | | | | 0.00 |
| $[Fe^{3+}]_{eq}$ | 0.00 | | | | | | | | |
| $[SCN^-]_{eq}$ | | | | | | | | 0.00 | 0.00 |
| $\varepsilon$ | XX | | XX | XX | XX | XX | XX | | XX |
| $K_f$ | XX | XX | | | | | | XX | XX |

Average $\varepsilon$ (from tubes 2 & 8) = _____

$K_f$ average = _____

Standard deviation and relative standard deviation of the calculated values of $K_f$:

$$\text{Standard deviation (S)} = \sqrt{\frac{\sum_1^n \left(x_i - \overline{x}\right)^2}{n-1}} = \underline{\hspace{3cm}}$$

$$\text{Relative standard deviation} = \frac{100S}{\overline{x}} = \underline{\hspace{3cm}}$$

Calculate from your experimental results the difference between the "rounded ratio" of $[Fe^{3+}]_i/[SCN^-]_i$ and "experimental ratio". Calculate the "theoretical ratios" for complexes of the formula $Fe(SCN)_n^{(3-n)}$ with n = 3, 4, 5, 6, or 7. Assume that the difference "rounded" to "experimental ratio" from your experiment is the same for the complexes with n = 3 through 7. Can Job's method under these conditions distinguish the complexes with n = 3 and 4, 4 and 5, 5 and 6, 6 and 7? Explain your answer on an attached sheet.

Date _____ Signature _____

# Investigation 32

## COLORIMETRIC DETERMINATION OF COPPER AS [Cu(ethylenediamine)$_2$]$^{2+}$

## INTRODUCTION

Every day 250 million analyses are carried out in the United States. Heavy metals, such as cadmium and lead, are determined in vegetables and drinking water; glucose is determined in blood and urine of patients suspected of suffering from diabetes; drinking water is analyzed for nitrate to protect children who are sensitive to this ion. A variety of methods are used for these determinations. Color is one of the characteristics of a substance that may be used to qualitatively detect its presence. When the intensity of the color can be ascertained, the concentration or quantity of the colored substance in a sample can be determined. For instance, one uses the "darkness" of the brown color of iced tea to judge whether the tea is weak or strong. If a substance does not have a color of its own, the possibility exists of reacting the colorless substance with an appropriate reagent to give a colored product. Many such color-producing reactions are in use as clinical tests, for instance, for glucose in blood (Investigation 30).

Although the types of color (blue, yellow, green, red) can be easily distinguished by the human eye, the intensity of a color is much more difficult to determine with the naked eye. Therefore, spectrophotometers (colorimeters) were developed that electronically measure the intensity of a light beam before it passes through a solution and after it has passed through a solution. A device that is part of a photometer provides light of defined wavelengths. With such a device (a monochromator or a series of filters) the wavelengths of light that are absorbed by a solution can be determined.

In this experiment the very weakly-colored hydrated Cu$^{2+}$ ion, [Cu(H$_2$O)$_n$]$^{2+}$ is reacted with ethylenediamine to produce the deeply-colored [Cu(H$_2$NCH$_2$CH$_2$NH$_2$)$_2$]$^{2+}$ complex. The wavelengths of light that are absorbed by solutions of this copper complex (determination of the absorption curve) are identified and the dependence of the intensity of the color or the concentration of the complex is elucidated (establishment of a calibration curve). The concentration of two unknown solutions containing Cu$^{2+}$ are then determined.

## CONCEPTS OF THE EXPERIMENT

"White light" is light that contains all wavelength of electromagnetic radiation in the wavelength region (400-800 nm) visible to the human eye. When such white light passes through a colored solution, light of a certain narrow range of wavelengths will be absorbed and light of all other wavelengths will pass through the solution. The solution will have the color complementary to the color absorbed.

When monochromatic light (light of only one wavelength) is passed through the solution, the light will be strongly absorbed, weakly absorbed, or not absorbed at all. When one measures the intensity of the incident and the emerging light at various wavelengths, an absorption curve is obtained. The monochromator in the spectrophotometer separates the white light into its color (wavelengths). The white light comes from a lamp.

**Cuvet**

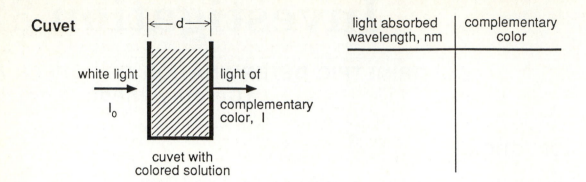

| light absorbed wavelength, nm | complementary color |
| --- | --- |
| | |

The light intensity is measured electronically. If the incident light has an intensity of $I_0$ and the transmitted light has an intensity I, then the transmittance is defined as the fraction $I/I_0$. Because I can take on all values from 0 (no light coming through the solution) to $I_0$ (all incident light passes through), the transmittance can take on values from 0 to 1. Spectrophotometers have scales of "percent transmittance" (%T) defined as $100\ I/I_0$ with possible values between 0 and 100. Most spectrophotometers also have an absorbance (A) scale. Absorbance is defined as the negative logarithm of transmittance (equation 1).

$$A = -\log\frac{I}{I_0} = -\log\frac{\%T}{100} = \log\frac{100}{\%T} \tag{1}$$

The absorbance is found to be proportional to the concentration C of the absorbing substance in solution and the distance d, through which the light travels in the absorbing solution (equation 2).

$$A = \varepsilon Cd \tag{2}$$

$$\varepsilon = A\ /\ Cd \tag{3}$$

The proportionality constant $\varepsilon$ is known as the molar absorptivity. $\varepsilon$ has the units $mol^{-1}\ L\ cm^{-1}$ (equation 3) and is the absorbance of a one centimeter thick layer of a 1.0 M solution. $\varepsilon$ changes with wavelength and is generally given for the wavelengths corresponding to the maxima in the absorption curve.

The absorption curve of a colored solution is obtained when its absorption is measured at various wavelengths and the absorbance plotted versus wavelength. Recording spectrophotometers vary the wavelength continuously and draw a continuous absorption curve. As an example, the absorption curve of an aqueous 0.10 M solution of copper sulfate is shown.

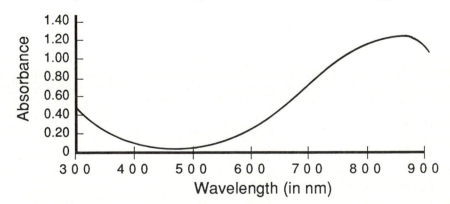

When absorbance measurements are to be used for the quantitative determination of a substance, the wavelength at which the absorption curve has a maximum is often chosen. Absorption curves may have more than one maximum. However, the absorption must not be too strong because low light intensities cannot be measured accurately. The best range is 15 to 65 percent transmittance (0.19 to 0.82 absorbance). A determination in this range can be performed by picking an appropriate wavelength (not necessarily at a maximum) or by dilution of a solution that absorbs too strongly. For precise measurements at a wavelength chosen, the absorbance should not change much with a change in wavelength. This requirement is especially important when the spectrophotometer is not equipped with a good monochromator and provides a band of wavelengths instead of monochromatic light. If light of all the wavelengths in this band are absorbed nearly to the same extent, only a small error is introduced by the non-monochromaticity of the light.

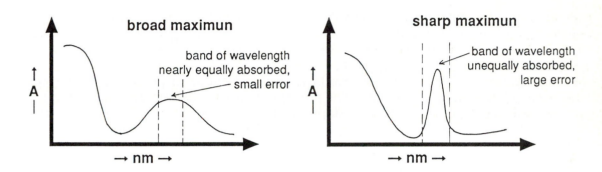

Beer's Law (equation 2) is used for the quantitative determination of a concentration. In principle, the absorbance of a standard solution of known concentration and the absorbance of the unknown could be measured. The standard and the unknown must have the same absorbing species. Then the molar absorptivity is the same. The resulting equation 6 has the concentration of the unknown as the only not-known quantity (equations 3-6).

$$A_S = \varepsilon C_S d \qquad (3) \qquad\qquad A_{UK} = \varepsilon C_{UK} d \qquad (4)$$

$A_S$: absorbance of standard

$C_S$: concentration of standard

$\varepsilon$: molar absorptivity

$d$: thickness of absorbing layer

$A_{UK}$: absorbance of unknown

$C_{UK}$: concentration of unknown

$$\varepsilon d = \frac{A_S}{C_S} = \frac{A_{UK}}{C_{UK}} \qquad (5) \qquad\qquad C_{UK} = \frac{A_{UK} C_S}{A_S} \qquad (6)$$

Because the measurement of the absorbance of a single standard could be seriously in error, a calibration curve is usually established by measuring the absorbances of at least three standards with different concentrations. Absorbance is then plotted versus concentration and a straight line drawn through the experimental points. Measurements that are in error (e.g., point E) are easily recognized in such a plot and can be eliminated.

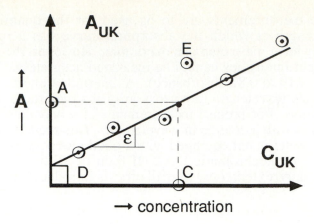

→ concentration

The concentration of an unknown is obtained by drawing a parallel to the concentration axis at the absorbance value of the unknown until the parallel intersects the calibration line. At the intersection a parallel is drawn to the absorbance axis. The intersection of this parallel with the concentration axis gives the concentration of the unknown.

In a more rigorous alternate way, the absorbance and concentration values can be entered into a calculator with a least-squares program. The program will calculate the best straight line for the experimental points entered and report the slope and the intercept of the best line represented by a linear equation (equation 7). The program will also calculate statistical parameters for the fit of the

$$A = \varepsilon C + D \tag{7}$$

A: absorbance

C: concentration (mole $L^{-1}$)

$\varepsilon$:  molar absorptivity (slope of calibration curve)

D: intercept on absorbance axis

straight line to the experimental points. In the ideal case the intercept D should be zero. After $\varepsilon$ and D are calculated, the absorbance of the unknown is entered into equation 7 and the equation solved for the concentration of the unknown.

Copper ions in aqueous solution have a blue color of low intensity. The molar absorptivity is low. When ethylenediamine is added to such a $Cu^{2+}$ solution, the deeply colored complex of $Cu^{2+}$ with ethylenediamine forms (equation 8).

$$Cu^{2+}(H_2O)_n + 2\ H_2NCH_2CH_2NH_2 \longrightarrow \left[ \begin{array}{c} CH_2\!-\!N \\ | \\ CH_2\!-\!N \end{array} \overset{H_2}{\underset{H_2}{\phantom{x}}} Cu \overset{H_2}{\underset{H_2}{\phantom{x}}} \begin{array}{c} N\!-\!CH_2 \\ | \\ N\!-\!CH_2 \end{array} \right]^{2+} \tag{8}$$

The molar absorptivity of the complex is much higher than the molar absorptivity of the hydrated $Cu^{2+}$ ion. Therefore, much smaller concentrations of copper can be measured using the ethylenediamine-$Cu^{2+}$ complex. At sufficiently high concentrations of ethylenediamine, almost all of the $Cu^{2+}$ ions are present as the complex.

The detection limit of a photometric determination, the lowest concentration that can still be determined, depends on the molar absorptivity and the ability of the spectrophotometer to measure small absorbances at a constant pathlength of the cuvet. Beer's law (equation 2) can be used to calculate detection limits when $\varepsilon$ and the smallest measurable absorbance are known.

## ACTIVITIES

- Complete the PreLab Exercises before coming to the laboratory.

- Prepare 100 mL of a 0.010 M solution of $CuSO_4$.

- Prepare standard solutions of $[Cu(H_2NCH_2CH_2NH_2)_2]^{2+}$.

- Determine the absorption curve of the Cu-ethylenediamine complex, the wavelength of the maximum, and the molar absorptivity at the maximum.

- Establish a calibration curve for the determination of copper as the copper-ethylenediamine complex.

- Determine the concentration of two unknowns.

- Complete the Report Form.

## SAFETY

*You will be working with some potentially dangerous reagents. Wear approved eye protection at all times. In case of contact with chemicals, wash the affected areas immediately with lots of water.*

## PROCEDURES

### A.  PREPARATION OF SOLUTIONS

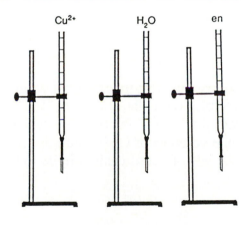

**A-1**.  Select a partner or work with the partner assigned to you by your instructor. Set up three clean burets and label them "$Cu^{2+}$", "$H_2O$", and "en" (ethylenediamine).

**A-2**.  Place a piece of weighing paper on the pan of an analytical balance. Weigh to the nearest 0.1 mg the amount of $CuSO_4 \cdot 5H_2O$ needed to prepare 100 mL of a 0.010 M solution of $CuSO_4$. A mass within $\pm 10$ mg of the theoretical amount is acceptable. Record all masses. Avoid spillage of $CuSO_4$ during the weighing process. Clean up if spillage did occur.

**A-3**.   Place a funnel into a clean but not necessarily dry 100-mL volumetric flask.  Transfer the copper sulfate quantitatively from the weighing paper into the flask.  Rinse the last traces of copper sulfate from the funnel into the flask with a stream of distilled water from your washbottle.  Fill the flask half full with distilled water.  Swirl the flask until all the crystals have dissolved.  Then fill the flask with distilled water to the 100-mL mark.  Stopper the flask snugly and invert it several times to make the solution homogeneous.  Calculate the concentration of the $Cu^{2+}$ solution.

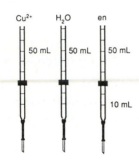

**A-4**.   Rinse the "$Cu^{2+}$" buret twice with 5-mL portions of the $Cu^{2+}$ solution.  Fill this buret with the $Cu^{2+}$ solution.  Rinse and fill the second buret with distilled water.  Rinse the third buret with a small volume of the 1.0 M ethylenediamine (en) solution provided by your instructor.  Fill the "en" buret only to the 10-mL mark.  Drain enough liquid from each buret into a "waste" flask to expel air from the buret tips.

| Test Tube | mL $Cu^{2+}$ |
|:---------:|:------------:|
| 2 | 0.5 |
| 3 | 1.0 |
| 4 | 2.0 |
| 5 | 3.0 |
| 6 | 4.0 |
| 7 | 5.0 |

**A-5**.   Clean and dry nine test tubes with a volume not smaller than 8 mL.  Label the test tubes "1", "2", ..."9".  Drain from the "$Cu^{2+}$" buret into test tubes "2" through "7" the volumes of $Cu^{2+}$ solution given in the Table to the left.  Record each volume to the nearest 0.01 mL.

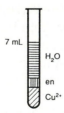

**A-6**.   To each of the test tubes "1" through "7" add 0.5 mL of the 1.0 M solution of ethylenediamine and a sufficient volume of distilled water to bring the total volume of liquid in each test tube to 7.0 mL.  Record all volumes to the nearest 0.01 mL.

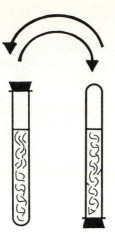

**A-7.** Stopper each test tube tightly with a clean and dry stopper. Invert each tube several times to thoroughly mix the reagents. Set the test tubes into a test tube rack. Calculate the concentrations of copper in each of these solutions.

## B.  SETUP OF THE SPECTROPHOTOMETER

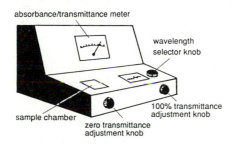

absorbance/transmittance meter

wavelength selector knob

sample chamber

zero transmittance adjustment knob

100% transmittance adjustment knob

**B-1.** Turn on the spectrophotometer assigned to you and allow the instrument to warm up for several minutes. Obtain a pair of cuvets from your instructor. Set the wavelength selector knob to 400 nm.

**B-2.** Fill one of the cuvets with distilled water. Check that no air bubbles adhere to the sides of the cuvet. If air bubbles are visible, shake the cuvet gently to dislodge the bubbles. Rinse the second cuvet with a few drops of the solution in test tube "7". Then fill the cuvet with this solution.

**B-3.** Having allowed the spectrophotometer to warm up, with the wavelength set at 400 nm and the empty sample chamber closed, adjust the "zero transmittance" knob until the dial reads zero transmittance. Place the cuvet of distilled water into the sample chamber and adjust the "100% transmittance" knob until the dial reads 100% transmittance (absorbance = 0).

Repeat the zero-transmittance and 100% transmittance adjustments until the dial remains at zero (chamber closed, empty) and at 100 (cuvet of distilled water in the chamber).

**B-4.** Place the second cuvet (from test tube 7) into the sample chamber. The mark on the cuvet must be positioned toward the front of the instrument.

## C. ABSORPTION CURVE OF THE $[Cu(en)_2]^{2+}$ COMPLEX

**C-1.** Read the percent transmittance and absorbance obtained in Procedure B-4. Record both values in your notebook.

**C-2.** Move the wavelength selector knob to 420 nm. Repeat Procedures B-3 and B-4, except this time the wavelength is set to 420 nm. Read the percent transmittance and absorbance. Record both values in your notebook.

**C-3.** Repeat Procedure C-2, but this time advance the wavelength selector knob another 20 nm.

**C-4.** Repeat Procedure C-3 until you have reached 600 nm.

**C-4.** Plot absorbance (y-axis) versus wavelength (x-axis). Locate the maximum of the absorption curve of $Cu(en)_2^{2+}$. Calculate $\varepsilon$ at the wavelength that gave the maximal absorption. Assume that all the $Cu^{2+}$ was present as the en-complex.

## D. CALIBRATION CURVE

**D-1.** Set the wavelength selector at the wavelength corresponding to the maximum of the absorption curve. Using the cuvet containing the distilled water, adjust "0" and "100%" transmittance (Procedure B-3). Bring the $Cu(en)_2^{2+}$ solution (test tube "7") into the beam and measure the transmittance and absorbance. Record the test tube number and the transmittance and absorbance.

**D-2.** Remove the cuvet with the $Cu(en)_2^{2+}$ solution from the sample chamber and place it on a clean towel. Connect a hose to an aspirator and a piece of plastic tube drawn into a fine point to the hose. Turn the aspirator on and draw the solution from the cuvet. While handling the cuvet, never touch the optical surfaces of the cuvet with your fingers. Place a small volume of solution from test tube "6" into the cuvet. Swirl the solution in the cuvet. Then draw the solution from the cuvet. Fill the cuvet with solution from test tube "6" and place the filled cuvet into the sample chamber. Check "0" and "100%" transmittance (Procedures B-3), place the cuvet with the $Cu(en)_2^{2+}$ solution into the light beam, and read and record the transmittance and absorbance.

**D-3**. Repeat Procedure D-2 with the solutions from test tubes "5", "4", "3", "2", and "1".

**D-4**. Construct a calibration curve by plotting absorbance versus copper concentrations in the solutions. If you have a calculator with a least-squares (straight line) program, obtain the slope and intercept and write an equation for the calibration curve in the form $A = \varepsilon[Cu^{2+}] + D$.

## E.  DETERMINATION OF UNKNOWN COPPER CONCENTRATIONS

**E-1**. Take two clean, dry test tubes to your instructor and exchange them for two test tubes containing unknown $Cu^{2+}$ solutions. The test tubes will be labeled with a code "U-#" and a volume.

**E-2**. To each of your two unknowns add 0.5 mL of the 1.0 M ethylenediamine solution and a volume of distilled water to bring the total volume of solution to 7.0 mL. Stopper the test tubes and invert them several times.

**E-3**. Measure the transmittance and absorbance of these solutions at the wavelength corresponding to the maximum in the absorption curve (Procedure C-4).

**E-4**. Obtain the copper concentrations in the two solutions (7 mL) from the calibration curve or from the straight line equation. Complete the Report Form.

**E-5**. Clean the cuvets, burets, and all other glassware. Switch off the power to the spectrophotometer.

# PRELAB EXERCISES

## INVESTIGATION 32: COLORIMETRIC DETERMINATION OF COPPER AS [Cu(ethylenediamine)$_2$]$^{2+}$

Name:_____

Instructor:_____          ID No.:_____

Course/Section:_____      Date:_____

1. Define:

   Molar absorptivity

   Light transmittance

   Absorbance

   Spectrophotometer

2. Calculate the mass of CuSO$_4 \cdot$5H$_2$O required for preparation of 100 mL of 0.010 M Cu$^{2+}$ solution.

(over)

3. A solution known to contain 0.0205 M $Cu^{2+}$ forms a complex with ethylenediamine that yields an absorbance of 0.27 in a cell of 1.0 cm pathlength at a selected wavelength. Calculate the concentration of an unknown copper solution with an absorbance of 0.32 under the same conditions.

4. Calculate $\varepsilon$ from the data given in question 3.

5. From the absorption curve of hydrated $Cu^{2+}$ (Section: Concepts of the Experiment) deduce:

   the wavelength of maximum absorption: _____ nm

   the molar absorptivity $\varepsilon$ at the maximum: _____ $mol^{-1}L\ cm^{-1}$

   the molar absorptivity 50 nm past the
     maximum (toward longer wavelength): _____ $mol^{-1}L\ cm^{-1}$

   the molar absorptivity 50 nm before the
     maximum (toward shorter wavelength): _____ $mol^{-1}L\ cm^{-1}$

   the molar absorptivity at the short wave-
                    length minimum: _____ $mol^{-1}L\ cm^{-1}$

(Keep a record of your answers for question #5. You will need to refer to this question and your answers when you complete the Report Form.)

# REPORT FORM

## INVESTIGATION 32: COLORIMETRIC DETERMINATION OF COPPER AS $[Cu(ethylenediamine)_2]^{2+}$

Name: _____    ID No.: _____

Instructor: _____    Course/Section:_____

Partner's Name (if applicable): _____    Date:_____

---

Show all calculations on attached pages.

$CuSO_4 \cdot 5H_2O$ + weighing paper: _____ g          Concentration of standard $Cu^{2+}$

weighing paper: _____ g          solution: _____ M

$CuSO_4 \cdot 5H_2O$: _____ g

| Tube No. | 1 | 2 | 3 | 4 | 5 | 6 | 7 | U#__ | U#__ |
|---|---|---|---|---|---|---|---|---|---|
| Volume Copper | 0.00 | | | | | | | | |
| Volume "en" reagent (mL) | | | | | | | | | |
| Volume distilled water (mL) | | | | | | | | | |
| Transmittance | | | | | | | | | |
| Copper(II) concentration (M) | | | | | | | | | |
| Absorbance | | | | | | | | | |

Wavelength of maximum absorption (Proc. C-4): _____

$\varepsilon$ obtained in Procedure C-4: _____

Calibration curve: $\varepsilon$ = _____

(Optional) Intercept of calibration curve: D = _____

(Optional) Equation for calibration curve: _____

## Concentrations of Unknowns

U# _____ (based on 7 mL volume): _____ M (from graph)

_____ M (from equation)

U# _____ (based on volume marked       _____ M (from graph)

on test tube)       _____ M (from equation)

## Detection Limits

Use the data from Question 5 of the PreLab Exercises to calculate the detection limit for the determination of copper using the color of the hydrated $Cu^{2+}$ ion.

Smallest absorbance measurable with the spectrophotometer in your laboratory (provided by your instructor):  $A_{smallest} =$ _____

      Detection limit  _____  mol $L^{-1}$

                               _____  g $Cu^{2+}$ per liter

Use the data from the absorption curve of the $[Cu(H_2NCH_2CH_2NH)_2]^{2+}$ complex to calculate the detection limit for the determination of copper using the color of the $Cu^{2+}$-ethylenediamine complex.

      Detection limit  _____  mol $L^{-1}$

                               _____  g $Cu^{2+}$ per liter

A solution of copper sulfate contains 1.0 microgram of copper per milliliter of solution. Can the color of either the hydrated $Cu^{2+}$ or the $Cu^{2+}$-"en" be used for the determination of copper in this solution?  Justify your answer.

Date _____ Signature _____

# Investigation 33

## MODELS FOR ORGANIC COMPOUNDS

### INTRODUCTION

Before man had a written history, he had learned to extract useful substances from sources provided by nature. Herbs provided flavors for foods or medicines for treating the sick. Bark extracts helped in tanning animal hides, and volatile chemicals in wood smoke helped in preserving his food. He could protect himself with arrows dipped in concentrated poisons from natural sources, celebrate his victories with alcoholic beverages, or seek temporary escape of questionable value from the realities of his life with a variety of drugs. Often the arts of extraction and concentration of physiologically active compounds were carefully guarded secrets.

Today man no longer needs to rely primarily on natural sources for useful chemicals. He can isolate an active substance, purify it carefully, determine its complete structure, and, frequently, make it synthetically for wide-scale use much more economically than he could extract it from the natural source. The arts of the chemist are no longer secret. A general understanding of the concepts involved and the techniques employed can be gained rather easily. Such knowledge should help in the wise determination of the uses for which these arts will be employed.

Organic chemistry is primarily the study of covalent compounds of carbon. These compounds include a fantastic variety of substances ranging from the simple hydrocarbons used as fuels and lubricants to the complex nucleic acids of the genetic material.

Carbon forms more different chemical compounds than any other element because carbon-carbon bonds are exceptionally stable, permitting the formation of long chains or rings of carbon atoms. In addition, carbon forms stable covalent bonds with hydrogen and with many other elements such as oxygen, nitrogen, sulfur, and the halogens.

Organic compounds may be categorized according to certain arrangements of heteroatoms, the functional groups, which give characteristic properties to the molecules in which they occur. Such a classification is important in systematizing the study of the enormous number of organic compounds.

Because carbon normally forms four covalent bonds, carbon atoms can be connected many different ways. The existence of isomers, molecules having the same molecular formulas but different structures, is another reason for the immense number of different organic compounds.

In an attempt to produce more effective drugs with fewer adverse side effects, pharmaceutical chemists have synthesized thousands of compounds with structural variations designed to reveal the molecular characteristics essential for optimal biological activity.

As an example, quinine ①, the antimalarial drug that was extracted from the bark of Cinchona trees, has been largely replaced by synthetic drugs developed over years of research involving syntheses, structural elucidation, and biological evaluation. The synthetic antimalarial drug ② has the methyl group next (in ortho- or 2- position) to the diethylaminoethyl group in the benzene ring. Compounds ③ & ④ that differ from the active drug only by the position of the methyl group are inactive. Such minor variations in structure often have profound influences on biological activities.

quinine (1)

1-[2'-(diethylamino)ethyl]-2-methylbenzene
synthetic antimalarial drug (2)

1-[2'-(diethylamino)ethyl]-4-methylbenzene
inactive (3)

1-[2'-(diethylamino)ethyl]-3-methylbenzene
inactive (4)

Carbon atoms, hydrogen atoms, and heteroatoms form many organic compounds that differ with respect to the number of the various atoms present in the molecules. Formulas that indicate only the kinds of atoms and their numbers in molecules are called "molecular" formulas. Examples of molecular formulas are:

| $CH_4$ | $C_{16}H_{34}$ | $C_8H_{18}$ | $C_6H_6$ | $C_{20}H_{24}N_2O_2$ |
|---|---|---|---|---|
| methane | hexadecane | octane | benzene | quinine |
| (a gas) | (a waxy solid) | (a liquid) | (a liquid) | (antimalarial drug) |

| $C_{13}H_{21}N$ | | $C_2H_4O$ | $C_2H_6O$ | $C_6H_{12}O_6$ |
|---|---|---|---|---|
| 1-[2'-(diethylamine)ethyl]-2-methylbenzene | | acetic acid | ethanol | glucose |
| (synthetic antimalarial drug) | | (a liquid) | (a liquid) | (a sweet solid) |

These molecular formulas are not very informative. They do not tell much about how the atoms are connected in the molecules.

The deduction of the structure of quinine from its molecular formula alone is next to impossible. Only for very simple molecules such as methane ($CH_4$), methanol ($CH_4O$), and ethane ($C_2H_6$) can the structures be deduced unequivocally from the molecular formulas. For the more complex molecules too many possibilities exist to connect the various atoms. Even for the simple molecular formula $C_2H_6O$ two possibilities exist to connect the atoms in a chemically reasonable way:

ethanol
(a liquid)

dimethyl ether
(a gas)

The structural formulas written for these two isomers with the same molecular formula are much more instructive than molecular formulas. They show how the atoms are connected in the plane of the paper. However, molecules are three-dimensional entities and, therefore, three-dimensional models are the best way to see how molecules are built and to picture the various isomers of a compound. The three-dimensional representations are particularly useful to understand the principle of optical isomerism.

This experiment provides the opportunity to explore the three-dimensionality of organic compounds and to deepen the understanding of the various types of isomers.

Figure 1: Classification of Isomers.

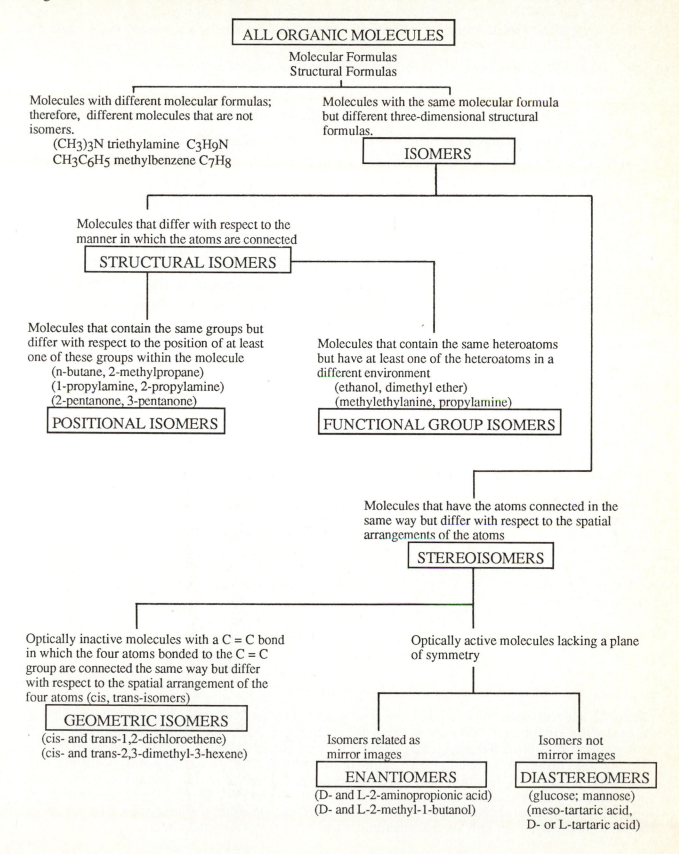

## CONCEPTS OF THE EXPERIMENT

For three-dimensional representations of organic molecules the following rules must be considered:

- A carbon atom bonded to four other atoms is in a tetrahedral environment.
- Free rotation exists about carbon-element single bonds.
- All four atoms bonded to a $\diagup C=C \diagdown$ group and the two carbon atoms are in the same plane. Rotation about the $\diagup C=C \diagdown$ double bond does not occur.
- The two atoms bonded to the carbon atoms in a $-C\equiv C-$ group are in one line with the two carbon atoms.

These rules are helpful in representing the various types of isomers discovered in organic molecules. Figure 1 categorizes the isomers.

Positional, functional group, and geometric isomers can be represented rather well by two-dimensional structural formulas in the plane of the writing paper. However, the rotation about single bonds and the conformers that are generated by such rotations are best explored with three dimensional models. Pentane, for instance, can adopt several conformations that interconvert rapidly and, therefore, are not isomers.

$$CH_3 \diagup {}^{CH_2} \diagdown {}_{CH_2} \diagup {}^{CH_2} \diagdown CH_3$$

"Stretched" Conformation

$$CH_3 - {}^{CH_2} \diagdown {}_{CH_2} \diagup {}^{CH_2} \diagdown {}_{CH_3 - CH_2}$$

"Closed" Conformation

$$CH_3 \diagup {}^{CH_2} \diagdown {}_{CH_3} \\ CH_3 - {}_{CH_2}$$

"Half-Closed" Conformation

When one is searching for all possible isomers of a compound, one must watch out for these conformations because they are not different compounds.

Enantiomers and diastereomers almost require three-dimensional models for their representation, particularly at the beginning of their detailed study. After familiarity with enantiomers and diastereomers has been acquired, various two-dimensional presentations will suffice.

When four different atoms or groups are bonded to one carbon atom, two tetrahedral arrangements about the carbon atom are possible. These arrangements are related to each other as an object to its mirror image. Such a pair of compounds – called enantiomers – has the same molecular formula and has the atoms connected in the same way. The object and its mirror image

mirror

cannot be superimposed on each other no matter how their three-dimensional models are turned. Because the difference between a pair of enantiomers lies only in the three-dimensional arrangement of four different groups bonded to a carbon atom (such a carbon atom is called asymmetric carbon or a chiral center), almost all the properties (melting point, boiling point, solubility, most reactions) are the same. However, they differ in two important respects: they rotate the polarization plane of plane-polarized light in opposite directions by the same amount and their biochemical properties are usually very different because the enzymes that serve as biochemical catalysts act only on one enantiomer and leave the other untouched.

A molecule may contain more than one asymmetric carbon atom. Such molecules may have several pairs of enantiomers; in addition, isomers exist that differ only with respect to the placement of atoms or groups about the asymmetric carbon atoms, are optically-active, but are not mirror images of each other. Such non-mirror image, optically-active isomers are called diastereoisomers.

$$CH_3 - \overset{\overset{\displaystyle CO_2H}{|}}{\underset{\underset{\displaystyle OH}{|}}{\overset{*}{C}}} - OH$$

lactic acid molecule with
one asymmetric carbon atom;
enantiomeric

$$H - \overset{\overset{\displaystyle O}{\|}}{C} - \overset{\overset{\displaystyle OH}{|}}{\underset{\underset{\displaystyle H}{|}}{\overset{*}{C}}} - \overset{\overset{\displaystyle H}{|}}{\underset{\underset{\displaystyle OH}{|}}{\overset{*}{C}}} - \overset{\overset{\displaystyle OH}{|}}{\underset{\underset{\displaystyle H}{|}}{\overset{*}{C}}} - \overset{\overset{\displaystyle OH}{|}}{\underset{\underset{\displaystyle H}{|}}{\overset{*}{C}}} - \overset{\overset{\displaystyle H}{|}}{\underset{\underset{\displaystyle H}{|}}{C}} - OH$$

glucose molecule with four asymmetric
carbon atoms (marked with *);
enantiomers and diastereomers exist

Organic molecules may have combined within a molecule all of the possible types of isomers giving rise to a large number of different compounds all with the same molecular formula. The ot very complex molecule with the molecular formula $C_{13}H_{17}NO_2$ may serve as an example for a molecule capable of forming positional isomers, functional group isomers, geometric isomers, and a pair of enantiomers.

2-(1'-oxopent-3'-en-1'-yl)-
4-(1'-amino-1'-ethyl)phenol

positional isomer with
respect to the OH group

functional group isomer with
respect to C = O, double bond

geometrical isomers

asymmetric
(chiral)
carbon atom

pair of enantiomers are possible

The number of possible isomers increases exponentially with increasing molecular mass. The hydrocarbon decane, $C_{10}H_{22}$, has 136 and the hydrocarbon, $C_{20}H_{42}$, has 3,395,964 isomers. A truly enormous number of positional isomers is possible with high-molecular-mass proteins that are built up from approximately 20 different L-amino acids. If only one molecule of each of the positional isomers of such proteins were made, the entire volume of the cosmos would be insufficient to hold them.

## ACTIVITIES

- Review the chapters on organic chemistry in your textbook.

- Complete the PreLab Exercises before coming to the laboratory.

- Construct models of positional, functional-group and geometric isomers, and of a pair of enantiomers.

- Classify the isomers belonging to a compound assigned to you.

- Complete the Report Form.

## SAFETY

*If you are working in a laboratory in which chemical reagents are set out, wear approved eye protection.*

## PROCEDURES

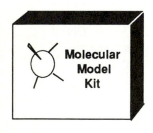

1. A stick-and-ball model kit will be available for this experiment. Check your kit to assure that all atoms and connectors that should be in the kit are present. Find out how multiple bonds are modeled with the kit available in your laboratory. More complex molecules might require more atoms than are available in your kit. Join forces with your neighbors and their kits to complete these assignments. Your instructor will identify the molecules that cannot be built with one kit.

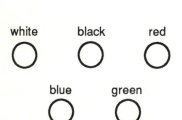

2. Find out which atoms are present in your kit. You should find color-coded atoms such as hydrogen atoms, carbon atoms, oxygen atoms, nitrogen atoms, and chlorine atoms. Each spherical ball will have holes in the correct positions to accept the sticks required for forming bonds. The univalent hydrogen and chlorine balls will have only one hole,

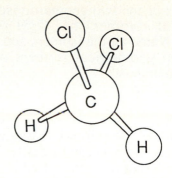

the oxygen atom two holes at an angle of approximately 109°, the carbon atoms four holes required for a tetrahedral arrangement (bond angle 109° 28'), and the nitrogen atom may have three or four holes.

**3.** Build a model of the tetrahedral molecule of carbon tetrachloride, $CCl_4$. Observe the Cl-C-Cl bond angles. Draw the molecule and show that the four chlorine atoms are positioned at the corners of a tetrahedron and the carbon atom is located at the center of the tetrahedron.

**4.** Construct a model of dichloromethane, $CH_2Cl_2$. Convince yourself that no second isomer of this compound is possible in a tetrahedral arrangement. Two isomers are expected in the case of a planar arrangement of the two hydrogen and two chlorine atoms around the carbon atom. The fact that two isomers of dichloromethane never could be isolated is chemical evidence for the tetrahedral structure of this molecule.

Cl - - - Cl
C
H - - - H
**One of the isomers
of planar $CH_2Cl_2$**

$C_5H_{12}$   pentane

CH_3 — C(H_2) — C(H_2) — C(H_2) — CH_3
**n-pentane**

**5.** Construct models of all possible isomers of the hydrocarbon pentane, $C_5H_{12}$. Draw two-dimensional representations of these positional isomers. Make sure you are not confused by conformers and isomers of this hydrocarbon.

$C_4H_9OH$

CH_3 — C(H_2) — C(H_2) — C(H_2) — OH
**n-butanol**

**6.** Construct models of all alcohols of the molecular formula $C_4H_9(OH)$. Draw two-dimensional representations of these butanols and write their systematic names.

$C_2H_2Cl_2$

one of the isomers
of 1,2-dichloroethene

1,2-dichlorobenzene
(one of the possible
isomers of $C_6H_4Cl_2$)

$C_4H_8O$

2-butanone or
2-oxobutane
(one of the isomeric
oxobutanes)

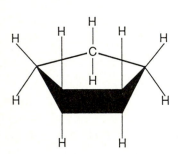

**7.** Construct models of the two geometric isomers and the positional isomer of $C_2H_2Cl_2$, 1,2-dichloroethene. Name all the isomers and draw their structures. Do geometrical isomers of the second positional isomer exist?

**8.** Draw all possible positional isomers of dichlorobenzene. These isomers can be easily drawn on paper and models do not have to be built. Name all the isomers.

**9.** How many isomeric oxobutanes, $C_4H_8O$, exist? You do not have to build models to find these isomers. The structural formulas can be easily drawn on paper. Name all the isomers.

**10.** Build a model of cyclopentane, $C_5H_{10}$. Describe the shape of this molecule. Then build a model of hydroxycyclopentane, $C_5H_{10}O$. How many isomers of hydroxycyclopentane exist? Then build a model of 1,2-dihydroxycyclopentane. Do isomeric 1,2-dihydroxycyclopentanes exist? If yes, classify these isomers with reference to Figure 1.

**11.** Construct models of the two enantiomers of alanine, 2-aminopropionic acid ($C_3H_7NO_2$). Identify the asymmetric carbon atom. Attempt to superimpose the two enantiomers upon each other. You may turn the models any way you want, but you may not move the sticks (the bonds). The proof that enantiomeric pairs are not superimposable is more easily carried out with four different atoms (balls) on the asymmetric carbon atom instead of full representation of the $CH_3$, COOH, and $NH_2$ group.

**12.** Your instructor will assign you a compound for which you have to find one or several isomers. The isomers have to be classified (Figure 1). A model of one of the isomers must be built. Your efforts will be graded as announced by your instructor.

# PRELAB EXERCISES

## INVESTIGATION 33: MODELS FOR ORGANIC COMPOUNDS

Name:_____

Instructor:_____  ID No.:_____

Course/Section:_____  Date:_____

1. Draw the tetrahedral molecule $CH_4$, methane. Show the C-H bonds, the H-C-H bond angles, and the center of the tetrahedron.

2. Give an example of a pair of positional isomers. Do not use the examples given in this experiment.

3. Give an example of a pair of functional group isomers.

4. Does a positional isomer of phenol, $C_6H_5OH$, exist? Justify your answer.

5. How many positional isomers of dichloronaphthaline are possible?

1,2-dichloronaphthaline

6. Identify the asymmetric carbon atoms (if any) in the following molecules.

$$CH_3C \equiv CH, \quad CH_3 - \underset{\underset{Cl}{|}}{CH} - COOH, \quad H - \underset{\underset{Cl}{|}}{\overset{\overset{Br}{|}}{C}} - COOH, \quad CH_3 - COOH, \quad C_{10}H_{21} - \underset{\underset{OCH_3}{|}}{CH} - CH_2 - CN$$

# REPORT FORM

## INVESTIGATION 33: MODELS FOR ORGANIC COMPOUNDS

Name: _____     ID No.: _____

Instructor: _____     Course/Section:_____

Partner's Name (if applicable): _____     Date:_____

Drawing of $CCl_4$:

Drawing of tetrahedral $CH_2Cl_2$ and planar isomers of $CH_2Cl_2$:

Functional formulas and names for the isomers of pentane:

Structural formulas and names for the isomeric butanols:

Drawings and names for the geometric and positional isomers of $C_2H_2Cl_2$:

Structural formulas and names for the isomeric dichlorobenzene:

Structural formulas and names for the functional group isomers of oxobutanes:

Drawings, structural formulas, names and classification for the isomers of 1,2-dihydroxycyclopentane:

Drawings of the enantiomeric pair of alanine:

Structural formulas, names, and classification for the isomers of the assigned compound:

Date _____ Signature _____

# Investigation 34

## IDENTIFICATION OF ORGANIC COMPOUNDS

## INTRODUCTION

There are many reasons why we might wish to determine the identity of an organic compound. In natural product chemistry, for example, we might find some exotic herb that contains something of therapeutic value. If we could isolate from the complex mixture a single compound possessing the useful properties and find out what it is, we might discover a valuable new drug. Quinine, the first good antimalarial drug, was found this way.

The identification of a biologically active component in a complex natural mixture is not a simple task. First, the compound must be isolated from the mixture in a pure form. Then structural information must be accumulated to permit an unambiguous assignment of chemical identity. If the compound of interest is one that is already known, this may involve relatively simple comparisons of properties to establish the identity. The determination of the structure of a new compound is much more difficult, especially if it is a relatively complex molecule. Isoamyl acetate, an alarm pheromone of the common honeybee, was isolated and identified by comparison with the known chemical in a fairly short time. On the other hand, many research teams all over the world have been investigating the complex venom system of the honeybee for more than 40 years and, although some 40 different compounds have been identified, the total composition of honeybee venom is not yet determined.

There are, of course, many reasons for identifying organic compounds in addition to those of the natural product chemist. The forensic chemist may be required to determine the nature of a confiscated drug to be used as evidence in a trial or to identify some exotic poison. The environmental chemist may have to detect an undesirable organic pollutant. The chemical oceanographer is interested in the identity of organic compounds in the marine environment, and the "astrochemist" may use very special techniques for detecting organic compounds in the atmosphere of a far planet or even in the vast reaches of interstellar space.

The determination of the composition of a chemical mixture, or even the isolation and identification of a single component from a mixture, is a time-consuming task requiring special knowledge and skills. It is considerably easier to identify a single pure compound, especially if we have narrowed the possibilities to a fairly small list.

In this experiment two pure organic compounds are to be identified on the basis of characteristic chemical reactions and physical properties.

## CONCEPTS OF THE EXPERIMENT

In this experiment the number of possible unknowns will be limited to only five classes of compounds: alcohols, aldehydes, amides, carboxylic acids, and ketones. The following Table 1 lists organic compounds that you may receive as unknowns to be identified. Your instructor may add to or delete from this list.

You will be assigned two unknowns, a liquid and a solid. By testing the properties of a known from each class, you will gain insight into characteristic properties of each class of compounds. By testing the chemical and physical properties of your unknowns, you will be able to identify your unknowns by comparision of their properties with the properties of the compounds listed in Table 1.

**Table 1**: Organic Compounds That May be Assigned as Unknowns and Some of Their Properties

| Class | Compound | m.p. (ºC) | b.p. (ºC) | Derivative* (m.p.) | Solubility in water** | Comments |
|---|---|---|---|---|---|---|
| alcohols | ethanol | -117 | 78 | - | ++ | active ingredient of "alcoholic" beverages |
| | cylcohexanol | 25 | 161 | - | + | a common solvent for resins and insecticides |
| aldehydes | benzaldehyde | -26 | 179 | 237 | - | almond flavoring agent |
| | salicylaldehyde | -7 | 197 | 252 | + | used in perfumes, has a "bitter-almond" odor |
| | p-anisaldehyde | 0 | 248 | 254 | - | odor reminiscent of vanilla beans |
| | cinnamaldehyde | -8 | 252 | 253l | - | cinnamon flavoring agent |
| | vanillin | 80 | 285 | 269 | + | vanilla flavoring agent |
| amides | acetamide | 82 | - | - | ++ | used as an additive to plastics and often to "denature" alcohol |
| | formamide | 2 | 210 | - | ++ | used as a softener for paper and animal glues |
| | phenylacetamide | 157 | - | - | - | used in manufacturing Penicillin |
| | salicylamide | 139 | - | - | - | an aspirin substitute |
| carboxylic acids | acetic | 16 | 118 | - | ++ | acid component of "vinegar" |
| | butyric | -19 | 163 | - | ++ | responsible for odor of rancid butter |
| | palmitic | 63 | 390 | - | - | found, as the glyceryl ester, in many vegetable oils and fats |
| | phenylacetic | 76 | 265 | - | - | used in production of some synthetic perfumes |
| ketones | acetone | -95 | 56 | 128 | ++ | common solvent for varnishes, cements, etc. |
| | cyclohexanone | -16 | 156 | 162 | + | valuable solvent and a reagent for preparing adipic acid for nylon synthesis |
| | 2-butanone | -86 | 80 | 111 | ++ | valuable solvent and used in preparing "smokeless" gunpowder |
| | acetophenone | 20 | 202 | 237 | - | used in perfumes to produce an orange-blossom aroma |
| | camphor | 179 | 204 | 177 | - | a moth repellant and a mild antiseptic |
| | 3-pentanone | -40 | 101 | 156 | + | useful starting material in the syntheses of more complex compounds |

*The derivatives used will be 2,4-dinitrophenylhydrazones for aldehydes and ketones
**++: high solubility; +: somewhat soluble; -: insoluble

In the first part of the experiment you will familiarize yourself with some of the properties of compounds representative of each class of organic derivatives considered in this experiment:

- alcohols: $R-OH$

- carboxylic acids: $R-\overset{\overset{\displaystyle O}{\|}}{C}-OH$

- carboxylic acid amides: $R-\overset{\overset{\displaystyle O}{\|}}{C}-NH_2$

- aldehydes: $R-\overset{\overset{\displaystyle O}{\|}}{C}-H$

- ketones: $R-\overset{\overset{\displaystyle O}{\|}}{C}-R'$

"R" in these formulas represents an aliphatic organic group, such as H, $CH_3$, $C_2H_5$, or an aromatic organic group such as $C_6H_5$.

Properties that are easily determined are the physical state (solid? liquid?) and the odor of a sample. Visual inspection classifies a sample either as a solid or a liquid at room temperature and, thus, considerably reduces the number of compounds, one of which is identical with your sample. Many compounds have characteristic odors. You might have encountered the memorable odor of moth balls (naphthalene), the unpleasant stink of rotten eggs (hydrogen sulfide), or the life-saving smell of the odorants added to natural gas (butanethiol). Carefully test the odors of the known compounds as described in Procedure A-2. If a liquid does not have any odor, it cannot be acetone, formic acid, or ethanol. Most organic liquids and some organic solids have characteristic odors that provide welcome clues to their identity.

Another property sometimes useful for the identification of a compound is its solubility in water or organic solvents. In this experiment ethanol, a polar organic solvent, is used. In general, substances with polar groups such as carboxyl $\left(-\overset{\overset{\displaystyle O}{\|}}{C}-OH\right)$, hydroxyl $(-OH)$, carbonyl $\left(-\overset{\overset{\displaystyle O}{\|}}{C}-\right)$ or amide $\left(-C\overset{\nearrow O}{\underset{\searrow NH_2}{}}\right)$ will dissolve in water provided that the organic part of the molecules consisting only of carbon and hydrogen atoms is not too large. If the hydrocarbon part of the molecule becomes predominant, the compound will not be soluble in water, but will dissolve in ethanol. For instance, formic acid, $H-C\overset{\nearrow O}{\underset{\searrow OH}{}}$, and acetic acid, $CH_3-C\overset{\nearrow O}{\underset{\searrow OH}{}}$, are easily soluble in water, whereas stearic acid, $C_{17}H_{35}-C\overset{\nearrow O}{\underset{\searrow OH}{}}$, is not. In this manner solubility will aid in the identification of a compound.

Of prime importance for the identification of an organic compound is the recognition of the functional groups $(-OH,\ -COOH,\ -\overset{\overset{\displaystyle O}{\|}}{C}-,\ -\overset{\overset{\displaystyle O}{\|}}{C}-NH_2)$. These groups give characteristic reactions that are used to establish their presence.

Solutions of carboxylic acids in water or aqueous ethanol react with sodium hydrogen carbonate liberating carbon dioxide. Bubbles of carbon dioxide rise to the surface of the solution indicating the presence of an acid (equation 1). Carboxylic acids are neutralized by sodium

$$R-\overset{\overset{\displaystyle O}{\|}}{C}-OH(aq) + NaHCO_3(aq) \rightarrow R-\overset{\overset{\displaystyle O}{\|}}{C}-ONa(aq) + CO_2(g) + H_2O(\ell) \qquad (1)$$

hydroxide (equation 2). When a weighed amount of a carboxylic acid is dissolved in water or

$$R-\overset{\overset{O}{\|}}{C}-OH \ + \ NaOH \ \rightarrow \ R-\overset{\overset{O}{\|}}{C}-ONa \ + \ H_2O \tag{2}$$

aqueous ethanol and the resulting solution titrated with an aqueous solution of sodium hydroxide of known molarity, the molecular mass of carboxylic acids with one carboxylic acid group in the molecule can be calculated as described in Investigation 18.

Carboxylic acid amides liberate ammonia when solutions of the amide are heated with aqueous sodium hydroxide solution (equation 3). The ammonia can be detected by its characteristic odor or by the color change of red litmus paper to blue on contact with the vapors of ammonia.

$$R-\overset{\overset{O}{\|}}{C}-NH_2(aq) \ + \ NaOH(aq) \ \rightarrow \ R-\overset{\overset{O}{\|}}{C}-ONa(aq) \ + \ NH_3(g) \tag{3}$$

Ketones and aldehydes have very characteristic odors unless their molecular mass is too high and concomitantly their volatility too low. However, a time-honored test for carbonyl groups is the formation of highly crystalline, orange to red 2,4-dinitrophenylhydrazones (equation 4). The melting points of the crystalline hydrazones can be readily determined and compared with known derivatives.

$$R-\overset{\overset{O}{\|}}{C}-R' \ + \ NH_2-NH-\!\!\!\bigcirc\!\!\!-NO_2 \ \rightarrow \ \overset{R}{\underset{R'}{C}}=N-NH-\!\!\!\bigcirc\!\!\!-NO_2 \ + \ H_2O \tag{4}$$

An easily quantifiable property of a solid is its melting point; of a liquid its boiling point and its melting point (obtained by cooling the liquid below room temperature).

Although specific tests for alcohols do exist, they cannot be carried out easily in an introductory chemistry laboratory. Therefore, the alcohols must be identified on the basis of negative results during the tests for other functional groups and melting points and boiling points. When the tests with the known compounds are completed, the work for the identification of the unknowns can begin. The flowchart in Figure 1 will assist in performing the tests in the proper sequence.


# ACTIVITIES

- **Complete the PreLab Exercises before coming to the laboratory.**

- **Study the techniques identified in the Procedure section and described in earlier experiments before coming to the laboratory.**

- **Perform the test for carboxyl, amide, and carbonyl functional groups on known compounds.**

- **Complete the Report Form.**


# SAFETY

*Wear approved eye protection. Be very careful with open flames. Most of the compounds used are highly flammable. Under no circumstances should you taste any of the chemicals. Although some are used as flavoring agents, many can be toxic in undiluted form.*

*In case of contact with chemicals, wash the affected area thoroughly with soap and water. Notify your instructor. Be very careful when working with the oil bath. Spillage of hot oil may cause serious burns.*

Figure 1:  Flowchart for the Identification of Unknown Organic Compounds.

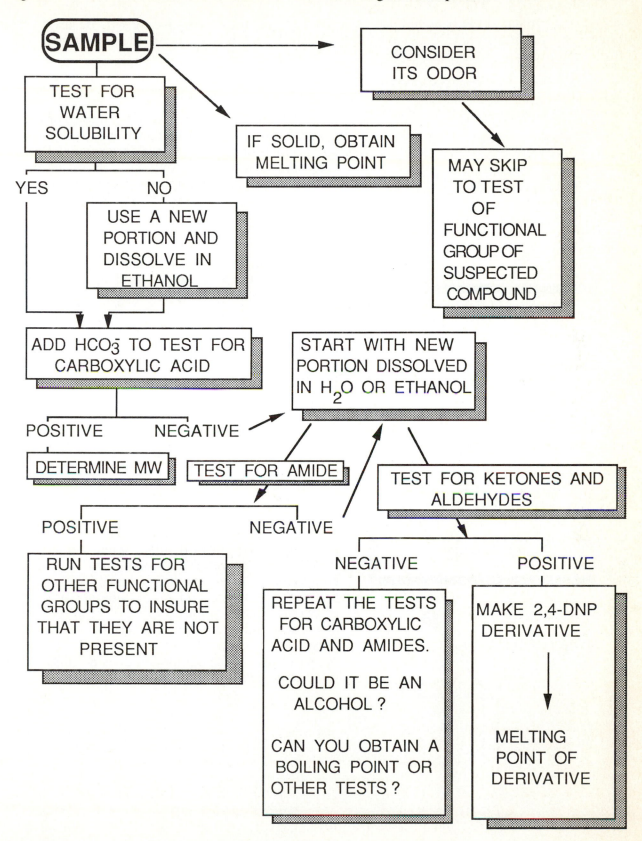

## PROCEDURES

In this experiment several techniques described in previous experiments will be used.   The description of these techniques must be reviewed before coming to the laboratory.   The investigations in which the descriptions are given are identified in this section.

### A.   TESTS WITH KNOWN COMPOUNDS

**A-1.**  Obtain from your instructor samples of known compounds.  Your instructor will provide you with the formulas of these reference compounds.  Practice the tests described below with these reference compounds.  It is very important that detailed records are kept of the procedures, observations, and results.

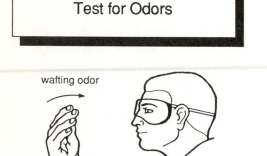

**Test for Odors**

wafting odor

**A-2.**  Unstopper the test tube and carefully waft the vapors from the tube toward your nose.  If you do not smell anything, bring your nose closer to the test tube.  *Do not inhale the vapors.*  Record the odor as precisely as possibly (smells like vanilla extract, like rancid butter, like ...).

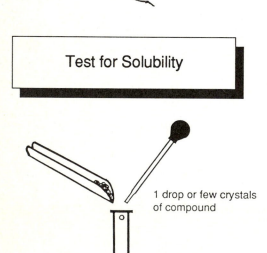

**Test for Solubility**

1 drop or few crystals of compound

20 drops water

**A-3.**  Add 20 drops of distilled water to a clean test tube.   To the water add one drop of a liquid compound or a few crystals of a solid compound.  Shake the test tube and check whether or not the compound has gone into solution.  If the compound is soluble, add two more drops or a few more crystals and shake the tube again.   Record your observations.  If the compound does not dissolve in water, repeat the test with ethanol as solvent.

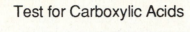

## Test for Carboxylic Acids

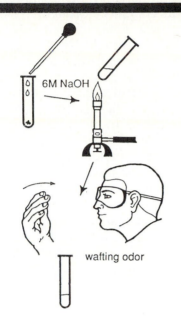

6M NaOH

wafting odor

**A-4.** Add 20 drops of distilled water and 2 drops of a water soluble carboxylic acid to a clean test tube. Swirl the mixture until the solution has become homogeneous. If the carboxylic acid is not soluble in water, use ethanol as the solvent. Add 20 drops of 95% ethanol to a clean and dry test tube; add 2 drops or a few crystals (20 mg) of the carboxylic acid. Shake the test tube. If the acid does not dissolve, heat the test tube with a burner flame. Then add two drops of water and homogenize by shaking. To the solution of the carboxylic acid add 15 drops of a 5-percent aqueous solution of sodium hydrogen carbonate. Observe carefully to see the rising bubbles of carbon dioxide.

## Test for Amides

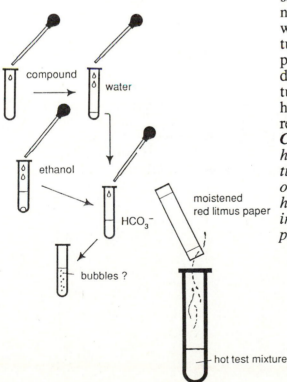

compound

water

ethanol

HCO₃⁻

bubbles ?

moistened red litmus paper

hot test mixture

**A-5.** Place a small sample (50 to 100 mg) of a carboxylic acid amide in a clean test tube. Add 15 drops of a 6 M aqueous sodium hydroxide solution. Swirl the mixture to mix its constituents. Gently heat the mixture with a burner flame and after 30 seconds sample the odor of the vapor for ammonia with your nose or hold a strip of red litmus paper moistened with distilled water close to the opening of the test tube. If the litmus paper turns blue, ammonia is present in the vapor. Make sure that the litmus paper does not come in contact with the inside of the test tube or with liquid in the test tube. The sodium hydroxide in the reaction mixture will certainly turn red litmus paper to blue.

*CAUTION: Before you waft the vapors from the hot reaction mixture toward your nose, shake the test tube gently to establish that the solution is not overheated and will not suddenly vaporize propelling hot caustic liquid into your face. Consult your instructor if you are unsure about this precautionary procedure.*

## Test for Ketones / Aldehydes

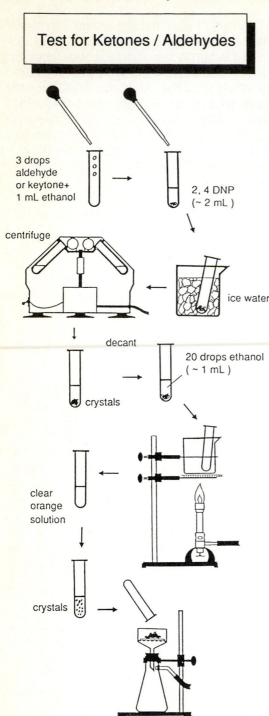

3 drops
aldehyde
or keytone +
1 mL ethanol

2, 4 DNP
(~ 2 mL)

centrifuge

ice water

decant

20 drops ethanol
(~ 1 mL)

crystals

clear
orange
solution

crystals

**A-6.** To three drops of a liquid or approximately 50 mg of a solid ketone or aldehyde in a clean and dry test tube add 20 drops of 95% ethanol and then 2 mL of the 2,4-dinitrophenylhydrazine solution. Swirl the tube to mix its contents thoroughly. Heat the test tube for 5 minutes in a beaker half-full with water that is maintained at 60 to 70°C (not warmer). Take the tube from the warm water and place it in a beaker containing tap water at room temperature. Red to orange crystals may form at the time the reagents are mixed, during the time the mixture is heated, or on cooling the mixture to room temperature. Should crystals not have formed when the reaction mixture is at room temperature, scratch the inside of the test tube covered by the solution with a glass rod to induce crystallization. If necessary, cool the test tube in a beaker containing crushed ice and water.

Centrifuge the test tube (Appendix E) and decant the liquid. Add 1 mL of 95% ethanol to the crystals and heat the mixture in a water bath. Should the crystals not have dissolved after several minutes in the water bath, add additional ethanol in small increments until a homogeneous solution is obtained.

Cool the tube and isolate the crystals by vacuum filtration (Appendix F). Allow air to pass through the filter for several minutes to dry the crystals as much as possible. Transfer the crystals from the filter into a clean and dry test tube. Place the test tube into a boiling water bath until all the ethanol has evaporated from the crystals. When the crystals are dry, determine their melting point.

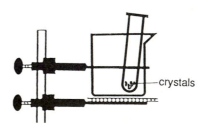

crystals

### Molecular Mass of a Carboxylic Acid

**A-7.** Weigh a clean and dry 125-mL Erlenmeyer flask with stopper on the analytical balance to the nearest 0.1 mg. Then add approximately 0.2 g of the acid, stopper the flask, and reweigh it. Record the mass of the acid to the nearest 0.1 mg. Add 15 mL water or 15 mL ethanol to the flask, shake the flask to dissolve the acid, and titrate the solution with standardized 0.1 M sodium hydroxide to the phenolphthalein endpoint (see Investigation 18). Calculate the molecular mass of the carboxylic acid from the titration data.

### Determination of a Melting Point

**A-8.** Melting points of compounds that are solid at room temperature are determined by placing a small amount of the solid in a capillary, securing the capillary to a thermometer, immersing the lower ends of the thermometer and the capillary into an oil bath, and slowly heating the oil bath. The temperature is read at the moment when the solid is completely melted (see Investigation 16 for details).
*CAUTION: Watch out for the oil bath! Handle carefully. The oil may still be hot. Do not spill hot oil. If you do not know how to handle an oil bath, consult your instructor.*

Liquids with melting points between 0° and room temperature can be determined by placing a small amount of a liquid into the capillary, immersing the thermometer-capillary assembly into a beaker containing ice cubes and water, and allowing the liquid to freeze and the water to reach 0°. Then remove the ice cubes and heat the water slowly to find the melting point.

## Determination of a Boiling Point

**A-9.** A boiling point can be easily determined with a few milliliters of a liquid when a microdistillation apparatus is available. Your instructor will supply you with the necessary information and equipment should you have to determine a boiling point. Boiling points vary with pressure. If a boiling point is given without pressure, the listed temperature refers to atmospheric pressure (760 torr).

## B.  IDENTIFICATION OF TWO UNKNOWNS

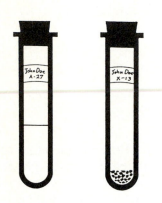

**B-1.** Obtain from your instructor approximately 2 g each of one solid and one liquid unknown in clean, dry test tubes. Label the test tubes with a specified code and your name. Keep these tubes stoppered and open them only to perform the odor test or to remove a sample.

**B-2.** With the help of the Flowchart (Fig. 1) and your experience identify the two unknowns. Compare the properties of your unknowns with the properties of the reference compounds listed in Table 1.

**B-3.** If you have established the identity of your solid, you should determine the melting point of a 1:1 mixture of your solid and its reference compound (if it is available). If the melting point of the mixture is the same as the melting point of its components, the two substances are the same. Mixed melting points should also be taken for the 2,4-dinitrophenylhydrazones prepared from the unknown and its reference compound.

**B-4.** In the absence of boiling points, alcohols must be identified by their odor, melting point, and the negative results for the tests specific for amide, carboxyl and carbonyl groups.

# PRELAB EXERCISES

## INVESTIGATION 34: IDENTIFICATION OF ORGANIC COMPOUNDS

Name:_____

Instructor:_____  ID No.:_____

Course/Section:_____  Date:_____

1. Give the functional group class and the names for each of the following compounds:

   $CH_3CH_2OH$

   $CH_3CONH_2$

   $CH_3CHO$

   $CH_3COCH_3$

2. Which class of organic compounds liberates $CO_2$ from $NaHCO_3$? Liberates ammonia when boiled with aqueous sodium hydroxide? Forms a precipitate with 2,4-dinitrophenylhydrazine? Describe the reactions by balanced equations.

3. An unknown liquid carboxylic acid (60 mg) with two oxygen atoms in the molecule is titrated with 0.05 M sodium hydroxide solution. The phenolphthalein endpoint was reached when 20.2 mL of the base had been added. Calculate the molecular mass of the acid and find its formula.

4. A solid unknown had a melting point of 126°. Potential reference solids melted at 124° and 127°. The mixed melting point with the 124°-solid was 110°, with 127°-solid 125°. Which reference substance is identical to the unknown?

# REPORT FORM

## INVESTIGATION 34: IDENTIFICATION OF ORGANIC COMPOUNDS

Name: _____     ID No.: _____

Instructor: _____     Course/Section:_____

Partner's Name (if applicable): _____     Date:_____

---

Data for Unknown

| Properties or behavior | Unknown code _____ | Unknown code _____ |
|---|---|---|
| odor descriptions | | |
| odor similar to that of known | | |
| water solubility (Procedure A-3) | | |
| behavior with $HCO_3^-$ (Procedure A-4) | | |
| behavior with 6 M NaOH (Procedure A-5) | | |
| behavior with 2,4-DNP reagent (Procedure A-6) | | |
| melting point °C | | |
| mixed melting point with reference compound | | |
| boiling point | | |
| functional group present | | |
| name and formula of compound | | |

Date _____ Signature _____

# Investigation 35

## AMYL ACETATE, A NATURAL FLAVORING AGENT

### INTRODUCTION

Carboxylic acids are a very important class of acidic organic compounds. Carboxylic acids and their derivatives occur widely in nature. Amino acids, the building blocks of proteins (enzymes), are amino-substituted carboxylic acids. Fats (butter, margarine, oils) have carboxylic acids and glycerol as their constituents. The unpleasant odor of sweat is caused by butyric acid (butanoic acid), $CH_3CH_2CH_2COOH$. Most low-molecular mass carboxylic acids have characteristic unpleasant odors, whereas the high-molecular mass compounds are odorless. However, the products of the reactions between malodorous carboxylic acids and alcohols (hydroxyalkanes) usually have sweet, "fruity" odors. These products – called esters – are found as natural flavoring agents in many fruits (Table 1) and are, separately or mixed, the constituents of several synthetic flavoring formulations.

Organisms synthesize carboxylic acid esters by relatively complex, stepwise processes that are catalyzed by enzymes. Laboratory and industrial processes are considerably simpler, but often less efficient in terms of energy utilization and yield. Several routes to esters are available. The *direct esterification* of a carboxylic acid with an alcohol is a simple procedure, but the yields of ester are limited to 50 to 75 percent because the water formed during the reaction hydrolyzes the ester. An equilibrium is established between the reactants and products (equation 1). The equilibrium constant for the formation of ethyl acetate from acetic acid and ethanol is approximately 2. When the concentrations of the starting materials are the same, only half of the reactants will have been converted to the product at equilibrium.

$$CH_3C\overset{O}{\underset{OH}{\diagdown}} + H-OCH_2CH_3 \rightleftharpoons CH_3C\overset{O}{\underset{O}{\diagdown}}CH_2CH_3 + H_2O$$

$$K = \frac{[CH_3COOC_2H_5][H_2O]}{[CH_3COOH][C_2H_5OH]} \cong 2 \ \ (\text{at } 25^\circ C) \tag{1}$$

According to Le Chatelier's principle, the yield of an ester can be increased by use of an excess of alcohol or acid. Which of the two reactants will be used in excess is determined by availability and cost. Alternatively, one of the products, ester or water, can be removed. Depending on the boiling points of ester and water, the reaction may be performed at a temperature at which one of the products distills from the reaction mixture.

An alternate method for the preparation of esters starts with a carboxylic acid chloride and an alcohol (equation 2). These reactions are performed in non-aqueous solvents from which hydrogen chloride can be easily expelled. The yields of esters are generally higher than 90 percent. Carboxylic acid chlorides are rather expensive substances and are, for this reason, only rarely employed in large-scale industrial processes for the manufacture of esters.

**Table 1.** Formulas, Trivial Names, and IUPAC Names for Carboxylic Acid Esters Serving as Flavoring Agents.

| Formula | Trivial name | IUPAC name | A flavoring agent and odor component of |
|---|---|---|---|
| $HC(=O)-O-CH_2CH_3$ | ethyl formate | ethyl methanoate | rum |
| $H_3CC(=O)-O-CH_2(CH_2)_3CH_3$ | n-amyl acetate | pentyl ethanoate | pears, bananas |
| $H_3CC(=O)-O-CH_2CH_2CH(CH_3)_2$ | isoamyl acetate | 3-methylbutyl ethanoate | pears, bananas |
| $H_3CC(=O)-O-CH_2(CH_2)_6CH_3$ | n-octyl acetate | octyl ethanoate | oranges |
| $H_3CCH_2C(=O)-O-CH_2CH(CH_3)_2$ | isobutyl propionate | 2-methylpropyl propanoate | rum |
| $H_3CCH_2CH_2C(=O)-O-CH_3$ | methyl butyrate | methyl butanoate | apples |
| $H_3CCH_2CH_2C(=O)-O-CH_2CH_3$ | ethyl butyrate | ethyl butanoate | pineapples |
| $H_3CCH_2CH_2C(=O)-O-CH_2(CH_2)_2CH_3$ | n-butyl butyrate | butyl butanoate | pineapples |
| $H_3CCH_2CH_2C(=O)-O-CH_2(CH_2)_3CH_3$ | n-amyl butyrate | pentyl butanoate | apricots |
| $H_3C(CH_2)_3C(=O)-O-CH_2CH_2CH(CH_3)_2$ | isoamyl valerate | 3-methylbutyl pentanoate | apples |
| (benzene ring)$-C(=O)OCH_3$, $-OH$ | methyl salicylate, oil of wintergreen | methyl 2-hydroxybenzoate | wintergreen plant |

$$RC\overset{O}{\underset{Cl}{\diagdown}} + H-OR' \longrightarrow RC\overset{O}{\underset{OR'}{\diagdown}} + HCl \qquad\qquad (2)$$

This experiment demonstrates the preparation of an ester from an alcohol and an acid and provides the opportunity to become familiar with laboratory glassware frequently used in organic syntheses.

## CONCEPTS OF THE EXPERIMENT

The conversion of a carboxylic acid/alcohol mixture to an ester and water (equation 1) is an acid-catalyzed, multi-step process. Experiments with oxygen-18 labelled acids or alcohols established that the oxygen atom in the water formed during esterification comes from the carboxylic acid. On the basis of these experiments, the mechanism for the formation of an ester involves several steps (equation 3).

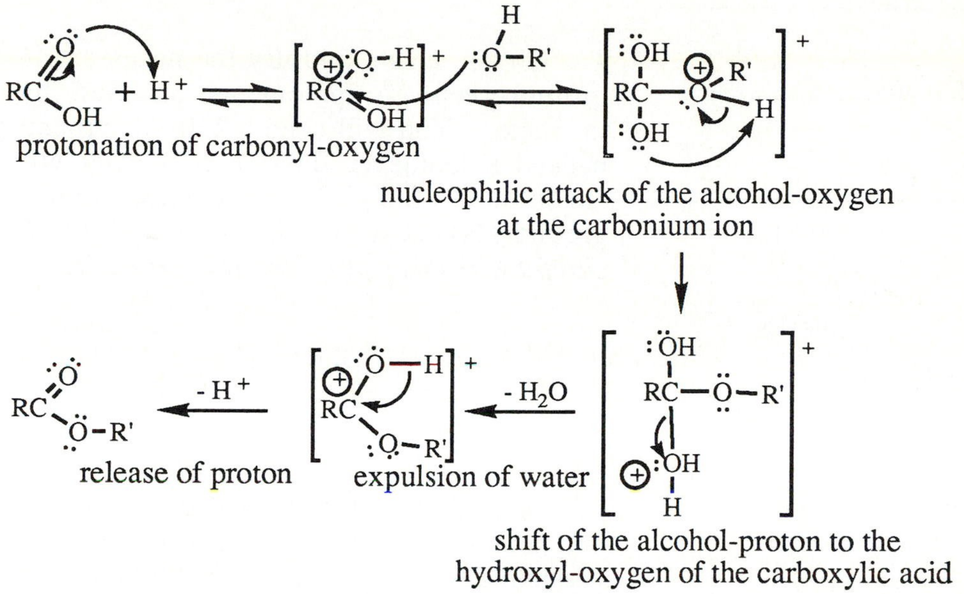

protonation of carbonyl-oxygen

nucleophilic attack of the alcohol-oxygen
at the carbonium ion

(3)

shift of the alcohol-proton to the
hydroxyl-oxygen of the carboxylic acid

expulsion of water

release of proton

The proton that initiates the reaction is regenerated in the last step. Any Lewis acid (an ion or molecule that can accept an electron pair; the proton is an example of a Lewis acid) may serve as a catalyst in the esterification reaction in place of a proton. The catalyst for a particular esterification reaction is chosen with consideration of cost, efficiency, solubility in the medium in which the reaction is performed, and ease of purification of the product. In the preparation of amyl acetate, the subject of this experiment, 1-pentanol and acetic acid are used as starting materials and sulfuric acid as the catalyst. The product is separated from the reaction mixture and purified by distillation.

## ACTIVITIES

- Complete the PreLab Exercises before coming to the laboratory.
- Assemble a ground-glass reflux apparatus.
- Prepare amyl acetate from 1-pentanol and acetic acid.
- Separate the ester from the reaction mixture.
- Purify the ester.

## SAFETY

*Wear eye protection at all times and use proper care to avoid ignition of the flammable vapors from the organic compounds used. Be particularly careful with ground-glass equipment to avoid breakage.* Never *heat or* reheat *a liquid before fresh boiling chips have been added to the* <u>cool</u> *liquid.*

## PROCEDURES

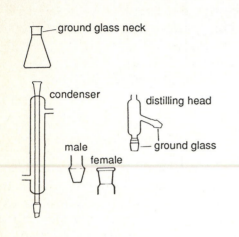

ground glass neck

condenser

distilling head

male

female

ground glass

1. Obtain, clean, and dry the necessary items of ground-glass equipment as directed by your instructor. You will need a 250-mL ground-glass-necked Erlenmeyer flask, a condenser with both "male" and "female" ground-glass joints, and a ground-glass distilling head. *Caution: Be very careful with this equipment. It is expensive.*

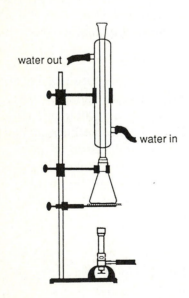

water out

water in

2. Moisten the tube connections on the condenser and attach the rubber condenser tubing snugly. Assemble the apparatus for a *reflux* setup as shown clamping the flask snugly in position with the flask supported by a wire gauze on a ring clamp. Do not clamp the condenser *tightly*. Attach the condenser clamp loosely to avoid a possible strain on the glass-to-flask connection. Have your instructor check the apparatus. Then remove the condenser and lay it in a safe place.

3. Pour from the storage bottle into a clean, dry graduated cylinder the volume of glacial acetic acid corresponding to 0.80 mole. *Carefully* test the odor of acetic acid. Describe this odor in your notebook.

4. Place 4 mL of concentrated sulfuric acid into the dry ground-glass-necked Erlenmeyer flask. Then pour the acetic acid from the graduated cylinder into

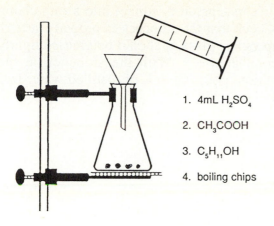

1. 4mL $H_2SO_4$

2. $CH_3COOH$

3. $C_5H_{11}OH$

4. boiling chips

this flask.  Do not rinse the graduated cylinder. *Caution: Glacial acetic acid and sulfuric acid are corrosive chemicals*.  When you pour the reactants into the Erlenmeyer flask, do not allow the reactants to come in contact with the ground-glass surface of the neck.  The use of a funnel will keep the neck free of chemicals.

Pour into your unrinsed graduated cylinder from the pentanol-storage bottle the volume of 1-pentanol corresponding to 0.40 mole.  Note and describe the odor of 1-pentanol in *its original container*.  Pour the measured alcohol into the flask.  Add two or three fresh boiling chips.

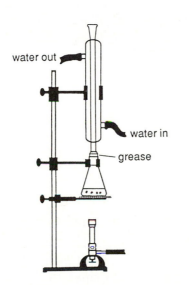

water out

water in

grease

**5**.  Very lightly grease the "male" joint of the condenser with a thin film of silicone grease.  Insert the condenser into the flask and rotate the condenser to distribute the grease uniformly over the ground-glass joint.  Secure the condenser with a clamp *just* tight enough to hold the condenser in place but *not* tight enough to place any strain on the apparatus.  Connect the lower condenser tubing to a cold-water tap and place the end of the other condenser tubing (water outlet) in the sink or drain trough.  Turn on the water *carefully* and adjust the flow rate so that a steady stream of water *runs* (not "gushes" or "trickles") out the drain tubing.  Have your instructor check the apparatus.

**6**.  Light the burner and adjust the flame to achieve a slow rate of reflux (boiling liquid with condensed vapor dripping back into the flask).  Continue adjusting the flame until the reflux rate gives about two drops of condensate per second.  Reflux the mixture at this rate for 20 minutes.

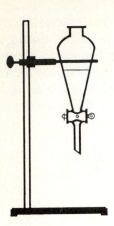

**7**.   Set a clean separatory funnel into a ring clamp support.  Check that the stopcock is properly seated. (If the stopcock is glass, it should be greased lightly with silicone grease.)  With the stopcock in the closed position, pour 100 mL of distilled water into the funnel.

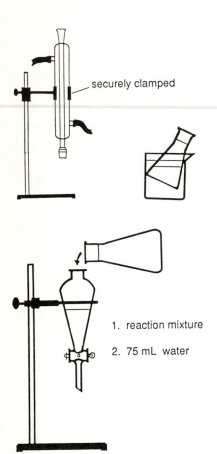

securely clamped

**8**.   After the reaction mixture has refluxed for 20 minutes, turn off the burner.  Continue running water through the condenser until the liquid in the flask has  stopped boiling.  Then turn off the water. Holding the neck of the flask securely (with split rubber tubing or folded towel to protect your hand), loosen and remove the flask-clamp. *While holding the neck of the flask*, tighten the condenser clamp so that it will support the condenser.  Lower the ring clamp and turn it out of the way.  Hold the condenser with one hand and with the other gently twist and remove the flask.  Place the flask in a 600-mL beaker half filled with cold tap water.

1. reaction mixture

2. 75 mL  water

**9**.   When the flask has cooled to about room temperature, pour the liquid from the flask into the separatory funnel.  Pour 75 mL of distilled water into the flask, swirl the flask to rinse the interior, and pour this rinse water into the separatory funnel. Stopper the separatory funnel.

vacuum hose

**10**. Wash the flask thoroughly with soap and water; rinse it with tap water and then with distilled water. Wipe the inside of the joint clean with a towel. Carefully shake the last of the rinse water from the flask; then rinse the flask with a small portion of acetone. Use an aspirator hose to pull air through the flask to evaporate all the acetone-water film. *Caution: Flammable vapors.*

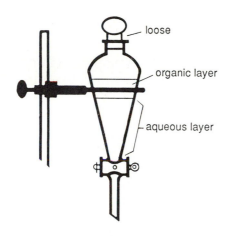

loose

organic layer

aqueous layer

**11**. *Using techniques described by your instructor,* extract the reaction mixture thoroughly with the distilled water. Remember to hold the stopper and stopcock securely and to vent the funnel through the stopcock after its initial inversion and periodically during the shaking procedure. Then set the funnel back into the ring support, loosen (but do not remove) the stopper and allow the funnel to stand until the two liquid phases have clearly separated.

**12**. While waiting for the phase separation, assemble a distillation apparatus as shown. All ground-glass connections should be greased lightly and the clamps should be arranged to hold the flask and condenser securely but not to place any strain on the apparatus. Lightly grease the opening through which the thermometer will be inserted. Carefully insert the thermometer with a twisting motion until the bulb is just opposite the side-arm opening. Have your instructor check your setup.

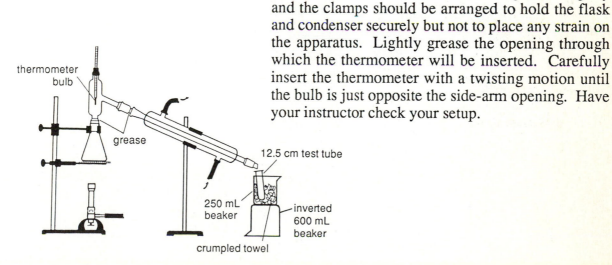

thermometer bulb

grease

12.5 cm test tube

250 mL beaker

inverted 600 mL beaker

crumpled towel

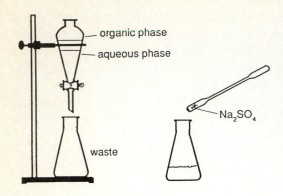

organic phase

aqueous phase

waste

$Na_2SO_4$

**13.** When your distillation setup has been approved and the liquid phases have separated in the funnel, remove the stopper from the separatory funnel and carefully drain all the lower liquid phase (aqueous) into a waste flask. Close the stopcock as soon as the organic phase has reached the stopcock bore. Discard the separated aqueous phase; transfer the organic phase into a 125-mL Erlenmeyer flask. Add four small spatula loads of anhydrous $Na_2SO_4$ to the liquid in this flask and swirl the mixture gently. The salt crystals will "cake up" as they absorb moisture. Continue adding (up to four more) spatula loads of $Na_2SO_4$ until additional increments no longer "cake up."

loose cotton plug

**14.** Carefully loosen the condenser clamp and disconnect the condenser and distilling head from the flask. Place a funnel containing a small, loose cotton plug into the neck of the distilling flask. Decant the organic phase from the $Na_2SO_4$ into the distilling flask through the funnel. Take care to keep the bulk of the $Na_2SO_4$ in the original flask. Add two or three fresh boiling chips to the distilling flask and reassemble the distilling apparatus. Properly adjust the water flow through the condenser. Have your instructor check your setup.

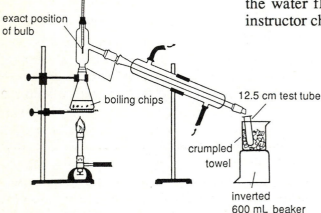

exact position
of bulb

boiling chips

12.5 cm test tube

crumpled
towel

inverted
600 mL beaker

**15.** Check that the test tube receiver is properly held by the towel in a position to catch drops falling from the condenser. Check the water flow rate again; then light the burner and begin to heat the distilling flask carefully.

1 drop every
2 - 3 seconds

After boiling starts, you will observe a film of condensate moving gradually up the interior of the distilling head. Watch the thermometer reading carefully as the vapors approach the thermometer bulb. Record the temperature reading when the thermometer bulb is bathed with a thin film of condensate and liquid just begins to collect inside the condenser. Adjust the heating rate so that drops fall from the condenser into the test tube receiver at the rate of about one drop every 2 to 3 seconds. Continue this heating rate and check the thermometer regularly. When the temperature remains constant during collection of 10 to 12 drops, carefully remove the first test tube and replace it with another clean test tube. Continue the slow heating rate as long as the temperature reading remains within $\pm 3^\circ C$ of the "constant" reading. If the temperature changes by more than about $\pm 3^\circ C$ or if the receiver test tube is more than three-quarters full, change receivers. Keep a record of temperature ranges for all fractions collected and label the test tubes in the order used. Continue distillation until nearly all of the ester has distilled. Under no circumstances should you heat the flask to dryness or allow the vapor temperature to exceed $160^\circ C$. (*n*-Amyl acetate boils around $149^\circ C$ at 1.00 atm. The 1-pentanol boils around $137^\circ C$.) When distillation is complete, turn off the burner and water and let the apparatus cool to room temperature.

**16.** Weigh a 250-mL beaker on the triple-beam balance. Place your first receiver test tube in the beaker and record the mass of the test tube and contents. Note and describe the odor of this fraction. Empty the test tube into the specified container and reweigh the empty tube and beaker. Repeat for each receiver test tube, recording the masses of each fraction collected. (Your instructor may specify separate containers for various fractions.)

**17.** After the distillation apparatus is cool, carefully disassemble the equipment. Wipe the grease film from all joints with a towel. Wash the distilling flask and the emptied test tubes thoroughly with soap and water. Rinse with tap water, then with distilled water (and, if directed, with acetone). Do not clean the distilling head or condenser unless you are given appropriate directions by your instructor.

**18.** Use your notebook records and calculations to complete the Report Form. Attach separate pages containing details of your calculations.

# PRELAB EXERCISES

## INVESTIGATION 35: AMYL ACETATE, A NATURAL FLAVORING AGENT

Name:_____

Instructor:_____    ID No.:_____

Course/Section:_____    Date:_____

1. The density of 1-pentanol at room temperature is 0.814 g mL$^{-1}$. Calculate the volume of 1-pentanol that corresponds to 0.40 mole of the alcohol.

2. The density of acetic acid at room temperature is 1.05 g mL$^{-1}$. Calculate the volume of pure acetic acid that contains 0.80 mole of the acid.

3. Define the following:

   Esterification

   Hydrolysis

   Lewis Acid

   Acid Chloride

   Decant

(over)

411

4. Identify the carboxylic acids and alcohols from which each of the esters listed in Table 1 can be prepared.

# REPORT FORM

## INVESTIGATION 35:  AMYL ACETATE, A NATURAL FLAVORING AGENT

Name: _____     ID No.: _____

Instructor: _____     Course/Section: _____

Partner's Name (if applicable): _____     Date: _____

---

Attach separate pages for calculations and problems.

## Calculated Quantities

0.40 mole of 1-pentanol corresponds to: _____ g  or  _____ mL

0.80 mole of acetic acid corresponds to: _____ g  or  _____ mL

The limiting reagent for the reaction was: _____

The theoretical yield of *n*-amyl acetate: _____ g

The mass of amyl acetate collected: _____ g

The percentage yield for this preparation: _____ %

## Experimental Results

Volume of acetic acid used:     _____ mL

Odor of acetic acid: _____

Volume of 1-pentanol used:     _____ mL

Odor of 1-pentanol: _____

Odor of *n*-amyl acetate: _____

| Property | Fraction number from distillation | | | | |
|---|---|---|---|---|---|
| | | | | | |
| boiling range (°C) | | | | | |
| mass of fraction | | | | | |
| odor | | | | | |

413

On the basis of boiling range and odor, the fractions that principally appear to be the product ester were: _____

Total mass of fractions corresponding to the product ester: _____ g

The obtained percentage yield was: _____.

**PROBLEMS** (If additional space is needed, answer these on an attached sheet of paper.)
1. Describe the function of the sulfuric acid in the synthesis of amyl acetate.

2. Why was the reaction performed under "reflux conditions"?

3. Why was excess acetic acid used?

4. On the basis of the relationship between molecular structure and water solubility, outline the purpose of the extraction procedure. Which components of the reaction mixture were probably transferred into the aqueous phase during the extraction?

5. Did you observe any evidence of 1-pentanol in the liquid placed into the distilling flask? Account for your answer in terms of the nature of the esterification reaction and the predistillation procedures.

6. (Optional) Based on your experience with the distillation of the amyl acetate reaction mixture and the handbook-acquired boiling points of ethanol, formic acid, ethyl formate, and water, could the excess of the reactants and products be separated during the preparation of ethyl formate from ethanol and formic acid? Which effect would the removal of one of the products have on the yield of the ester?

Date _____ Signature _____

# Appendix A

## WEIGHING WITH A TRIPLE–BEAM BALANCE

The triple-beam balance is a rugged balance. It consists of a pan, a supporting base, a fulcrum, a zero adjustment screw, a beam split into three sub-beams on the pointer side of the fulcrum, three mass riders, a pointer, and a fixed scale. A triple-beam balance may have a magnetic dampening device.

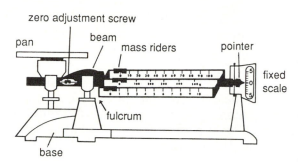

The middle of the three sub-beams carries the 100-g rider and has notched positions in 100-g steps from 0 to 500 grams. The second sub-beam carries the 10-g rider and has notched positions in 10-g steps from 0 to 100 grams. The third sub-beam, equipped with a 1-g rider, has marks at 0.1-g steps from 0 to 10 grams

The mass of an object can be read to 0.1 g from the positions of the riders when the balance is at equilibrium. The position of the 1-g rider can be easily determined to 0.05 g and can be estimated with some difficulty to 0.01 g. Depending on the requirements of an experiment, the mass of an object may be found to the nearest 0.1, 0.05, or 0.01 gram. Examine a triple-beam balance in your laboratory and identify its parts. Push the 1-g rider to an arbitrary position and practice reading the indicated mass to 0.1 g, 0.05 g, and 0.01 g.

Before you use the triple-beam balance, fix the following rules in your mind:
- **NEVER PLACE CHEMICALS DIRECTLY ON THE BALANCE PAN!!!**
- **KEEP THE BALANCE AND THE AREA AROUND THE BALANCE SPOTLESSLY CLEAN!!!**

If you are not familiar with the use of a triple-beam balance, weigh 7.80 g of sand on a balance without magnetic dampening. If the triple-beam balances are magnetically dampened, you will receive modified instructions from your instructor.

1. Check the positions of the three riders. If they are not in the zero position, slide the 100-g and the 10-g riders into the zero notches and the 1-g rider exactly to the zero position. Then *lightly* push the balance beam to cause the pointer to swing between the +5 and -5 positions on the fixed scale. If the balance is properly zeroed, successive pointer turning points should be at +4, -4; +3, -3; +2, -2. When such or very similar readings are obtained, the balance is at equilibrium. If readings such as +5, -3; +4, -2 are obtained, ask your instructor to help you adjust the balance with the zero-adjustment screw.

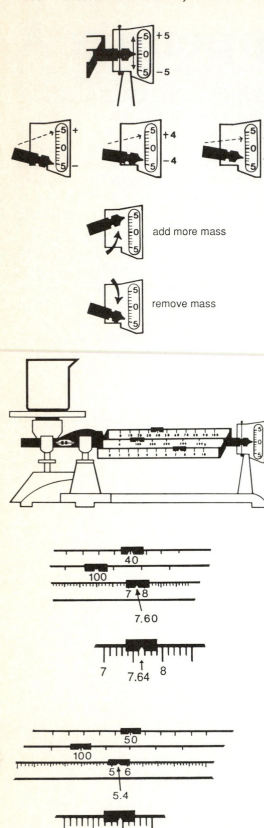

**2.** Place a clean and dry 100-mL beaker on the pan. The pointer will go to the top of the scale. Slide the 100-g rider into the 100-g notch. If the pointer moves down and remains down, the beaker is lighter than 100 g. Move the rider back to the zero notch. If the pointer moves up and remains up, the beaker is heavier than 100 g. In this case advance the rider one notch at a time until the pointer goes down. Then set the rider back one notch.

**3.** Repeat this procedure with the 10-g rider until the pointer goes down. Then set the rider back one notch. Advance the 1-g rider to the 5-g position. If the pointer moves up, set the rider to the 7.5 position; if it moves down, to the 2.5-g position. Continue moving the rider to half-point positions until the swinging beam appears to be in equilibrium.

Stop the beam; then push it lightly to have the pointer swing between the +5 and -5 scale position. If the turning points are not as described in Procedure 1, move the 1-g rider in very small increments until the beam executes the proper swings.

**4.** Read the rider positions and record the mass of the beaker. Read the 1-g rider position first to 0.05 g, then to 0.01 g. The positions of the riders in the illustration give the mass of the beaker as approximately 147.6 g. The position of the 1-g rider in the enlarged illustration shows the mass to be 147.65 or - with estimation to 0.01 g - 147.64 g.

**5.** Change the rider positions to make them correspond to the mass of beaker plus sample (147.64 g + 7.80 g = 155.44 g). Set the 10-g rider into the 50-g notch and slide the 1-g rider to the 5.44-g position.

**6.** With your clean and dry scoopula take some sand from the storage bottle and add it to the beaker in small increments. Do not spill any sand during this operation. From time to time push down gently on

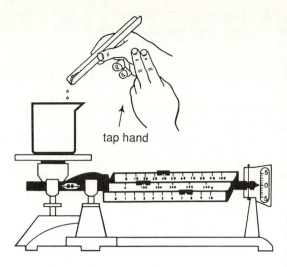

tap hand

the balance pan. If the beam is reluctant to swing, you are not yet close to the desired mass of the sample. If the beam starts to swing easily, add the sample in very small increments by gently tapping the hand holding the scoopula over the beaker with a finger from your other hand. When the beam begins to swing freely, observe the turning points of the pointer on the fixed scale. Add tiny increments of the sand until the point has turning points symmetrically located with respect to the zero-point on the scale. This operation is referred to as "returning the balance to equilibrium".

If you added too much of the sand, you may remove some of it from the beaker and repeat the incremental addition with care. Removal of sample from the beaker is a more difficult operation than careful incremental addition. When you judge the balance to be at equilibrium, stop the beam. Then push the beam gently and observe the turning points of the pointer to check whether or not the balance is at equilibrium. If necessary, adjust the amount of sample in the beaker and repeat the check procedure.

**7**. Read the rider settings and record the mass of beaker plus sand to 0.05 g and also to 0.01 g. Remove the beaker gently from the pan. Return all riders to the zero-position. Before leaving the balance, check the balance and the surrounding area. If they are not completely clean, clean up!

**8**. Return the sample of sand to the designated container. Calculate the mass of the sand to the nearest 0.05 g and also to the nearest 0.01 g.

# Appendix B

## WEIGHING WITH AN ANALYTICAL BALANCE

An analytical balance is a delicate instrument for the determination of masses to 0.1 of a milligram. Such a balance is probably the most expensive instrument you will encounter in your laboratory. These balances will be used in most of the experiments. You must be thoroughly familiar with the proper use of these balances and the recommended weighing process to prevent damage, avoid costly repair, prevent delays in performing an experiment, and obtain reliable results. Proper weighing on the analytical balance takes time and frequently causes "weighing jams". You will be able to use your laboratory time most efficiently when you know the balance procedures and have become an "expert weigher".

Two types of analytical balances are now common in laboratories: the mechanical analytical balance and the electronic analytical balance. The electronic balances are much easier to use than the mechanical balances. The following procedures are general guidelines for the proper use of these balances. Your instructor may provide additional information about the type of balance that is in use in your laboratory.

### B-1. THE MECHANICAL ANALYTICAL BALANCE

The single-pan, mechanical analytical balance has a housing for the balance pan with a window in front and sliding glass doors on the left and right. All dial and readout devices are generally located on the front panel of the balance. The *arrest knob* has three positions. When the balance is not in use, the pointer on the arrest knob must always be in the arrest position. Turning the arrest knob to the right stop will partially release the balance beam. In this "partial release position" mass standards may be added or removed. This position may not be used when objects are to be placed onto or removed from the balance pan; for this operation the balance must be fully arrested. Turning the arrest knob to the left stop will place the balance into the fully-released position and allow the balance beam to swing freely. The balance may be placed in the fully released position only when the mass of the object on the pan is within 1 gram of the masses dialed with 1-g, 10-g (100-g) mass dials.

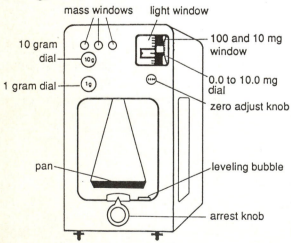

The *mass dials* (1-g dial, 10-g dial, perhaps 100-g dial) must be in the zero position when the balance is not in use. The mass dials may be turned cautiously to the left or to the right, but only when the balance is fully arrested or partially released. The dials may never be turned when the balance is released and the beam is free to swing. When, contrary to strict instruction, the mass dials are turned with the balance in the released position, the knife-edges may be damaged and the balance may become useless. The *mass windows* display the masses dialed with the mass dials. When the mass windows show zeros only, all the mass standards will rest on the balance beam when the

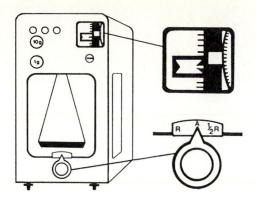

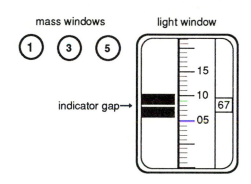

mass windows          light window

indicator gap→

balance is fully released. In this condition the balance is at equilibrium. When an object is placed on the pan, the load on the beam is too high. To restore equilibrium, mass standards are removed from the beam by turning the mass dials until the balance is again close (within 1 g) to equilibrium. In this way, the object on the pan is a substitute for the mass standards removed from the beam. The mass windows display the total mass removed from the beam and, thus, the mass of the object on the pan to the nearest 1 gram.

The *light window* becomes illuminated when the arrest knob is in the partial-release or the release position. Projected on the window is a mass scale with numbers from 00 to 100 or 120. One scale division represents 0.01 g (10 mg). When the mass of an object has been dialed to the nearest 1 g, the arrest knob is turned to the release position. The beam is free to swing and is allowed to reach a new equilibrium position. Because the beam is dampened, this position will be reached in a few seconds. The deflection of the beam from the horizontal position is correlated with the additional mass of the object to be weighed in excess over the 1-g units dialed. This additional mass can be read from the projected scale using the scale division that is just below the indicator gap. The additional 1-mg and 0.1-mg masses are found by centering the scale division below the indicator gap into the gap. This operation is achieved by turning the *0.0-to-10.0-mg dial*. When the scale division is in the center of the gap, the 1-mg and 0.1-mg-units are read from the small window. The mass of the object is then read from left to right beginning with the mass windows, the projected scale division centered in the gap, and the window for the 0.0-to-10.0-mg dial. The illustration corresponds to an object with a mass of 135.0867 g. Before you weigh an object on an analytical balance, always check the balance first. You should find the balance in the following condition:

- Both balance doors closed.
- The arrest knob in the arrest-position.
- All mass windows with zeros.
- No light in the light window.
- The window for the 0.0-to-10.0-mg dial set at 00.
- The pan and interior of the balance spotlessly clean.
- The surroundings of the balance spotlessly clean.

If any of these conditions are not met, contact your instructor, who will inform you how to correct the situation.

You will encounter three types of weighing procedures:

1. Weighing an object (beaker, watchglass, piece of paper) to ±0.1 mg.
2. Weighing an amount of a reagent to ± 0.1 mg within a given mass range. For instance, for the preparation of a potassium hydrogen phthalate solution, an amount of the reagent between 1.4 and 1.6 g is needed. Any mass within this range is acceptable. However, the mass must be determined to ±0.1 mg. This type of procedure is called for with statements such as "weigh out approximately 1.5 g ±10%", or "weigh out an amount between 0.6 and 0.8 g".
3. Weighing an exactly prescribed amount of a reagent to ±0.1 mg such as an amount of a primary standard of which 0.7564 g are needed. Fortunately, this rather complex procedure is not needed very frequently.

## BEFORE YOU USE ANY ANALYTICAL BALANCE, REMEMBER THE FOLLOWING RULES:

- **Never place any hot object inside the balance or on the balance pan.**
- **Never place any chemical directly on the balance pan. Protect the pan with a watchglass, a beaker, or a piece of paper.**
- **The analytical balance is a delicate, expensive instrument. Work with it cautiously and gently.**

## B-1.1.  Weighing an Object to ±0.1 mg

The object to be weighed on the analytical balance (watchglass, small beaker, sugar package) must be clean and dry. The object must not be heavier than the maximal load for which the balance was designed. Check with your instructor to learn about the maximal loads of the balances in your laboratory.

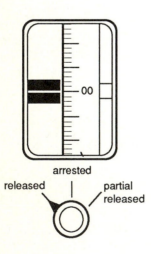

arrested

released | partial released

**1.** Check the balance: doors closed? mass dials all on zero? 10 to 0.1 mg dial on zero? balance pan clean? inside of balance clean? If any of these conditions are not met, check with your instructor.

**2.** With the balance doors closed, nothing on the pan, and all mass dials at zero, turn the arrest knob to the fully released position. Watch the scale in the light window and wait until the scale has stopped moving. The "00" scale mark will probably not be centered in the indicator gap. Use the *zero-adjust knob* to center the "00" line in the gap. This procedure, called "zeroing" the balance, must be performed before weighing an object. When the balance is zeroed, do not touch the zero-adjust knob until you have completed your weighing.

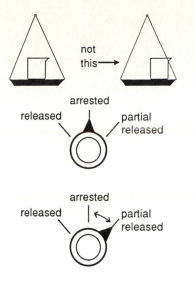

**3.** Arrest the balance by turning the arrest knob from the fully released position to the arrest position. Open one of the balance doors and *gently* place the object (Is it clean and dry?) on the middle of the balance pan. Close the balance door.

**4.** Turn the arrest knob slowly to the partially released position. The light will come on in the light window. Should the balance pan begin to swing like a pendulum, gently move the arrest knob several times between the partial-released position and the arrest position until the pan stops swinging. Bring the arrest knob to the partial-released position.

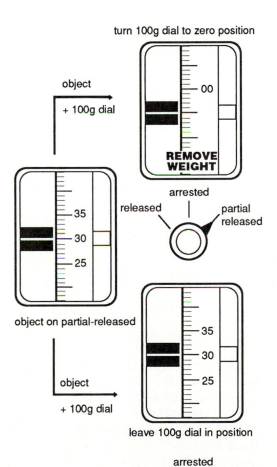

**5.** Check the scale in the light window. Because the load on the beam is too high for the beam to be close to equilibrium, a scale mark between 25 to 35 will be inside the indicator gap. If you do not know the approximate mass of your object, you will have to proceed systematically.

**6.** If your balance has a 100-g dial, turn it and watch the scale in the light window. Should the scale move to show the "00" mark, a mark below "00", or the sign "REMOVE WEIGHT", your object is lighter than 100 g. In this case, turn the 100-g dial to its zero position. Should the scale remain in the position it was in before the 100-g dial was turned, the object is heavier than 100 g. In this case, leave the 100-g dial in its position.

**7.** Repeat this procedure with the 10-g dial. Gently dial in 10 g, 20 g, 30 g, . . . , until the scale moves and the "REMOVE WEIGHT" sign appears. Then back up one unit.

**8.** Repeat this procedure with the 1-g dial. Gently dial in 1 g, 2 g, 3 g, . . . , until the scale moves and the "REMOVE WEIGHT" sign appears. Then back up one unit.

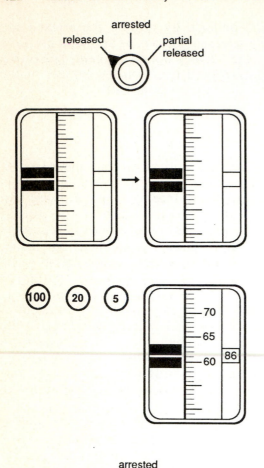

**9.** This procedure can be shortened considerably if you know that your object weighs less than 10 g or less than 100 g. If your object weighs less than 10 g, only the 1-g dial has to be used (Procedure 8). If your object weighs less than 100 g, the 10-g dial (Procedure 7) and the 1-g dial (Procedure 8) must be used. If you know the mass of your object to the nearest 1 g, dial this mass using the appropriate dials while the balance is arrested. Then turn the arrest knob to the partially released position.

**10.** Turn the arrest knob to the arrested position and then gently to the fully released position. When the scale in the light window comes to rest, turn the fractional mass dial until the line just below the indicator gap is exactly centered in the gap.

**11.** Read the mass of your object from left to right in the sequence 100-g window − 10-g window − 1-g window − scale mark in light window − 10 to 0.1 mg window. The illustration given corresponds to a mass of 125.6186 g. Record the mass of your object.

**12.** Turn the arrest knob to the arrest position. Open one of the balance doors. Gently remove the object from the pan without touching the pan with your fingers. Close the door. Return all dials to the zero position.

## B-1.2. Weighing a Chemical Within a Given Mass Range to ±0.1 mg

Because a chemical *must not* come in contact with the pan of the balance, a watchglass, beaker, or piece of paper is first placed on the pan and weighed to ±0.1 mg. Then the required amount of the chemical is added to the watchglass, the beaker, or the paper, and the mass of the container + chemical is determined to ±0.1 mg. The mass of the chemical is obtained as the difference between the two masses. The choice of container is a matter of convenience and of the properties of the chemicals to be weighed. For corrosive solids, the watchglass or the beaker is suitable; for liquids the beaker, and for inert solids a piece of paper. Pieces of paper shorten the weighing process because they often weigh less than one gram and do not require the use of the mass dials.

**1.** Determine the mass of your container to ± 0.1 mg as described in Section B-1.1. The following directions are given for an amount of solid between 1.4 and 1.6 g.

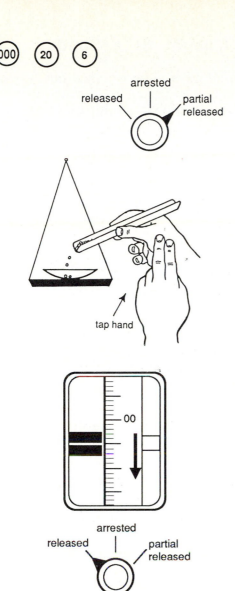

**2.** Add the mass of your container (let us assume 25.3672 g) and the smallest amount of your solid you need (1.4 g). Make sure the balance is arrested. Turn the appropriate dials to achieve 26 g.

**3.** Open the door of the balance. Turn the arrest knob to the partial-release position. Load your scoopula with the finely powdered chemical. Hold the end of the scoopula not more than 2 cm above the center of the watchglass. With the other hand lightly tap the hand holding the scoopula. The chemical will fall in small portions from the scoopula onto the watchglass. Watch the scale in the light window. At the beginning of this operation a scale mark below "00" will be near the indicator gap or the "REMOVE WEIGHT" sign will be visible. Keep adding your chemical in small increments until the scale begins to move down. Turn the arrest knob to the arrest-position and then to the fully released position. Reduce the size of your added portions, but keep adding your chemical until the scale has moved to bring the "76" mark to the indicator gap. Because the range of acceptable masses includes the "96" mark, you can overshoot the "76" mark slightly. It is much easier to very carefully add small increments than to remove some of the chemical from the watchglass. Close the door of the balance. Arrest the balance. Then turn the arrest knob to the fully released position and determine the exact mass of the watchglass and your chemical as described in Procedure 10 of Section B-1.1. Record the mass.

**4.** Should you have added so much of the chemical that the upper limit of the mass range is exceeded, you must remove some of the chemical from the watchglass. **NEVER REMOVE ANYTHING FROM THE BALANCE PAN WHILE THE ARREST KNOB IS IN THE FULLY RELEASED OR PARTIAL-RELEASED POSITION.** Arrest the balance. With an empty scoopula or spatula remove some of the chemical from the watchglass. Then repeat Procedure 3 of this section with great care to avoid exceeding the mass range.

**5.** Leave the balance clean, with doors closed, and all dials at zero.

## B-1.3. Weighing an Exactly Specified Amount of a Solid

Fortunately, exactly specified amounts of solids do not have to be weighed out very often. Weighing out an amount specified to ±0.1 mg is tedious at best and often impossible. With solids that are not finely powdered and drops of liquids, there is a physical limit to the size of the achievable increment. Therefore, in the experiments that you will have to perform an "exactly specified amount" (specified at most to ±0.1 mg) is achieved when the amount you weighed out is within ±2 mg of the specifications. In the following directions the exactly specified amount is 7.532 g.

**SEE SECTION B-1.1.**

**1.** Weigh a watchglass to ±0.1 mg as described in Section B-1.1. Assume the watchglass weighed 12.2122 g. Record the mass.

**2.** Calculate the sum of the mass of the watchglass and the specified mass of the solid (12.2122 g + 7.532 g = 19.7442 g).

**SEE SECTION B-1.2.**
**PROCEDURE 3**

**3.** Add the chemical as described in Procedure 3 of Section B-1.2. When the scale in the light window begins to move, be especially careful to add very small increments until the "70" mark (four marks before the target "74") is reached. Then arrest the balance. Close the door. Turn the arrest knob to the fully released position and determine the mass to ±0.1 mg. When the balance door is open, air currents might influence your weighing. The result of your weighing with the door closed will tell you whether you have reached the exactly specified mass (reading between 72 and 76 mark) or whether you need to add more solid. When more solid is needed, open the door, carefully add a tiny amount of the solid, arrest the balance, close the door, and determine the exact mass. Repeat this procedure until you are within the acceptable ±2 mg. Record the mass.

**4.** Leave the balance clean, doors closed, and all dials at zero.

## B-2. THE ELECTRONIC ANALYTICAL BALANCE

An electronic or digital analytical balance is much easier to use than a mechanical analytical balance. An electronic balance has no mass dials. The operationally important parts of the electronic balance are the *Control Bar*, the *Calibration Lever*, and the *Balance Pan*. The directions given here apply to the Mettler Model AE 200 balance. If a different model is in your laboratory, ask your instructor for any required modifications to these procedures.

A: brief operating instructions

B: control bar

C: level indicator

D: calibration lever

E: pan / windshield ring

F: leveling screw

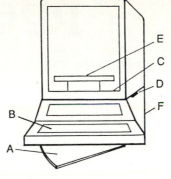

## TURN ON THE BALANCE

## CALIBRATE

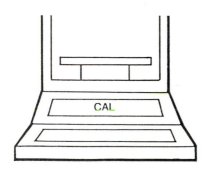

## WEIGHING AN OBJECT

## WEIGHING A CHEMICAL

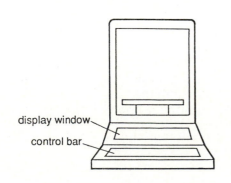

display window

control bar

**1.** The electronic balance must be kept spotlessly clean. Chemicals shall never be placed directly on the balance pan. Use a watchglass, a beaker, or a piece of paper to hold the chemicals. Do not put objects heavier than the maximal load on the pan. Keep the doors of the balance closed unless the weighing operation requires an open door. The electronic balance should be switched on one hour before the first use to allow for a sufficient warm-up time. The balance is switched on by briefly depressing the control bar. After a few seconds the display will show only zeros.

**2.** After the balance has warmed up for at least one hour, the calibration can be performed. Make sure the balance doors are closed and the pan is spotlessly clean. Depress and hold the *Control Bar* until the display shows "-CAL-". Then release the bar. The display will change to a blinking "CAL 100". Move the calibration lever all the way to the rear. The display will change to "CAL----", then to "100.0000", and then to a blinking "CAL 0". As soon as the the display blinks, move the calibration lever all the way to the front. The display changes to "----" and then to "000000". The balance is now ready to be used for a weighing.

**3.** Open one of the doors of the balance. Place your clean and dry object on the balance pan. Close the door. Wait until the display has stabilized. Read and record the mass.

**4.** The electronic balance has a convenient taring option. Place the container (beaker, watchglass, folded paper) on the balance pan. Close the door. Press the control bar briefly. The display changes to zero. The mass of the container is now tared. Open the door and add your chemical as described in Procedure 3, Section B-1.2. Watch the display telling you the mass of chemical in the container. Add the chemical in very small increments when you approach the desired mass. When you have reached this mass, close the balance door and wait until the display has stabilized. Read and record the mass. Remove the container from the balance.

**5.** Before leaving the balance, make sure the weighing compartment is spotlessly clean. Close the balance doors. Do not turn off the balance unless specifically instructed to do so.

# Appendix C

## TRANSFER OF SOLIDS OR LIQUIDS

Most laboratory chemicals are rather expensive. Many are corrosive or poisonous (or both). It is only good common sense, from both financial and safety standpoints, not to spill or otherwise waste chemicals. On numerous occasions you will have to transfer chemicals from stock bottles into beakers or flasks, or from one piece of equipment to another. The following suggestions will facilitate such operations.

Never take more of a chemical than you actually need. It is better to take smaller amounts and come back for more, if necessary, than to waste chemicals needed by other students. If you take too much, you may not return the excess to the stock containers except when specifically directed to do so. Even with the best of intentions, such returns often result in the contamination of reagents.

### C-1. TRANSFER OF SOLIDS

Solids are generally stored in wide-mouth, screw-cap reagent bottles. Before taking a sample from the storage bottle, inspect the physical state of the solid in the bottle. Ideally, the solid ought to be a fine, free-flowing powder. Many solids tend to "cake". In such cases the solid particles will be stuck together. To break up such "caked" chemicals you have several options.

**1.** Tap the closed bottle against the palm of your hand until the "cake" has been broken up.

**2.** Should tapping against the palm not produce the desired result, consult your instructor. The instructor will provide you with a completely clean and dry spatula, scoopula, or other appropriate tool with which to break up the caked chemical. This operation must be carried out with great care to avoid contamination of the chemical, breaking of the storage bottle, and injury to the hands. **Never use a glass rod to break up a solid in a storage container**.

**3.** A "caking" solid broken up into smaller pieces will often consist of particles too large to be used in a weighing process. In this case, place an amount of the chemical estimated to be slightly more than required into a clean mortar of appropriate size. With a clean pestle crush the solid into fine particles.

To avoid contamination of a solid in a storage bottle and subsequent problems for students using such a contaminated solid, **NEVER STICK A SPATULA OR SCOOPULA INTO A STORAGE BOTTLE.** To obtain a sample of the appropriate size, take a creased piece of paper, aluminum foil, watchglass, or beaker to the storage bottle, open the bottle, bring the mouth of the bottle over your container, incline the bottle appropriately, and gently turn the bottle. The solid will flow from the bottle into your container. To transfer only slightly more than the required amount of the solid, you may perform this transfer with the container on the pan of a triple-beam balance.

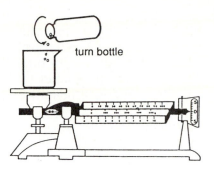

turn bottle

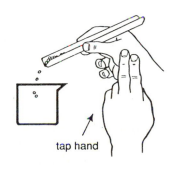

tap hand

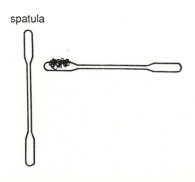

spatula

After you have transferred from the storage bottle to the container an amount of the solid sufficient for the experiment to be carried out, you will almost always be faced with weighing out an amount of the solid within a specified range. When the experiment calls for an amount to be weighed only to ±0.1 g or ±0.05 g, you can transfer the solid directly from the storage bottle to your container on the pan of a triple-beam balance using the "turn the bottle" method described above (see Appendix B for the weighing procedure). This transfer system will not work when the sizes of the receiving container and the storage bottle are not compatible. For instance, controlled transfer from a 5-kg storage bottle to a 100-mL beaker will be impossible and – when attempted – will cause unprecedented spillage. In such a case of mismatch, transfer first into a larger container (e.g., from the 5-kg bottle to a 500-mL beaker).

When you have to weigh out an amount of a solid on an analytical balance to ±0.1 mg, a creased piece of paper, a watchglass, or a beaker is placed on the balance pan. A scoopula is loaded with the solid from the container into which the solid was transferred from the storage bottle. Carefully move the scoopula over the paper, watchglass, or beaker to position the scoopula centrally about 1 cm over the container receiving the solid. Tap the hand holding the scoopula lightly with your other hand to make the solid slide off the scoopula. For the proper weighing procedure see Appendix B-1.2. When you have transferred the required amount and some solid is still left on the scoopula, return the excess to your container (not the storage bottle). Then wipe clean the scoopula with a towel.

A spatula is much more difficult to handle in transfer operations than a scoopula. Chances are very good that an unsteady hand will cause the solid to slide off prematurely and "mess up" the surroundings of the balance, the weighing compartment, the balance pan, or your work place. Spatulas are, therefore, used only for transferring very small amounts of a solid. **WHENEVER YOU SPILL ANY CHEMICAL, CLEAN UP AS SOON AS POSSIBLE. IF YOU ARE NOT SURE HOW TO CLEAN YOUR SPILL, ASK YOUR INSTRUCTOR FOR DIRECTIONS.**

## C-2.  TRANSFER OF LIQUIDS

Because liquids usually flow easily, their transfer is generally quite simple provided a few precautions are followed.  Usually, you will have to pour the liquid from a large into a smaller container.  Such an operation becomes problematic when the sizes of the two containers are very different.  Pouring a liquid from a 5-liter jug into a 10-cm test tube is almost impossible.  In such a case, the liquid is first poured from the 5-liter jug into a 500-mL beaker, from the 500-mL beaker into a 100-mL beaker, and finally from the 100-mL beaker into the 10-cm test tube.

To pour from a large (for instance, 1 gallon) screw-cap bottle or a bottle with a flat-topped glass stopper, first remove the cap or the stopper and place it upside down on a clean towel.  Then, holding your collection vessel at the mouth of the bottle, carefully tilt the bottle until liquid flows slowly and evenly into the container.  Avoid rapid, jerky movements that will splatter the liquid. When you have finished, replace the cap or stopper.  If any liquid has run down the outside of the container, wipe it off carefully with a damp towel.

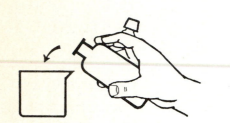

When pouring from a glass stoppered bottle with a "vertical disk" stopper, hold the stopper between two fingers as shown while pouring the liquid. Never place this type of stopper on the bench top.

Special care is needed when pouring concentrated acids, concentrated bases or solutions of corrosive or toxic reagents.  Such transfers should be performed in well-ventilated hoods, with the doors of the hood drawn down as far as possible, and – if necessary – your hands protected by appropriate gloves.  If you are careful, you will not spill any of these liquids. However, spills do occur.  If you have a spill, you must react quickly but without panic.  If you spilled some of the liquid on your hands, set the bottle and the container down in the hood and go immediately to the nearest sink.  Wash off the chemicals very thoroughly with water.  If you follow this procedure, you will at worst get a minor skin burn.  Should you panic and drop the bottle, you and your neighbors could suffer serious harm.

Consult your instructor for directions for clean-up of liquids spilled in the hood.

The experiments in this manual do not require the transfer of liquids with the help of pipets. Should you find it necessary to use a pipet, *never* fill the pipet with suction from your mouth. Always use a mechanical suction device such as a pipet bulb.

# Appendix D

## DELIVERING MEASURED VOLUMES OF LIQUIDS FROM A BURET

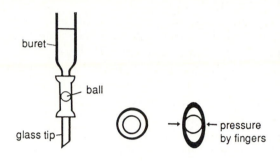

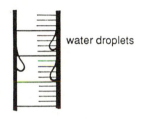

A buret is a glass tube that has been calibrated to deliver, in controlled amounts, known volumes of liquids. Burets are available in a wide range of sizes and with several different types of delivery control devices such as glass or Teflon stopcocks or a simple pinchcock. The pinchcock is the most easily maintained and operated delivery control. It consists of a short piece of rubber tubing with a glass sphere stuck inside. The rubber tubing connects the buret to the glass tip. To deliver liquid from the buret, the rubber tubing is squeezed between the thumb and middle finger to form an ellipsoidal shape allowing the liquid to flow past the glass sphere.

Before using a buret, you must check whether or not the buret is clean, paying particular attention to any dirt and grease that cause water droplets to remain when water is drained from the buret. Fill the buret with tap water, wipe dry the outside of the buret, allow the water to drain through the tip, and check for droplets clinging to the buret walls. If you observe droplets, you must clean the buret thoroughly.

Various special cleaning mixtures are sometimes employed, but most of these are dangerously corrosive and some ("chromic acid") may react violently with the rubber tubing. It is recommended that for cleaning a buret you use a long-handled buret brush and a detergent powder. Do not use abrasive cleansers that will scratch the interior of the buret and do not use soap that may leave a film on the glass. Wet the brush and sprinkle the bristles with the detergent powder. Insert the brush into the buret and slide it back and forth the full length of the buret 10 to 12 times. Remove the brush and rinse the buret with tap water several times. Fill the buret with tap water and drain the water through the tip. If droplets still cling to the wall, repeat the cleaning procedure. Once the buret drains cleanly, rinse it several times with distilled water. There is no need to use a lot of distilled water. With the plastic squeeze bottle squirt water around and around the buret top to let a thin film rinse down the entire inner surface. Drain the rinses through the pinchcock to rinse the pinchcock and the tip.

After the buret has been cleaned, rinsed, and clamped to a ringstand, use two successive 5-mL

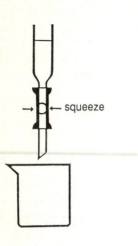

← ⊂ ← squeeze

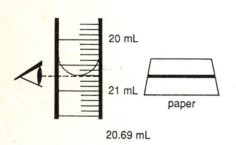

20 mL

21 mL

paper

20.69 mL

portions of the solution to be used in the experiment to rinse the buret again. Introduce these portions with a circular motion to have them flow as a thin film down the entire inner surface. Drain the solution through the tip. Then fill the buret with the solution to above the zero mark. Drain enough titrant into a waste flask to expel air from the buret tip and to drop the liquid level to just below the zero mark. Read and record the liquid level.

When you are ready to measure a certain volume of solution into a container, calculate from the initial buret reading and the needed volume the final buret reading. Then squeeze the rubber tubing at the position of the glass sphere between the thumb and one of the fingers to let the solution flow into the container. The harder you squeeze the more deformed the tubing will become and the faster the solution will drain into the container. With your squeeze you can adjust the flow from a steady stream to a drop at a time. When you come close to the final buret reading, slow down the delivery of solution. Add the last increments dropwise to avoid "overshooting" the required volume. When you titrate with the solution in the buret, you may add the titrant quickly at first and slow to a dropwise addition when the endpoint is near. Before reading the buret, wait a minute or two to allow liquid adhering to the walls to drain down. Then read the buret.

To read the buret, have your eye at the same level as the meniscus (curved surface) of the liquid. Record the position of the bottom of the meniscus with respect to the graduation marks on the buret. A black-and-white marked card held so that the dark line is just below (and reflected by) the bottom of the meniscus will help improve the precision of your readings. With practice you can learn to read a buret to ±0.01 mL.

When you have finished your work with the buret, drain out any remaining titrant and rinse the buret two or three times with distilled water as before. Either return the buret to the storage area or, if it is to be left clamped to the ringstand, fill it with distilled water and cover it with an inverted test tube or a small piece of aluminum foil.

If the burets in your laboratory are equipped with glass or Teflon stopcocks, ask your instructor for a description of their use and care.

# Appendix E

## USE OF A LABORATORY CENTRIFUGE

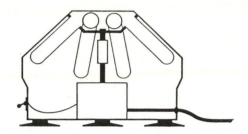

A centrifuge is useful for the rapid separation of an insoluble solid from a liquid medium. The centrifuge consists of a housing inside of which are located the electromotor and gear system needed to spin the rotor. Typical instructional laboratory centrifuges contain two, four, or six openings for 10-cm test tubes in the rotor head and have a power cord and an on-off switch. The centrifuge should be used in the following manner:

sample      match level with distilled water

1. Check that the centrifuge is plugged into a 110-V line.

2. Check the rotor head and remove any test tubes left in it.

3. Fill an empty test tube with distilled water to the level of the mixture in your sample tube. This will insure that the centrifuge is balanced.

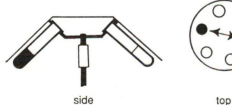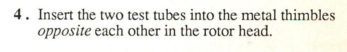

side      top

4. Insert the two test tubes into the metal thimbles *opposite* each other in the rotor head.

5. Turn on the centrifuge.

6. Wait until the rotor has attained full speed.

7. When full speed has been attained, switch off the centrifuge.

8. Let the rotor come to a *complete* stop on its own.

9. Remove both test tubes.

## SAFETY

- Do not touch the rotor while it is in motion.

- Do not attempt to stop the rotor motion with your hand after the motor has been turned off.

- Do not come close to the moving rotor with long hair or loose clothing.

# PRACTICE IN USING THE CENTRIFUGE

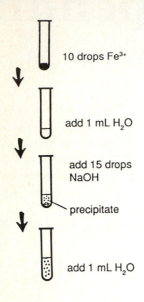

10 drops Fe³⁺

add 1 mL H₂O

add 15 drops NaOH

precipitate

add 1 mL H₂O

1. Place 10 drops of an $Fe^{3+}$ solution in a clean centrifuge tube.

2. Add 20 drops of distilled water.

3. Add dilute NaOH solution, dropwise, until a flocculent brown precipitate appears.

divide solution equally

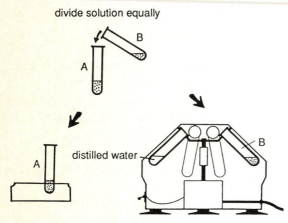

B

A

distilled water

A

B

4. Add 10 additional drops of the NaOH solution and 20 drops (1 mL) of distilled water.

5. Mix the contents of the tube thoroughly and pour half into a clean test tube. Set this tube in the test tube block.

6. Centrifuge the other tube using the procedures described.

A

B

7. Compare the appearance of the centrifuged and non-centrifuged tubes. Attempt with both tubes to decant (pour off) the clear solution without transferring solid $Fe(OH)_3$.

# Appendix F

## VACUUM FILTRATION

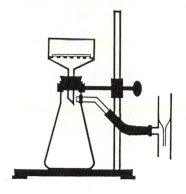

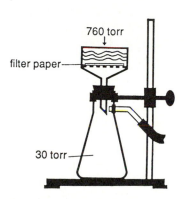

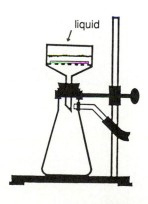

Although centrifugation is an effective separation method for many solid-liquid mixtures, it is limited (with typical instructional laboratory centrifuges) to small samples. Filtration is normally used for larger samples, and, except for gelatinous or very fine precipitates, the application of a vacuum speeds up the filtration process.

A simple vacuum filtration apparatus consists of a suction flask with a side arm and thick glass construction, a funnel (containing a rigid support for the filter) inserted through a rubber stopper, a thick-walled vacuum hose, and a water aspirator. When water rushes through the constricted tube of the aspirator into the wider pipe, air is pulled in through the side arm of the aspirator. This air flow creates the "vacuum" for filtration.

With a water aspirator, the theoretical limit of the "vacuum" is the vapor pressure of the water at the laboratory temperature. In practice, particularly when a number of aspirators are running from the same water line, this limit may not be approached very closely. Filtration is accelerated by the difference between the atmospheric pressure against the surface of the liquid in the funnel and the "vacuum" pressure within the flask. A pressure difference of 730 torr is equivalent to a "push" against the liquid surface of almost one atmosphere (0.96 atm). This pressure difference is sufficient to implode a thin-walled flat-bottomed flask (such as a 250-mL Erlenmeyer). Flat-bottomed flasks (unless they have special thick glass) that are larger than 125 mL must never be evacuated.

The following general procedure is recommended for vacuum filters.

**1.** Secure a clean porcelain or plastic funnel ("Buchner funnel") inserted through a rubber stopper to fit the vacuum flask. Set the funnel and stopper securely in the neck of the flask. Connect the flask to a water aspirator by a length of vacuum hose. To keep the flask stable, clamp it to a ringstand. Place a filter paper (of a size to fit easily in the funnel while

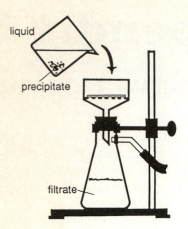

covering all the holes in the base of the funnel) into the funnel and moisten the paper. Turn on the aspirator to pull the moistened paper snugly against the filter support.

If the paper is not moistened, it might curl up when a mixture is poured into the funnel permitting part of the mixture to escape filtration.

**2**. Now decant as much liquid as possible from the settled solids into the funnel. Leave as much of the solid in the beaker as possible during the initial part of the filtration. The liquid will filter more rapidly before accumulated solid begins to clog the pores of the filter paper. Do not fill the funnel more than two-thirds at any time.

**3**. After most of the liquid has passed through the filter, transfer the remaining liquid and solid into the funnel. Rinse the original container with small volumes of the filtrate to transfer residual solid into the funnel. Keep the aspirator vacuum at least until no further liquid drops are coming from the funnel stem. The further treatment of the solid in the funnel will vary depending on particular experimental requirements. For example:

**a**. It is generally desirable to wash collected crystals with a small amount of liquid (usually the original solvent). To do this, first disconnect the vacuum hose carefully from the side arm of the flask. Then pour the desired amount of wash liquid into the funnel and stir the mixture with a clean spatula. Be very careful not to disturb or tear the filter paper. Reconnect the vacuum hose to pull the wash liquid through the filter. Washing may be repeated as necessary.

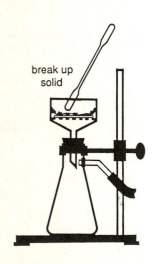

**b**. If it is desired to air dry the collected solid after all liquid has been removed, carefully break up the filter cake into finer pieces with a spatula or glass rod. The aspirator is allowed to pull air through the solid until it is dry.

**4**. When the desired filtration procedures have been completed, the vacuum hose is first disconnected from the flask. *Then* the aspirator is turned off. If the aspirator is shut off while the hose is still connected, water may back into the suction flask.

# Appendix G

## INSERTING GLASS TUBINGS INTO STOPPERS

One of the most common laboratory accidents involves the breakage of glass tubing or other glass equipment while attempting its insertion into a stopper. Three rules are *essential*.

- **Always protect your hands with a towel.**

- **Never apply force to a glass tubing at an angle.**

- **Never apply excessive force.**

Before attempting to insert glass tubing, be sure both ends of the tubing are smooth and rounded. If not, heat the end of the tubing at the edge of the hot inner cone of a burner flame until the glass softens to form a smooth, rounded edge. *Let the glass cool* before working with it again.

The following procedure is recommended for inserting a glass tubing into a stopper:

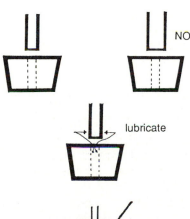

**1.** Only insert glass tubing (or other equipment) into a stopper having the proper size hole. The hole must be just *slightly* smaller than the tubing to give a snug fit, but not too small.

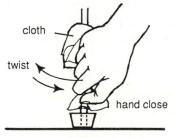

**2.** Lubricate the stopper hole and the part of the tubing to be inserted with a few drops of glycerin or water.

**3.** Protect both hands with a towel. Grasp the tubing *close to the end to be inserted* and with a twisting motion push the tube gently into the stopper. Never push at the end away from the stopper and never use excessive force. If you have problems, consult your instructor.

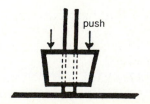

If you should have to remove tubing (or other glass equipment) from a stopper, protect your hands with a towel and push down *on the stopper* while the tubing is set at a 90 degree angle against a towel on the bench top. If the tubing does not loosen readily from the stopper, consult your instructor. *Do not use excessive force.*

# Appendix H

## PRECISION, ACCURACY, SIGNIFICANT FIGURES

During your study of chemistry you will employ a number of measuring devices such as a buret, a graduated cylinder, a thermometer, and a balance. Whenever you set out to make a measurement, you must use a device capable of doing the job properly. The particular measuring instrument selected will depend both on the accuracy required and on the detection limit of the instruments available.

If, for example, you wished to measure a 10-mg sample of a pharmaceutical chemical, you might think first of a triple-beam balance because such an instrument could easily give a two-place weighing accuracy. However, the triple-beam balance is not sufficiently "sensitive". At best, it could give a mass to within ±0.05 g (50 mg). You would, therefore, require the more "sensitive" analytical balance capable of weighing to ±0.1 mg. In general, the simplest measuring device having both the accuracy and detection limit required is selected. If you had needed to weigh "about 125 g" of a salt, for example, the triple-beam balance would have been a logical choice. For such a measurement, the analytical balance would have been unnecessary (and more time consuming).

The results of measurements are influenced by the limitations inherent in the instruments used. Each measurement will have a certain *error* (e.g., the ±0.05 g tolerance of the triple-beam balance) arising from this limitation. If you misread the mass settings on a balance, this is not referred to as an "error". We might call it a "blunder". It is assumed that your results will be limited only by instrument error; that is, "blunders" are not expected of you.

There are two types of errors: *systematic* and *random*. Systematic errors may result from incorrect calibration of instruments, from incorrect instrument design or manufacture, or from use of a method not well suited to the problem. Systematic errors affect all measurements the same way so that results may be precise, but inaccurate. Random errors result from the difficulty of exactly repeating a measuring procedure. Even though skill and practice can reduce random error, it is not possible to eliminate it completely. In some cases random errors occur for reasons beyond the control of the experimenter, as in a line voltage fluctuation during the use of an electric measuring device.

Every experimental result must be reported with an indication of its random error expressing the precision of the results. To obtain a measure of the precision, the experiment must be repeated several (n) times. The larger n is, the better is the calculated precision. As the first step toward finding the measure of precision, the average is calculated. The average result is calculated by summing the individual results and dividing this sum by the number (n) of individual values (equation 1) as shown for five weighings on an analytical balance.

$$\bar{x} = \frac{\sum_{n}^{i} x_i}{n} = \frac{x_1 + x_2 + x_3 + \ldots + x_i}{n} \quad (1)$$

$\bar{x}$ : average

$x_i$: individual results

n: number of measurements

| | |
|---|---|
| 1st mass | 1.6752 |
| 2nd mass | 1.6748 |
| 3rd mass | 1.6754 |
| 4th mass | 1.6747 |
| 5th mass | 1.6753 |
| Total | 8.3754 |

$$\text{Average mass} = \frac{8.3754}{5}$$
$$= 1.67508$$
(rounded to: 1.6751)

The precision of a result is measured by the standard deviation. To obtain the standard deviation, the differences between the average and every result in a series of measurements are formed; these differences are squared; the squares are added; and the sum of the squares is divided by the number of measurements reduced by one (n-1). The square root is extracted from this quotient to give the standard deviation, S (equation 2). The calculations leading to the standard

$$S = \sqrt{\frac{\sum_{1}^{n}(x_i - \overline{x})^2}{n-1}} \qquad (2)$$

Standard Deviation for the 5 weighings:

$$S = 0.0003$$

deviation become tedious when n is large. Fortunately, many electronic calculators have a program that calculates averages, standard deviations, and relative standard deviations. Only the individual results must be entered. However, you should perform at least one example by hand. The result of repeated measurements is correctly reported in the form:

AVERAGE ± STANDARD DEVIATION

1.6751 ± 0.0003 g (for the five weighings)

For the comparison of the precisions of several results, the relative standard deviation is often more convenient than the standard deviation. The relative standard deviation (RSD) expressed in percent is obtained by multiplying the standard deviation by 100 and dividing this product by the average (equation 3).

$$RSD = \frac{100S}{\overline{x}} \qquad (3)$$

$$RSD \text{ (5 weighings)} = \frac{100 \times 0.0003}{1.6751} = 0.018\%$$

As an example consider the results of two weighings: one series performed on the triple-beam balance, the other on the analytical balance.

| triple-beam balance | analytical balance |
|---|---|
| 126.2 ± 0.1 g | 0.0052 ± 0.0001 |
| RSD = 0.08% | RSD = 1.9% |

Consideration of the magnitude of the standard deviations (0.1 g, 0.0001 g) might lead to the erroneous conclusion that the result from the weighings on the analytical balance is more precise than the weighings on the triple-beam balance. Comparison of the relative standard deviations clearly shows the triple-beam result to be approximately twenty times more precise than the analytical balance result.

The accuracy of a result is expressed as the difference between the true value (T) and the average from a series of measurements (equation 4). True values are rarely known. The closest

$$\text{ACCURACY} = T - \overline{x} \qquad (4)$$

one can come to true values is through the use of certified standards issued, for instance, by the U.S. National Institute for Standards and Technology. For instance, the NIST standard "citrus leaves" is certified to contain 3.1 ±0.3 micrograms arsenic per gram of dry leaves. If you determine arsenic in this material several times and obtain 2.7 ± 0.3 µg/g, then you can calculate the accuracy of your result as +0.4 µg/g (neglecting the uncertainties in both results). When a result is not very accurate, the experiment is probably troubled by systematic errors.

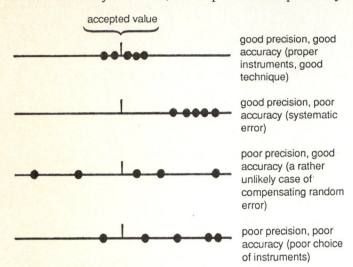

accepted value

good precision, good accuracy (proper instruments, good technique)

good precision, poor accuracy (systematic error)

poor precision, good accuracy (a rather unlikely case of compensating random error)

poor precision, poor accuracy (poor choice of instruments)

The calculations of precision and accuracy can produce four types of results shown in the diagram. All experimenters should do their best to obtain results with good precision and good accuracy. Because true values will rarely be known to you and in many cases will not be available to anyone, you must base your judgment on the precision of your results.

Another way of expressing precision uses the range by giving the highest and the lowest value or the difference between these values. A narrow range suggests a more reliable measurement than a broad range.

After the average and the standard deviation have been calculated for a series of measurements, the average result must be expressed using significant figures only. In a significant-figures-only result the last figure given is in doubt. This figure is identified with the help of the standard deviation. For instance, the average of the five weighings listed earlier is 1.67508 g with a standard deviation of ±0.0003 g. The standard deviation tells that the fourth decimal in the result (the 0.1-mg place) is in doubt by ±3 units. Carrying the "8" in the result is not meaningful because the "0" is already uncertain and the mass could, with high probability, be anywhere between 1.6753 and 1.6747 g. If a calculation as in the case of the five masses yields an average with more figures than are significant, the result must be rounded to the proper number of significant figures. When the first non-significant digit is 0 through 4, omit this digit and all the following digits. When the first non-significant digit is 5 through 9, add one unit to the preceding digit.

For example:     $\dfrac{1.6751}{5} = 0.33502$     (S = 0.0002)     round to 0.3350

$\dfrac{1.6751}{2} = 0.83755$     (S = 0.0003)     round to 0.8376

To avoid ambiguities in expressing results, use the scientific notation. Zeros to the left of the decimal point with no digits to the right of the decimal point may cause problems. When the result of a volume measurement is given as 100 mL, the two zeros are needed to fix the position of the "1" as a hundred-unit. If the two zeros were significant, for instance, in the case of a standard deviation of ±2 mL, then the result will be unambiguously expressed in scientific notation as (1.00 ± 0.02) x $10^2$ mL.

A numerical result should never be given without its standard deviation. Should a result appear without standard deviation, then the last digit of the result written in scientific notation can be assumed to be uncertain with a standard deviation of ±1. For instance, a temperature reported as 96°C can be assumed to be within 95 to 97° (96±1°C).

The numbers obtained from an experiment are frequently used to calculate the final results by addition, subtraction, multiplication, and/or division. Such calculations - particularly when carried out on an electronic calculator - almost always produce more figures than are significant. How does the uncertainty (the standard deviation) in the raw data propagate to the calculated, final result? The general rule states that no mathematical operation performed on data can increase the precision. The precision is best judged in terms of the relative standard deviation. The final result should have the same relative standard deviation as the least precisely known quantity used in the calculation. The following examples will clarify the application of the rule. It is advisable to carry more figures (that are generated by mathematical operation) than are significant until the final result is obtained. Only the final result should be rounded. Earlier rounding might cause serious errors (or blunders).

Calculation of a density from a mass and a volume measurement:

$$D = \frac{1.6751 \pm 0.0003 \text{ g}}{0.332 \pm 0.005 \text{ mL}} \qquad\qquad \text{RSD of mass: } 0.018\%$$
$$\text{RSD of volume: } 1.5\%$$

$D = 5.0454819 \text{ g mL}^{-1}$ (calculated result)

1.5% of the calculated result: 0.0756   round to 0.08

Calculated result with significant figures only: $5.0454819 \pm 0.08$  must be rounded to 5.05
$\pm 0.08 \text{ g mL}^{-1}$

Calculation of a sum of masses:

16.5 g + 21.2 g + 68.34 g + 40.1 g + 10.327 = 156.467 g

Assume every mass in this sum has a standard deviation of $\pm 1$ in the last digit. Find the mass with the highest relative standard deviation.

$16.5 \pm 0.1$      RSD = 0.6%

0.6% of 156.467 g = 0.94 g   rounded to 0.9 g

Calculated result with significant figures only: $156.467 \pm 0.9$ g  must be rounded to 156.5
$\pm 0.9$ g

These rules for error propagation are only approximate but suffice for the work during the first two semesters of a chemistry course. A more thorough treatment of statistical concepts applied to experimental data can be found in textbooks on analytical chemistry and monographs on statistical data treatment for natural scientists.

It is important in laboratory work to be aware of the limitations in the experimental results and the obligation of all experimenters to report their results with significant figures and standard deviations or relative standard deviations.